11

Springer Series on Chemical Sensors and Biosensors

Methods and Applications

Series Editor: G. Urban

For further volumes:
http://www.springer.com/series/5346

Springer Series on Chemical Sensors and Biosensors

Series Editor: G. Urban

Recently Published and Forthcoming Volumes

Solid State Gas Sensors – Industrial Application
Volume Editors: M. Fleischer, M. Lehmann
Vol. 11, 2012

Optical Nano- and Microsystems for Bioanalytics
Volume Editors: W. Fritzsche, J. Popp
Vol. 10, 2012

Mathematical Modeling of Biosensors
An Introduction for Chemists and Mathematicians
Volume Authors: R. Baronas, F. Ivanauskas, J. Kulys
Vol. 9, 2010

Optical Guided-wave Chemical and Biosensors II
Volume Editors: M. Zourob, A. Lakhtakia
Vol. 8, 2010

Optical Guided-wave Chemical and Biosensors I
Volume Editors: M. Zourob, A. Lakhtakia
Vol. 7, 2010

Hydrogel Sensors and Actuators
Volume Editors: Gerlach G., Arndt K. -F.
Vol. 6, 2009

Piezoelectric Sensors
Volume Editors: Steinem C., Janshoff A.
Vol. 5, 2006

Surface Plasmon Resonance Based Sensors
Volume Editor: Homola J.
Vol. 4, 2006

Frontiers in Chemical Sensors
Novel Principles and Techniques
Volume Editors: Orellana G., Moreno-Bondi M. C.
Vol. 3, 2005

Ultrathin Electrochemical Chemo- and Biosensors
Technology and Performance
Volume Editor: Mirsky V. M.
Vol. 2, 2004

Optical Sensors
Industrial, Environmental and Diagnostic Applications
Volume Editors: Narayanaswamy R., Wolfbeis O. S.
Vol. 1, 2003

Solid State Gas Sensors – Industrial Application

Volume Editors:
Maximilian Fleischer
Mirko Lehmann

With contributions by

O. Ahmed · M. Andersson · C. Baratto · T. Bürgler · E. Comini · I. Eisele · G. Faglia · N. Felde · P. Iskra · D. Kohl · F. Krogmann · C. Kübel · A. Lloyd Spetz · R. Moos · C. Peter · R. Philipona · C. Pijolat · J. Polak · G. Sberveglieri · H. Scherzinger · K. Schmitt · C. Senft · T. Tille · J.P. Viricelle · J. Wöllenstein

Springer

Editors
Maximilian Fleischer
Siemens AG
Corporate Technology
Corporate Research and Technologies
CT T DE
München, Germany

Mirko Lehmann
Innovative Sensor Technology (IST) AG
Wattwil, Switzerland

ISSN 1612-7617
ISBN 978-3-642-28092-4 ISBN 978-3-642-28093-1 (eBook)
DOI 10.1007/978-3-642-28093-1
Springer Heidelberg New York Dordrecht London

Library of Congress Control Number: 2012938651

Printed on acid-free paper

Springer is part of Springer Science+Business Media (www.springer.com)

Series Editor

Prof. Dr. Gerald Urban

IMTEK - Laboratory for Sensors
Institute for Microsystems Engineering
Albert-Ludwigs-University
Georges-Köhler-Allee 103
79110 Freiburg
Germany
urban@imtek.de

Aims and Scope

Chemical sensors and biosensors are becoming more and more indispensable tools in life science, medicine, chemistry and biotechnology. The series covers exciting sensor-related aspects of chemistry, biochemistry, thin film and interface techniques, physics, including opto-electronics, measurement sciences and signal processing. The single volumes of the series focus on selected topics and will be edited by selected volume editors. The *Springer Series on Chemical Sensors and Biosensors* aims to publish state-of-the-art articles that can serve as invaluable tools for both practitioners and researchers active in this highly interdisciplinary field. The carefully edited collection of papers in each volume will give continuous inspiration for new research and will point to existing new trends and brand new applications.

Preface

Stakeholders in Gas Sensing

The field of gas sensors is fascinating for people working in many different fields. Sensor researchers work in multidisciplinary teams with specialists in material science, physical chemistry, manufacturing technology, semiconductor physics and signal processing to build innovative gas sensors. The result of these efforts is key to gas sensor products in the areas of comfort, security, health, environment and energy savings. New applications in these fields have become tangible through the availability of sensing capabilities that were not available before. Moreover, gas sensors have become fascinating for an increasing number of companies which have no gas sensor-based product up to now, since their customers are more and more interested in gas sensor applications. Many of them are characterized by a significant economic potential.

However, the commercial success up to date is often found to be delayed compared to the expectations and the potential of the upcoming technological breakthroughs.

Development of a New Gas Sensing Technology

Researchers are inquisitive. They provide new creative ideas for sensing principles and readout mechanisms. However, the detection of a new physical effect eligible for sensing, by itself, is generally of limited value. Such detection principles often work well in a laboratory environment under simplified conditions, e.g. when gas X is diluted with clean and dry air. In a real application environment with changing temperature, humidity, various additional reactive gases and sometimes even strong corrosive effects, the result might be completely unsatisfactory. Such cases are often the reason that new effects are not followed up on and the economic benefit is lost.

However, the discovery, characterisation and cultivation of a new effect can be very promising, provided the existence of a strong belief of all participants in the success of this detection principle to detect gas X in a better way as well as an economically sound application scenario. The basis for success in a gas sensing application can usually be found in the choice of the appropriate physical–chemical effect that has the potential to function reliably under the given conditions of the application.

The follow-up of a new effect is even more reasonable when there is a chance to create a sensor platform that is based on this effect. Gas sensor markets are often niche markets. Due to the high cost of the commercialisation of a new sensor technology, it is beneficial to base the industrial launch on a multiple application target. The ability of a sensing principle to be adapted to different gases (e.g. by the exchange of the receptor surface, by the change of an electric control voltage or by the choice of operating temperatures) increases its chances for industrialisation. From an economic point of view, it makes much more sense to have a homogeneous but adaptable sensing technology for different applications than to approach different applications with a bundle of heterogeneous sensing principles. This rule retains its validity, even when it is a common knowledge amongst gas sensing specialists, that the heterogeneity of gas sensing applications does not allow for a "universal" sensing platform. As is the case with nearly every engineering development, we have to live with the best trade-off.

Implementation of a New Gas Sensing Application

To work efficiently, the scientist needs to be familiar with the specifications of the application. However, it also has to be made clear that the end user or financial investor needs to understand the capability of the technology as well as the timelines required to accomplish something profoundly new.

A person who is defining a gas sensor application might give a specification like: The sensor has to output a realistic gas value 1 s after power is supplied. For that request, the application engineer should scan the state of the art to find an answer for that specific problem. There might be a quick and straightforward existing solution but experience shows that this is rarely the case. Often the requirements cannot be met with an existing technology. At this point, the research scientist could step in and devise a new functional principle which fulfils the need. Unfortunately, the development of a new sensor technology lasts several years, and the application realisation according to the user or investor usually cannot wait that long.

If there is no technology available for a special application and it is not anticipated in the near future, the application scenario might be cancelled accompanied by financial losses. This can be avoided in many cases by using additional means which, in a particular way, lessen the requirements of the sensor itself in a confined way. Usually this is possible when application know-how is used to overcome the weaknesses of the sensor. For example, the existing drift of a sensor might be compensated by predicting and not measuring the gas concentration value in specific situations of the application. Such application information should be included in the evaluation of the sensor signal to make the sensor suitable for the application. We generally recommend in these situations to evaluate other gas sensor products/solutions and to learn from experts about how they realized a solution instead of giving up a promising product development. It must always be kept in mind that the gas sensor development can be a long, stony way but it is worth it if you want to stay ahead of competitors.

As a consequence we have to state that:

- Application restraints always require a pre-selection of appropriate operating principles.
- The selection of a sensor necessitates a compromise between application, performance, and cost.

For the successful implementation of a new gas sensing application, three groups have to collaborate. First, it is the application engineers who have to create a system that fulfils the requirements. They need intense exchange with the second group, the research scientists who know what a sensing principle is capable to deliver. The application engineers also need a strong exchange with the third group, the sensor suppliers to ensure that they get components with the sufficient performance at the required standard of quality. The sensor supplier needs to collaborate intensely with the research scientists to understand which production parameters are essential for a given sensing principle in order to be able to provide sensors with a sufficient quality standard at reasonable prices.

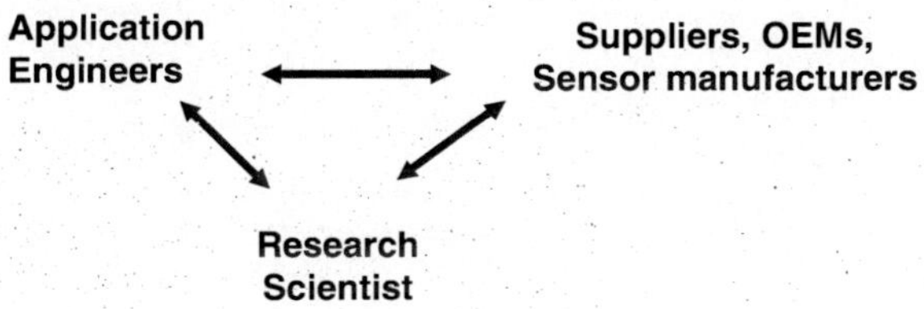

Better Communication Amongst the Stakeholders Is Needed

As stated above, a precondition for speeding up the development of a gas sensing application is strong interaction between all three stakeholders. The sometimes perceived suboptimal progress in the development of gas sensing technologies into real applications is coherent with the notion that at scientific conferences there is a remarkable participation of researchers but only a few users and application engineers can be seen. Vice versa when we visit trade fairs, we see many users but way too few scientists.

This compendium intends to contribute to open communication amongst the groups. It starts with a visionary view on how life can be improved in future buildings with the power of gas sensors. Then several contributions discuss the requirements for different application areas. The subsequent chapters address current trends in new sensing principles, followed by elaborations on how new sensing principles can be used for innovative applications. Renowned representatives report their experiences and expectations from research, applications and industrialisation in the most interesting fields of gas sensors.

This book focuses on what the research community labels as solid state gas sensors, where a gas directly changes electrical properties of a solid, which serves as a primary signal for the transducer. Therefore electrochemical, mass/viscoelastic, optical and calorimetric effects are out of the scope of this topic and have not been included.

The editors thank all the contributors of this book for their valuable chapters, and hope that it will leave the reader with a better understanding of gas sensors and encourage future research and investments in the fascinating world of gas sensors.

Munich, Germany Maximilian Fleischer
Wattwil, Switzerland Mirko Lehmann

Contents

Future Building Gas Sensing Applications

O. Ahmed

Abstract It is often thought that technology drives innovation – a process that creates great values from ideas. But historically, it is a great vision that drives innovation and seeks for technologies as enablers. Take computer as a technology. "There is no reason for any individual to have a computer in their home," said Mr. Ken Nelson, the then president of Digital Equipment Corporation or DEC in 1977 (Nova Online. "Traveling Through Time – Part 2". Public Broadcasting Service. http://www.pbs.org/wgbh/nova/time/through2.html). Although immensely successful in minicomputer business, DEC failed to imagine what Home PC can do, and hence, unfortunately, saw one of the most dramatic demise of a technology company. It was lack of vision and imagination and not technology that led to such demise. Currently, Apple's success in mobile industry is driven by Steven Jobs' vision of delivering consumer infotainment by bundling technologies with flair and sleek products.

Therefore, it is important to paint a vision of what gas sensing can do in the future (Hagleitner et al., Nature 414:293–296, 2001; Moos et al., Sensors 9:4323–4365, 2009). What are the possibilities? How it can touch and benefit environment and human lives? Below is an attempt to paint vision through common-life examples. The goal here is to inspire and motivate us to think or dream big as how sensing can change our daily life experience. The dream should not be bounded by what can be achieved or not. That shall ultimately be governed by the real-life constraints such as costs, available technology, etc. But such lofty sensing vision should set a roadmap by defining requirements and seeking solutions for future.

This chapter presents a vision of solutions, particularly with gas sensing, that are achievable today with various matured and emerging technologies. The landscape

O. Ahmed (✉)
Global Head, Strategic Collaboration- Smart Buildings, Siemens Industry, Inc., Building Technologies- BAU, 1000 Deerfield Pkwy., Buffalo Grove, IL 60089, USA
e-mail: Osman.ahmed@siemens.com

M. Fleischer and M. Lehmann (eds.), *Solid State Gas Sensors – Industrial Application*, Springer Series on Chemical Sensors and Biosensors (2012) 11: 3–12,
DOI 10.1007/5346_2011_13,
Published online: 11 February 2012

of sensing solutions is changing fast as more and more technologies are becoming commercially available at even faster rate. Hence, once again, it is our vision that should set the pace of what we can expect tomorrow. So, let us keep imagining.

Keywords Application, Building, Built-environment, Carbon footprint, Environment, Future, Gas sensing, Innovation, MEMS, Microsystems, Sustainability

Contents

1 Vision of Sustainable Built Environment

It was early in the morning, Adam wake up but not at the sound of alarm clock buzzer but by a voice that comes from his cell phone alerting him that his newly born baby's room has unusually high level of total volatile organic compounds (TVOC). Adam jumps from his bed and rushes toward that room. He takes a deep sigh of relief as he noted that his little daughter is sleeping well. He then checks on a new web-enabled sensor that not only measures TVOC, CO_2, CO, T, and relative humidity (RH) in that room but also transmits such data to his smart phone should there be any level of any gas that requires human attention. Adam checked the TVOC level and found that reading is somewhat high but nothing to be alarmed about. He then, using his smart phone, downloaded TVOC reading over last week and actually noticed that the TVOC level is slowly creeping up. Smart sensor on the wall really did its job by noticing gradual change in the TVOC value and sending an alarm message to the Smart Phone. This was great – Adam said to himself. He recalled that recently he painted this room with so-called Non-VOC type, but it seems that the paint is not completely free of volatile organic compounds (VOC). He decided to wait for few more days for the paint to dry before he puts his baby girl back into the room.

While Adam was on the train, he checked his Smart Phone which has a built-in sensor arrays for a variety of information besides the outside temperature and humidity such as any indication of any toxic substance or flu virus. He sat down and dialed a sensor address at an apartment where his 82-year-old mother lives by herself. He found there is no indication of alarm and all the readings from the sensor suites are normal.

Adam works in a hospital as a physician. As he enters through the hospital door, he gets a summary report of control and containment in the hospital with respect to any pathogens, biological, chemical, and radiological agents, and any disease that may be injurious to public health. The Smart Phone notifies Adam that the indoor environment quality or IEQ in the vicinity of Adam's office is good. There is no sign of any exposure of any gases or substance that may be injurious to Adam's health. It verified that the virus-monitoring system within the building is active and there is no report of any breakout of any virus within the building.

As Adam heads toward the lunch room, his Smart Phone continuously gets an update from the entire kitchen and food management systems notifying him about the freshness [1] of "Today's Catch" of Red Snapper but warns him about not to think about the frozen custard since the freezer temperature is not cold enough. The Smart Phone assures that he can drink from the faucet and does not need to buy any fancy expensive bottle water.

During the day break as Adam wants to use the toilet and he is close to it, he gets an alert that essentially asks to use another toilet because the smell inside is not acceptable due to malfunctioning of bathroom ventilation system.

What is described above may sound like a science fiction, but the description is a realistic projection of a home or building automation system of the future that is more integrated within our life to protect us and provide us a comfortable environment by using granular level, pervasive, and ubiquitous sensing.

2 The Role of Sensing Applications in Next Generation Building

The above vision can be only realized if sensing, particularly gases, become central to the future building and home automation systems. Sensing is fundamental in sustaining our life. With the advent of digital control systems in late 1970s, the sensing applications have seen exponential growth both in terms of its scope and utilization. Today, the basic sensing element is now coupled with a microprocessor that can process the sensed signal right within the sensor itself. The sensor communicates digital signal over building automation system's communication network. With the availability of inexpensive microprocessor and advancement in sensing element and sensor manufacturing technologies using latest complementary metal-oxide semiconductor (CMOS) and other integrated circuit (IC) processes, the overall sensor cost has come down significantly. The sensors are also integrating wireless communication chip and as a result offering more versatility and affordability by reducing the overall installed cost. Today's Micro-Electro Mechanical Systems technology (MEMS) allows to create an array of sensors (4–5) all into a single chip combined with the microprocessor, wireless communication, and power that can be harvested or scavenged from ambient sources, such as light, sound, or vibration, and thus not requiring plug-power that conventionally comes from fossil fuel source [2]. The power source can even come from a micro-fuel cell providing energy for years. Similar trend of converging functionalities into a single chip even to a smaller scale is evolving using nanotechnology. The technology platform [3] that can combine different functionalities using MEMS or Nano is known as Microsystems or Nanosystems, respectively. The fundamental technical advantages of such a system are better accuracy, precision, robustness, repeatability, and a smaller footprint of a sensor that is often a huge advantage to use sensors in an inconspicuous way. But what drives the advent of Microsystems or Nanosystems are compelling business reasons – a single chip that can sense multiple variables means lower manufacturing cost and combining wireless communication and self-powering means eliminating the need of field wiring altogether that can lower installed cost by as much as 70–80%. The rush is on to make the tiny microsensor or nanosensor to be almost self-sustaining with added features such as self-configuring, self-commissioning, self-healing, etc.

So, today's building automation systems (BAS) can measure, monitor, and communicate sensed values more easily and cost-effectively within a building. Using powerful computers and mass storage devices, the BAS can effectively process sensed values in creating rich and useful applications and knowledge. The building industry, which traditionally is very conservative, is slow to adapt, and first-cost conscious, finally having a major reawakening driven by the potential future that can be realized with the use of various types of gas sensors. Within buildings and outside, personal mobile device can sense, track, and monitor level of healthiness as captured in Fig. 1 depicted by different elements of sensing applications within the "Health" box or safety level as indicated by elements within

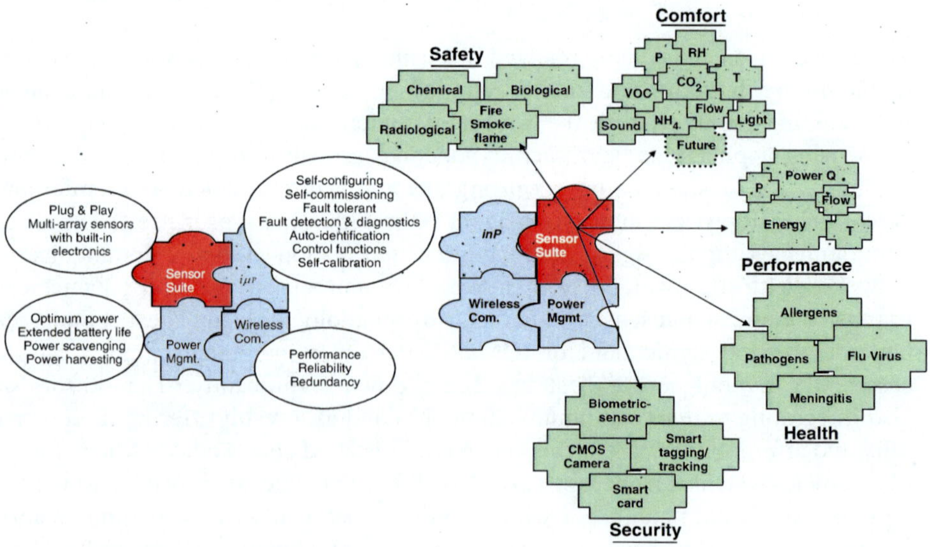

Fig. 1 Microsystems platform for sensing applications

the "Safety" box. Using same building blocks of Microsystems, various types of sensing applications can be packaged as shown in Fig. 1 [4].

While the future sensing applications are only limited by human imagination, the current scope is dictated by the legacy of the past. Although the drivers for rapid applications growth are plentiful, buildings are still lethargic and slowly overcoming the inertia of embracing new sensing technologies. However, this will change as soon as the industry reaches a threshold where demand for new applications exceeds past expectations. Following is an attempt to discuss such solutions and applications in today's buildings and beyond and more interestingly, what to expect in the future – only possible because of current and future sensor technologies.

Today, the most common sensing application in a building is dominated by space temperature control. Occasionally, relative humidity is added as a comfort variable. In a temperature centric building control system, a three-tier BAS is used. At the lowest or local level, the goal of sensing application is to maintain space temperature either by varying the flow rate of cooling or heating medium while keeping the temperature of that medium fixed or by varying the temperature of that medium while keeping the flow rate fixed.

In the future, sensing applications in a sustainable building will dramatically change. For the local comfort, additional variables will be added such as indoor air quality, often measured by CO_2 and TVOC. The result will be a complex mix and use of multiple variables that need to be sensed and controlled, and such application requires a new technology platform, for example, a Microsystems or a Nanosystem. Most dramatic change will emerge as a result of building's ability to sense the presence of individual occupant at a specific location and tailor a comfort, security, and safety solution that is suited for that particular individual via Smart Phone applications, for example. This will require execution of advanced algorithms such as adaptive learning by observing past behavior of an individual occupant. Imagine that in case of an actual fire breakout, the BAS activates floor lighting, just like in an aircraft, and guides people in the surrounding areas to the nearest fire exit. Imagine a simple building water cooler that has embedded Microsystems chips that measure, process, and communicate water flow rates, water temperature, and even water quality to a BAS for further system level processing. The results can detect problems with water quality or flow or even not having right temperature. Fixes can be made to provide fresh and good quality water throughout the building.

Building, in addition to its traditional role of providing comfort, shall be able to offer more granular knowledge-based solutions that are related to healthy environment, security, and life safety. The various embodiments of how the future building sensing solutions will look like are shown in Fig. 2 [5].

Using Microsystems as a technology platform [6], buildings will be able to offer three categories of solutions such as for healthcare, building environment that includes operation, and overall sustainable environment.

Based on vision and above examples, it is clear that in a future building and home, more advanced applications will be needed that can be categorized as

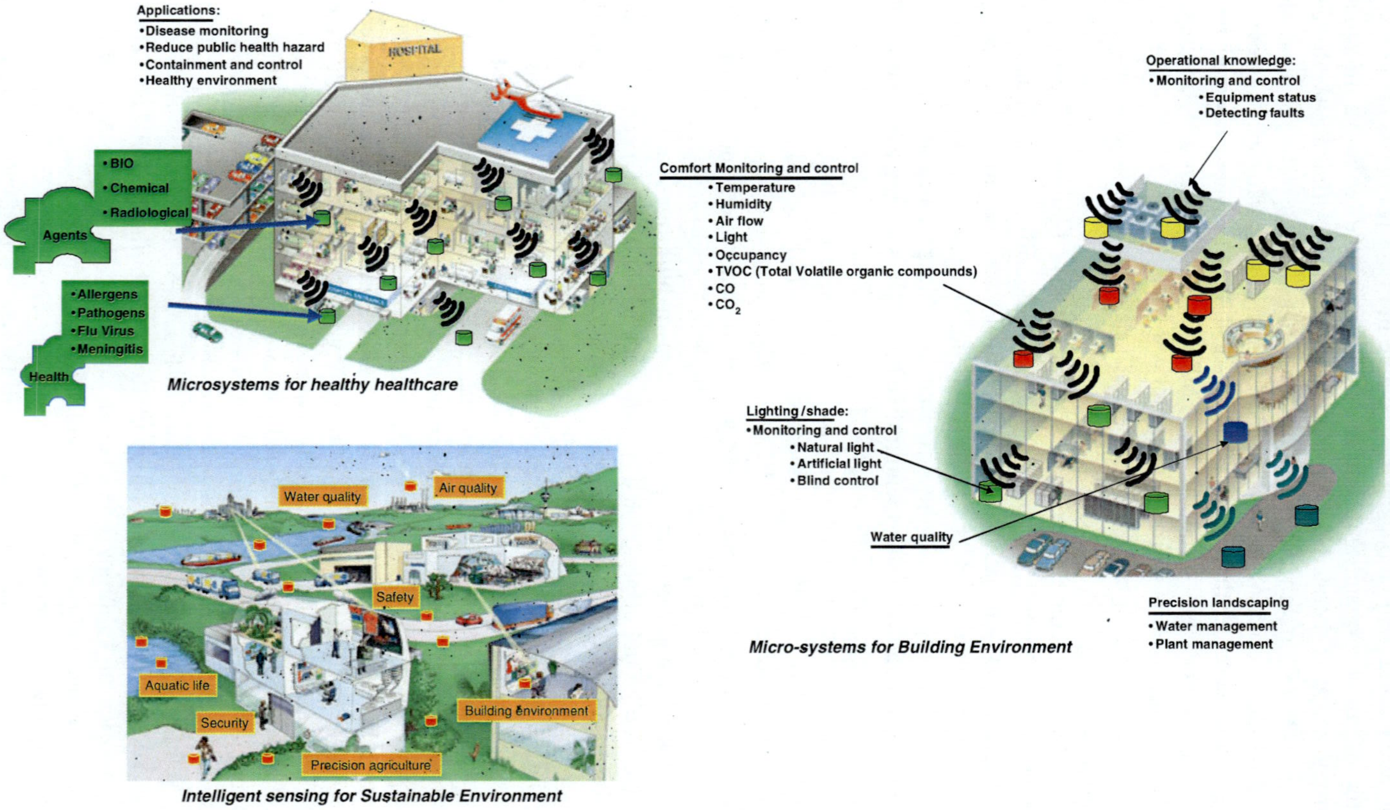

Fig. 2 Various embodiments of building microsystems environment

follows and includes both traditional domains such as comfort and energy management but new domains such as health and building that knows its occupants.

3 Sensing Need for Specific Applications

3.1 *Comfort*

Emphasis will be to measure comfort more routinely as compared to the present practice of measuring temperature. Sensing needs to calculate comfort shall include more variables such as relative humidity, TVOC, and CO_2. Mean radiant temperature (MRT) is also a variable that may be needed to measure in order to quantify "comfort" accurately. In some special cases, other chemicals may be needed to be measured such as CO or Ammonia (NH_3). The definition of comfort may even include light level, both natural and artificial, and sound.

3.2 *Health*

In certain facilities such as health care, educational, airports, train stations, etc., outbreak of air/water-borne diseases is common. In such facilities, health sensitive sensing may be needed for early detection of such viruses that may pose serious health risk to the population or even cold flu virus that may impact a company's productivity. So the sensing need for a healthy environment will become more prevalent. A future building management systems (BMSs) need to sense, monitor, analyze, and report on the overall health of the building environment.

3.3 *Energy Management Applications*

Sensing needs for energy management applications have seen tremendous growth in recent years. The building energy performance is a core functionality of such applications that require sensing, monitoring, and analyzing building's energy consumption and performance characteristics. For that building senses power consumptions at a main or primary power control panel but also more and more throughout the building at the equipment level such as chillers, fans, pumps, and lighting. A more granular tertiary level power monitoring is emerging and that is to measure power consumption for every electrical outlet. The level of sensing should match the overall goals of energy management applications in terms of what types of knowledge can be extracted and utilized by the applications in order to enhance overall building energy performance.

3.4 Security

Next generation buildings will require advanced security above and beyond what has been the norm today for basic property and asset and people protection such as burglary and thefts. Terrorism threat, protection of corporate intelligence, and need to safeguard people from violent criminals who are often driven by a social cause or personal grudge or revenge are driving the need for next generation building security. The scientists and technologists are working to secure a landmark building against a possible attack by a hijacked airline that closely approaches and endangers building by remotely taking over the control of such an airplane in order to avoid collision.

More incremental approach shall see wide usage of smart cameras, RFID or similar technology to track and tag people, biometrics, real-time data analytics, and extracting knowledge from video and picture streams. The overall results can then create several layers of security that are needed to protect people, asset, specific locations or zone, building, and the entire enterprise. Tracking and tagging people, without compromising privacy issue, is a key sensing requirement that can make significant improvement in overall security of company's intellectual intelligence and data by linking people to location, time of day, specific usage of a device, and activity.

3.5 Life Safety

Building life safety such as fire protection applications will continue to use core-sensing technologies such as photoelectric, ionization, CO, temperature, rise in temperature, and optical for fire detection. The applications fields in fire protection will become very interesting again because of the ability to combine sensing, processing, and communication functionalities in a single platform.

Examples are detecting or pinpointing the root cause of fire and its location or eliminating false alarms. Both goals often conflict to each other. For example, cooking smoke, excessive heat, welding smoke, fibers, dust, excessive moisture or heat, process exhaust gas all can falsely indicate fire, but at the same time smoke from cooking can also really indicate the presence of a fire. In order to make a distinction between false and real fires, real-time processing of advanced algorithms that is built into sensing is becoming more prevalent.

On top of things, life-safety sensing is going to include a building internal GPS in order to locate it. Not only that, advanced building modeling technology shall be able to create a virtual building that will provide information as what sorts of materials or processes are adjacent to the sensor. Hence, it will be possible not only to pinpoint the location of the fire but also to assess the fire hazard by knowing the surrounding environment.

The location of people and linking that information with the life-safety sensor reading and its location can create a dynamic building evacuation plan. Buildings floor like an airplane can literally light-up and instant message on mobile phone can provide direction to evacuate.

4 Building That Knows You and Your Physical Environment

A host of new applications shall emerge that will radically change the way the occupants are connected to their building environment. Using same core sensing that have been described above such as comfort, health, life safety, and security but by also sensing the location of a person and what surrounds him or her, BMSs can finally achieve the goal of knowing and serving its main clients or the occupants. The principal behind such integration between human occupant and building relies upon the fact that it is possible today to sense the surrounding, the needs of the occupant, and data that measures human comfort or health. It is possible to measure indoor environment quality, the location of a specific occupant, and also what surrounds that human occupant – carpet, building materials, exposure to any processes, equipments, external environment such as lights, sounds, etc. All these information along with the sensing can determine whether we are able to maintain the IEQ or not, and if not, what could be the main reason behind that. Could carpet be the cause based on TVOC? Or if we sense higher concentration of SO_x, then an adjacent industrial process should be investigated.

The building should have the ability to understand the behavior of its occupants and adapt to accommodate what occupants really need. Again, for example, using the basic sensing components, new applications will be able to adjust the blinds or shades in the late afternoon to provide the maximum natural light from the top portion of the window but block any direct sun on the computer screen or face by knowing the orientation of the office and seating configuration. In this example, lights need be measured – both natural and artificial and at the desk. The knowledge of occupant's preference of a cool office or warm environment in a late afternoon is also required. This knowledge can be extracted from a priori set of data that captures how occupant wanted space temperature or set comfort parameters in conjunction with the time of day and external environment conditions. The sensors, therefore, are required to measure localized external conditions. The rest will be to use advanced algorithms to map the comfort parameters that the occupant wanted and external conditions. Once such mapping is done, the algorithm then needs to optimize a solution that matches occupant's requirements but in a most cost-effective way. In our example, more natural lighting will lower the lighting cost but also increase cooling cost due to solar gain. So the BMS needs to reduce the overall energy cost but as an optimal combination of cooling and lighting energy.

5 Carbon and Sustainability Footprint

Perhaps, the most interesting areas where sensing applications will make most profound impact are related to carbon and sustainability in order to track, monitor, and manage them. A host of sensing technologies is emerging to measure carbon and green house gas (GHG) emissions. But an interesting factor that shall largely define the success of such technologies is not based on their performance but how affordably they can offer solutions in the market place. In this sense, Microsystems are already making its mark. Companies have developed low cost tiny power meters that can be embedded into an electrical outlet and at a fraction of a cost of regular meter. Such meters include wireless technologies or transmit measured power consumption at the outlet via Power line carrier or PLC. Besides electrical power, consumptions of other resources such as water, gas, and waste can also be measured at a granular level. Measurement of GHGs at the source will be possible. All these measurements will provide the ability to know and understand how, when, where, by whom, and what processes carbon is being generated within or outside (i.e., landscape) a building, and hence the chance of managing carbon and sustainability. Without a carbon and sustainability footprint infrastructure as sensing in its core, it is impossible to conceptualize a carbon-based society or economy.

6 Conclusion

The future building systems will largely depend on sensing as on of its core technologies. But the technology advancement in converging sensing with other functionalities such as communication, intelligent processing, and power management is the real key to imagine and implement a variety of applications affordably. And such applications are fundamental for a true smart or intelligent building. Microsystems and the future Nanosystems are the fundamental blocks to create applications for the future buildings.

References

1. Röck F, Barsan N, Weimar U (2008) Electronic nose: current status and future trends. Chem Rev 108:705–725
2. Ahmed O (2003) Future of building systems. Keynote speech. International conference on building systems and facilities management, Singapore, September 2003
3. Fleischer M (2008) Advances in application potential of solid state gas sensors. Meas Sci Technol 19:1–18
4. Ahmed O (2008) Microsystems for buildings – innovation challenges and opportunities. Keynote speech. 2nd Industry day conference on micro and nano science platform. ETH, Switzerland, May 2008
5. Ahmed O (2005) Intelligent sensing for sustainable building, energy, and environment. International conference on sustainable energy Asia, Singapore, November 2005
6. Fleischer M (2011) Solid state and spectroscopic chemical sensors and applications. Sens Lett 9(2):706–709(4)

Requirements for Gas Sensors in Automotive Air Quality Applications

T. Tille

Abstract The implementation of gas sensors in automotive environments aims to improve the air quality for vehicle occupants. These sensors provide output signals corresponding to the gas concentration of the prevalent pollutant and the degree of odor contamination. This output signal is primarily used for the automatic recirculation control of the vehicle's Heating, Ventilation, and Air Conditioning (HVAC) system.

This paper outlines the requirements of solid state gas sensors for use in automotive air quality applications. Implementing these sensors in an automotive environment poses a number of challenges, due to the wide range of possible temperature, atmospheric pressure, humidity, and vibration profiles. Additionally, the sensors must fulfill strict cost requirements and meet high standards of reliability and quality.

Based on the example of a metal oxide semiconductor gas sensor, the technical specification, data interpretation, application criteria, and automotive suitability are demonstrated. Furthermore, a brief overview of the use of solid state gas sensors for detecting the indoor cabin air quality via odor pattern recognition is given.

Keywords Air quality, Automotive, Gas sensors, Requirements

Contents

T. Tille (✉)
Research and Innovation Center, BMW Group, 80788 Munich, Germany
e-mail: Thomas.Tille@bmw.de

M. Fleischer and M. Lehmann (eds.), *Solid State Gas Sensors – Industrial Application*, Springer Series on Chemical Sensors and Biosensors (2012) 11: 13–34,
DOI 10.1007/5346_2011_1,
Published online: 21 January 2012

1 Introduction

Air quality sensors measure air quality in order to improve the comfort of a vehicle's occupants. These sensors provide an output signal corresponding to levels of pollution and smell, which is proportional to the gas concentration. To analyze the air quality outside the vehicle cabin, sensors based on the change in conductivity of metal oxide semiconductors are primarily used. These sensors have significant advantages compared to other sensors such as mass sensitive sensors or thermal conductivity sensors. The advantages of metal oxide semiconductor sensors are high sensitivity, and therefore a low detection limit for the gas to be detected, a low dependence on humidity and temperature, a long service life and relatively low product costs. However, disadvantages are a sensitivity drift with increasing life span and a limited reproducibility of the sensor properties. For this reason, metal oxide gas sensors are used to measure relative changes in gas concentration [1], for example in the detection of air quality changes for controlling vehicle air recirculation.

The operation of metal oxide gas sensors is based on the change in conductivity of a metal oxide semiconductor under the influence of reducing or oxidizing gases. In our case, these gases are the typical traffic exhaust fumes. As the temperature is an important factor influencing the behavior of the gas-sensitive metal oxide semiconductor, the metal oxide gas sensors is generally include a gas-sensitive layer with an additional heating element. The classical manufacturing of a metal oxide gas sensor uses powder metallurgy (pills or tubes form) or thick-film technology. In more advanced methods, a microstructured silicon substrate is applied to a metal oxide. Due to the limited space to be heated, and the shorter distance between heating layer and the sensitive surface layer, the necessary heating power of microstructured sensors is lower than that of conventional sensors. The design of a metal oxide gas sensor in microstructured silicon technology is illustrated in Fig. 1 [1].

As gas-sensitive materials metal oxides such as SnO_2, ZnO or WO_3 is mainly used, as well as mixed oxides of different metal oxides. The gas-sensitive metal oxide layer is contacted by a structured metallization, for example of platinum. The metallization and the gas-sensitive layer are electrically isolated from the heating layer (e.g., platinum meanders) by a passivation. Reducing gases (e.g., CO, C_xH_y)

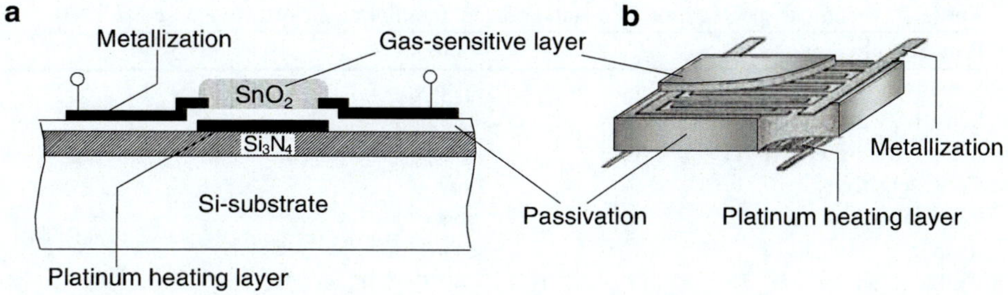

Fig. 1 Schematic illustration of a microstructured metal oxide gas sensor (**a**) Cross section, (**b**) Visualization of the metallization as interdigital structure, and the heating layer as platinum meander structure [1]

result in an increase in conductivity; oxidizing gases (e.g., O_2, NO_2) produce a reduction in the conductivity of the metal oxide, which acts electrically as an n-type semiconductor.

At the semiconductor surface (here SnO_2), oxygen is adsorbed from the ambient air. The adsorption energy level at the semiconductor surface is below the Fermi energy of the n-type semiconductor, so the electrons are absorbed by the semiconductor and negatively charged oxygen forms. With the release of electrons from the semiconductor, a depletion of electrons at the semiconductor surface is caused, resulting in a reduction of conductivity and an increase in electrical resistance R_S of the metal oxide semiconductor. If the semiconductor surface now comes into contact with molecules of a reducing gas (e.g., CO), these react with the adsorbed negatively charged oxygen to negatively charged CO_2. The negatively charged CO_2 has a higher energy level than the Fermi level, so electrons are supplied to the semiconductor, which results in an increase of conductivity. The now uncharged CO_2 desorbs from the semiconductor surface and the adsorption site can be reoccupied by new oxygen. The time available for the conductivity change is determined by desorption of CO_2 and the renewal of the adsorption sites by oxygen. In the following section, a technical specification for an air quality sensor for use in a vehicle is described.

2 Technical Specification

2.1 Functional Requirements

Table 1 gives an overview of the technical specification of a standardized automotive air quality gas sensor [2], which is explained in more detail in a follow-up.

For the detection of smell situations specific to traffic, emissions from petrol and diesel vehicles are primarily used. Representative substances in waste gases of gasoline vehicles are carbon monoxide (CO) and certain hydrocarbons (C_xH_y). For diesel vehicle exhaust emissions, nitrogen oxides (NO_x) are suitable as

Table 1 Technical specification of a standardized automotive air quality gas sensor [2]

Property	Value
Sensitivity 1st Gas (CO)	<1 ppm
Sensitivity 2nd Gas (NO_2)	<50 ppb
Sensitivity 3rd Gas (NH_3)	<50 ppb (optional)
Cross sensitivity	Low
Contamination	None
Response time t_{90}	<1 s
Ambient temperature ϑ_A	$-40°C \leq \vartheta_A \leq 85°C$
Storage temperature ϑ_S	$-40°C \leq \vartheta_S \leq 90°C$
Relative humidity H	$5\% \leq H \leq 95\%$
Operating voltage V_B	$9.0\ V \leq V_B \leq 16.5\ V$
Maximum voltage V_{Max}	18.5 V (1 h); 26.0 V (1 min)
Power consumption P_{Max}	<1 W
Interface	LIN 2.0 [3]
Plug	AMP2-967642-1
Pin connector	Terminal 1: Clamp 15, Terminal 2: Clamp 31 Terminal 3: Output
Plug frequency	≥10×
Mounting	Bayonet socket and dovetail guide
IP protection classes	IP64 + IP67 [4]
Identification marking	Interchangeable inserts or labeling with ink/laser
Dimensions	≤45 mm × 35 mm × 25 mm
Weight	<20 g
Recycling	End of life vehicle (ELV) directive [5]

indicator gases, particularly nitrogen dioxide (NO_2). In order to make an appropriate evaluation of the gas concentration, the CO sensitivity of the sensor should be less than 1 ppm and the NO_2 sensitivity less than 50 ppb. Furthermore, for air recirculation control, smell situations are also of interest, which arise, for example, through agricultural conditions (manure). The detection of these situations through a multigas air quality gas sensor is optional. To identify manure smell situations, ammonia (NH_3) can be used as an indicator gas, which is detected with a sensitivity of less than 50 ppb. Other odor-specific concentrations of indicator gases or total concentrations can also be used for the detection of agricultural odors.

The multigas air quality sensor should have a constant sensitivity at high life time without the occurrence of contamination effects; the sensor signal should exhibit a low humidity and temperature dependence. The cross sensitivity to other gases should be low and well defined. In order to achieve a prompt response of the recirculation flap in the application, the response time of the air quality gas sensor should be $t_{90} < 1$ s. The ambient temperature of the sensor is in the range of $-40°C \leq \vartheta_A \leq 85°C$. The storage temperature is in the range of $-40°C \leq \vartheta_S \leq 90°C$. The sensor needs to withstand a maximum temperature of $\vartheta_{Smax} = 90°C$ for a period of 2 h (in case of vehicle repainting). The air quality sensor must be designed for a relative ambient humidity range of $5\% \leq H \leq 95\%$.

The operating voltage of the sensor is in the range of $9.0\ V \leq V_B \leq 16.5\ V$. The sensor must withstand a maximum voltage of $V_{Max1} = 18.5$ V for a period of 1 h

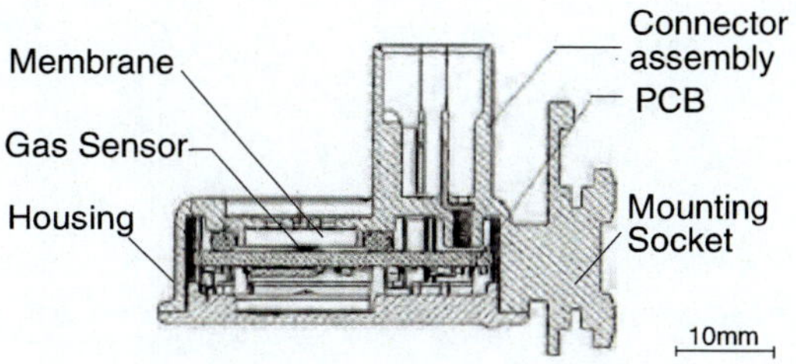

Fig. 2 Cross section of a typical automotive air quality sensor with embedded microstructured 2-channel metal oxide gas sensor and microcontroller-based evaluation circuit

and a maximum voltage of $V_{Max2} = 26.0$ V for a period of 1 min. The maximum power consumption of the sensor is limited to $P_{Max} < 1$ W. The electrical supply and exchange of data occurs over a 3-pin connector AMP 2-967642-1. Terminal 1 of the plug is assigned with clamp 15 (operating voltage), terminal 2 with clamp 31 (ground), and terminal 3 with the data output. The data interface is described in more detail in Sect. 2.2. The plug must be designed for a frequency of more than 10 plug socket attempts.

The geometry of the air quality gas sensor should be 45 mm × 35 mm × 25 mm at most, and the maximum weight of 20 g. The identification marking of the sensor (e.g., serial number) can be done either through an interchangeable insert in the housing tool or as an inkjet or laser marking. Figure 2 shows the cross section of a typical automotive air quality sensor with embedded microstructured 2-channel metal oxide gas sensor and microcontroller-based evaluation circuit.

The sensor needs to fulfill requirements for resistance against dust and splash water according to IP64 and IP67 [4]. A semipermeable membrane protects the sensor against dust and moisture. This membrane can be made from polytetrafluoroethylene (PTFE) and is built into the polyamide housing of the sensor. The materials of the sensor must be recycled in accordance with the ELV directive [5].

2.2 *Sensor Signal Conditioning*

Currently, air quality gas sensors are used in two basic forms. The first type of gas sensors is based on single elements to provide an output voltage signal, which is proportional to the conductivity change of the metal oxide. This analog signal is passed to a control unit for further processing (see Sect. 2.2.1). The second type of air quality sensors includes, in addition to the sensor element or more sensor elements, an internal preconditioning to prepare a classified sensor output signal of so-called air quality levels (AQLs) (see Sects. 2.2.2 and 2.2.3).

2.2.1 Conditioning of an Analog Output Signal

The resistance R_S of metal oxide gas sensors dependent on the gas concentration X_G is shown in Fig. 3a with the example of carbon monoxide normalized on the sensor

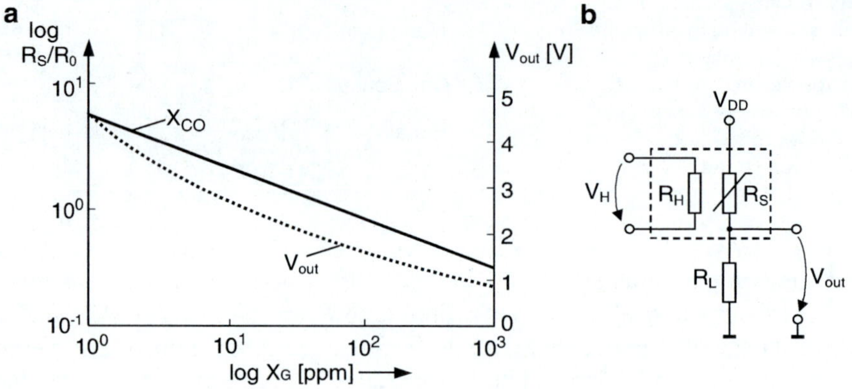

Fig. 3 (**a**) Typical characteristics of the normalized output resistance R_S/R_0 and the equivalent output voltage V_{out} of a CO gas sensor, (**b**) Simple control and evaluation circuit for a metal oxide gas sensor

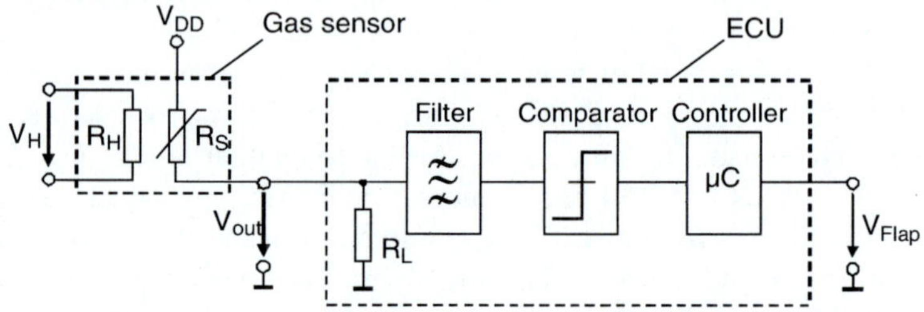

Fig. 4 Evaluation of the analog sensor output signal in a control unit (circuit diagram)

base resistance R_0. The gas concentration-dependent resistance is converted by the circuit in Fig. 3b into an equivalent output voltage signal V_{out}.

In the circuit shown in Fig. 3b, the gas sensor is supplied with a constant operating voltage V_{DD}. The heating resistor R_H of the gas sensor is operated with a heating voltage V_H. In series with the sensor resistor R_S, there is a load resistor R_L, needed to drop the output voltage V_{out}.

$$V_{out} = \frac{V_{DD}}{R_S(X_G)/R_L + 1}. \tag{1}$$

The output voltage V_{out} is passed to an electrical control unit (ECU), for example climate control unit for further processing, as shown in Fig. 4.

In traffic, sudden changes of gas concentration may occur. To evaluate these, the slope of increase or decrease of the sensor output signal V_{out} within a given unit of time is determined and compared with a predetermined threshold value. If the sensor signal exceeds this threshold in amount, then the recirculation mode of the vehicle air conditioning system is turned on. The determination of the slope can be realized by a filter (e.g., band pass). The comparison with the threshold value is done by a comparator. The digital comparator output signal is passed to a microprocessor, which outputs an activation signal V_{Flap} for the recirculation flap (see Fig. 4). In addition to the basic circuit shown, it is also possible to transfer the output signal V_{out} of the gas sensor to an integrated analog–digital converter of a microprocessor, and implement the evaluation of the sensor signal by a software algorithm.

2.2.2 Conditioning of a Pulse-Width Modulated Output Signal

The second class of air quality sensors contains one or more sensor elements, and it implements internal conditioning of sensor signals to a multilevel output signal. Here, we have a sensor system which realizes the allocation of air quality to a specific AQL through a corresponding evaluation logic which is implemented in a microprocessor. A typical characteristics of a pulse-width modulated (PWM) output signal V_{PWM} of an air quality sensor system is shown in Fig. 5. T is the period of an evaluation cycle, in which a current sensor signal value is assigned by a specific AQL. The illustrated example assumes a high-active signal. If, for example, the output signal V_{PWM} has the value high for 80% of the given time period, there is no contamination of the analyzed ambient air (AQL_0). If the output signal V_{PWM} has the value high for 60% of the period, there is a diminution of air quality based on the detectable indicator gases from the sensor system such as CO, NO_2, or C_xH_y (AQL_1). Another increment (e.g., in 20%-steps) then includes the AQLs proportional to the decreasing air quality. The example illustrated in Fig. 5 shows the change in AQLs from uncontaminated ambient air (AQL_0 for $0 \leq t < T$) to slightly polluted air (AQL_1 for $T \leq t < 2\,T$) to more polluted air (AQL_3 for $2\,T \leq t < 3\,T$). The classification of air quality is dependent on the strength of the change in concentration of the gas, or the rate of increase or decrease of the gas sensor signal.

The PWM output signal of the air quality sensor can be forwarded to a microprocessor of an ECU without major circuit measures. Here, it can be converted into an activation signal of the vehicle recirculation flap.

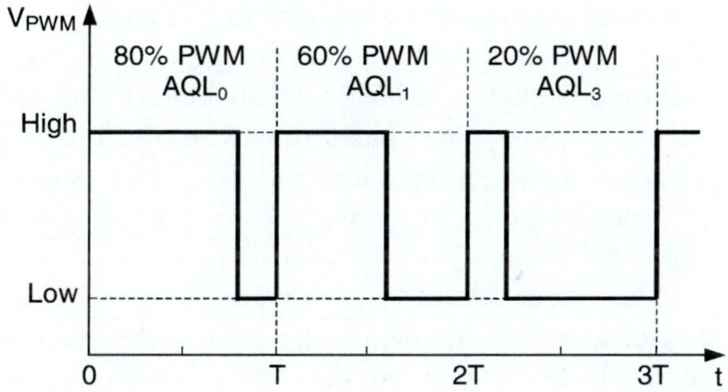

Fig. 5 Typical characteristics of the output voltage of an air quality sensor system with pulse-width modulated (PWM) output signal

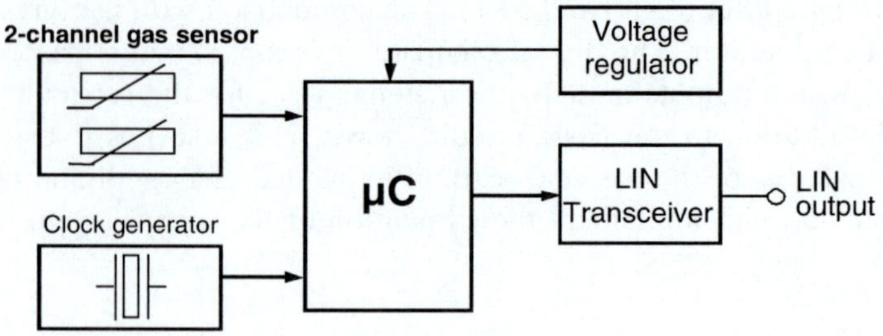

Fig. 6 Principle diagram of a 2-channel air quality gas sensor system with LIN output

2.2.3 Conditioning of an LIN Output Signal

In addition to the conditioning as a PWM output signal, the output signal can be generated as a Local Interconnect Network (LIN) signal. Figure 6 shows the principle diagram of a 2-channel air quality gas sensor system with LIN output.

The data communication between the air quality gas sensor and the vehicle electrical system is achieved via the single-wire bus system LIN, wherein the data protocol is realized in standard LIN version 2.0 [3]. The recommended data rates amount to 9,600 or 19,200 baud, with an automatic baud rate detection in accordance with [3]. The air quality sensor is a slave; the ECU assumes the master role. The slave cannot wake the master.

The following input data can be processed in the air quality sensor: outside temperature, vehicle speed, status of recirculation flap, status of fan speed level, navigation attributes, and basic sensor sensitivities. The data can be sent randomly. Depending on the signal availability (penetration rate of the vehicle), certain parts or the entire input data are processed. The more input data are available, the better the air quality sensor can determine the switching recommendation of the recirculation flap. The input data of the air quality sensor are described in greater detail in Tables 2 and 3.

The data source of the input data “status recirculation flap” is an ECU, which represents the percentage of the opening position of the recirculation flap in the vehicle air conditioning system. The “fan speed level” will also be provided in percentage by the ECU as an input for the air quality sensor. The “navigation attributes” allow the use of additional data as sensor input, which are valued in an ECU. The modes city, country, highway, and tunnel are differentiated based on different background levels dependent on odor localization between the environment scenarios. The “basic sensor sensitivity” is provided by the ECU and also serves as the set of input data that provides a target of the desired response behavior and the sensitivity of the gas sensor. If no selection of “basic sensor sensitivity” is done, the mean sensitivity as a standard in the sensor is valid.

The output data of the air quality sensor include the following messages: AQLs, sensor identification data, software and hardware level, LIN-supplier code and the node address. The output data “air quality level” is specified in detail in Table 4.

Table 2 (**a**) Input data "status recirculation flap", (**b**) Input data "status fan speed level" [2]

(a) Status recirculation flap	Data
Clean air	0001
Partial recirculation 10%	0001
Partial recirculation 20%	0010
Partial recirculation 30%	0011
Partial recirculation 40%	0100
Partial recirculation 50%	0101
Partial recirculation 60%	0110
Partial recirculation 70%	0111
Partial recirculation 80%	1000
Partial recirculation 90%	1001
Recirculation	1010
Initialization	1110
Invalid	1111
(b) Status fan speed level	Data
Fan off	0001
Level 10%	0001
Level 20%	0010
Level 30%	0011
Level 40%	0100
Level 50%	0101
Level 60%	0110
Level 70%	0111
Level 80%	1000
Level 90%	1001
Level 100%	1010
Initialization	1110
Invalid	1111

Table 3 (**a**) Input data "navigation attribute", (**b**) Input data "basic sensor sensitivity" [2]

(a) Navigation attribute	Data
Tunnel mode	0001
City mode	0010
Country mode	0011
Highway mode	0100
Initialization	1110
Invalid	1111
(b) Basic sensor sensitivity	Data
Maximum sensitivity	00
Medium sensitivity	01
Low sensitivity	10

The air quality sensor features an intelligent behavior: The sensor performs an adaptation of sensitivity to certain smells scenarios (e.g., city, country, tunnel, and highway). Furthermore, the implementation of a so-called "nose equivalent reaction" is provided, which emulates the human behavior of temporal olfactory adaptation.

Table 4 Specification of sensor's LIN output signal with air quality levels in 10%-steps

Air quality level	Data	Remark
Clean air	x0000	Normal range
Air quality level 10%	x0001	Normal range
Air quality level 20%	x0010	Normal range
Air quality level 30%	x0011	Normal range
Air quality level 40%	x0100	Normal range
Air quality level 50%	x0101	Normal range
Air quality level 60%	x0110	Normal range
Air quality level 70%	x0111	Normal range
Air quality level 80%	x1000	Normal range
Air quality level 90%	x1001	Normal range
Heavily polluted air	x1010	Normal range
Initialization	x1110	E.g., Heating phase, undervoltage, overvoltage
Invalid (malfunction)	X1111	Internal sensor error (generating DTC)
Communication OK	0xxxx	No check-sum error
Communication error	1xxxx	Check-sum error

Optionally, the use of certain input data is used to optimize the output data. Furthermore, a self-diagnosis of the sensor is required, which, for example, internally monitors the sensor elements, sensor heating, power supply and a plausibility check of sensor output values. In the case of an internal sensor error, an error entry (DTC: Diagnostic Trouble Code) in the ECU memory is generated. In addition, an algorithm for decontamination (e.g., by overheating) is provided, which prevents poisoning of the gas sensor surface.

3 Automotive Suitability

The automotive suitability of air quality sensors is verified by the following qualification tests: electrical qualification, electromagnetic compatibility (EMC), and environmental qualification. An excerpt from the specific test criteria is described below.

3.1 Electrical Requirements

Electrical stress tests are used to ensure the robustness against overvoltage, undervoltage, load dump, jump start, short interruptions, start pulses, etc. Transient overvoltages may occur in the electrical system due to the switching-off of loads and due to short accelerator tip-ins. Figure 7a shows the test pulse for overvoltage in the electrical system. One thousand cycles of this test profile are performed.

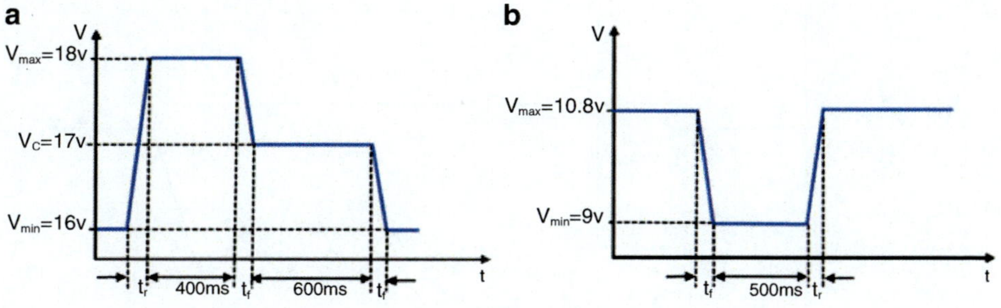

Fig. 7 (**a**) Test pulse for overvoltage, (**b**) Test pulse for undervoltage in the electrical system

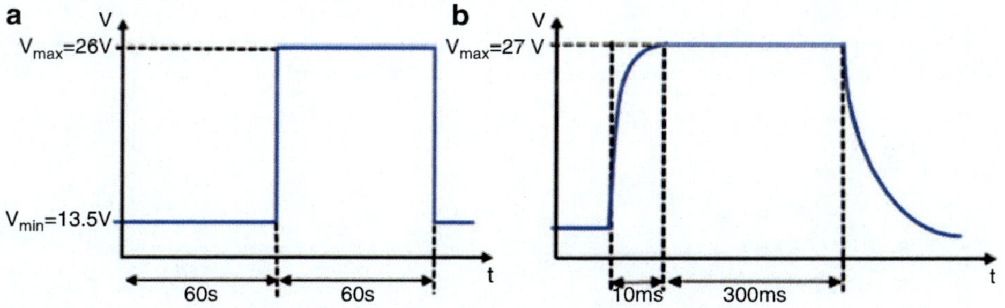

Fig. 8 Test profile for (**a**) Jump start and (**b**) Load dump

Transient undervoltages in the electrical system may occur due to switching-on of loads. Figure 7b shows the test pulse for undervoltage. One cycle of the test profile is performed.

In the jump start test, the external starting of a vehicle is simulated. The maximum test voltage V_{max} of 26 V results from commercial vehicle systems and their increased power supply voltage. Another test is the load dump test. This test simulates the dumping of an electric load, in combination with a battery with reduced buffering ability. It results in an energy-rich overvoltage pulse due to the generator characteristics of the vehicle. The resulting maximum test voltage V_{max} has a value of 27 V. Figure 8 shows the test profiles for jump start and load dump. For the jump start test, 1 cycle and for the load dump test 10 cycles of the test profile are performed.

Additionally, the robustness against the voltage characteristics of the start pulses "Cold Start" and "Warm Start" of a vehicle must be tested. A typical voltage characteristic of the start pulse "Cold Start" is shown in Fig. 9. The resulting minimum test voltage V_{min} has a value of 4.5 V for a time period of 19 ms. Ten cycles of the test profile are performed.

A typical voltage characteristic of the start pulse "Warm Start" is shown in Fig. 10. The resulting minimum test voltage V_{min} has a value of 6 V for a time period of 15 ms. Hundred cycles of the test profile are performed.

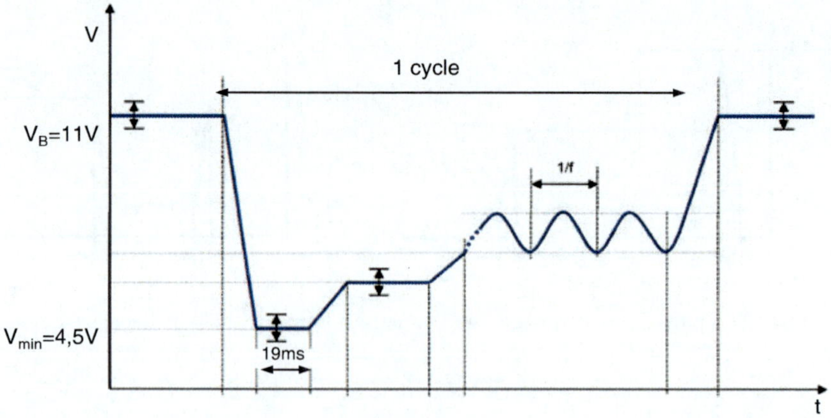

Fig. 9 Typical voltage characteristic of the start pulse "Cold Start" of a vehicle

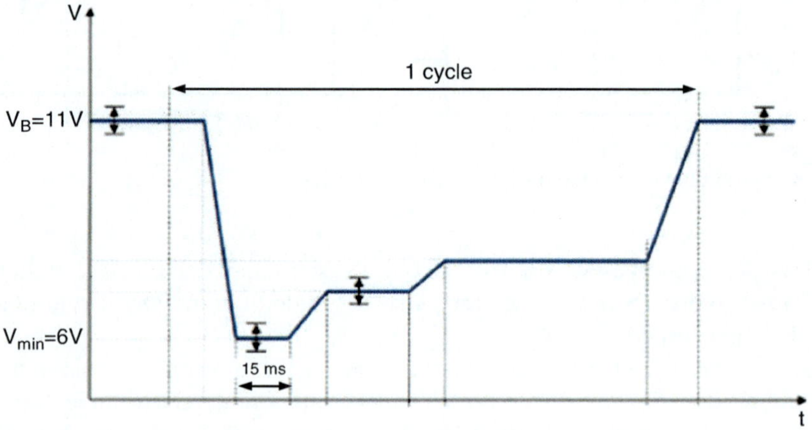

Fig. 10 Typical voltage characteristic of the start pulse "Warm Start" of a vehicle

3.2 EMC Qualification

For EMC emissions, the limit line for measurement of conducted interference from electronic components exemplarily in the frequency range of 5 MHz $\leq f \leq$ 1 GHz is shown in Fig. 11.

For EMC immunity, in the frequency range of 1 MHz to 3 GHz an electric field strength limit of 100 V/m (peak conservation) must be undershooting. Additionally, the robustness against electrostatic discharge (ESD) of the sensor must be verified. The test voltages are $\pm$8 kV for direct contact discharge and $\pm$15 kV for air discharge. The minimum number of discharges is three times each.

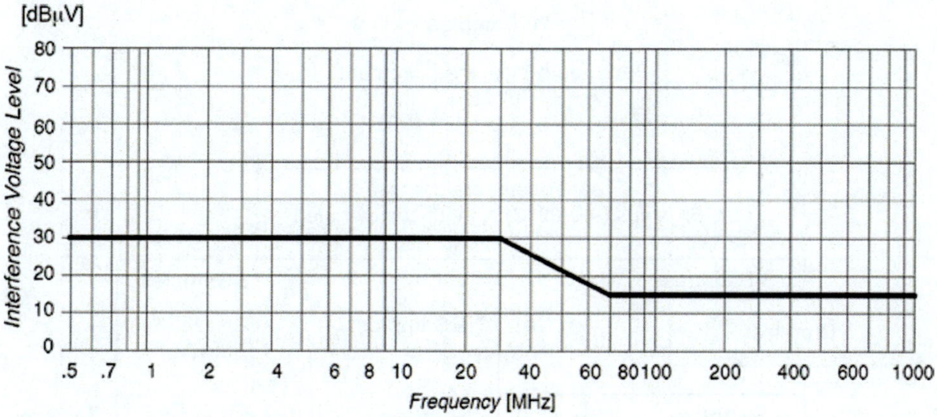

Fig. 11 Limit line for measurement of conducted interference from automotive sensors in a frequency range of 5 MHz $\leq f \leq$ 1 GHz

3.3 Mechanical, Climatic, and Lifetime Qualification

For the combined mechanical, climatic, and lifetime qualification test, 20 sensor samples are used for different tests, as related in Fig. 12.

Before and after the test procedure, a comprehensive full test of the samples is taken. During the procedure, simplified tests for functionality must be performed. The full test includes the testing of all electrical parameters and measured values at room temperature with a successively carried gas loading with NO_2 concentrations between 50 ppb and 10 ppm, with CO concentrations between 1 ppm and 50 ppm, and with a NO_2–CO gas mixture with a concentration of 500 ppb NO_2 and 10 ppm CO. The simplified function is a test at room temperature, electrical load, and a gas loading with 500 ppb NO_2.

The following shows in detail the tests that the sensors are submitted to: In a stepped temperature test, sensors are conditioned in temperature steps of 5 K in the range from ϑ_{min} of −40°C to ϑ_{max} of +85°C and subjected to a full function test. In the range from −10°C to 85°C, the samples are thereby exposed to the specified gas concentrations at the subsequent rise in temperature and rinsed with synthetic air. Figure 13 shows schematically the test profile of a stepped temperature test.

In the change of temperature test, the samples continuously pass the operating temperature range from −40°C and +85°C while being permanently tested for functionality. The tests are performed with 30 cycles per 8 h. After each cycle, a full test is required. Figure 14 shows the test profile of change of temperature test.

The use of the sensor in the exterior of vehicles requires them to be leak-proof to dust and splashing water, and the resistance to salt fog. The dust and splash water test will be performed in accordance with the required IP class according to [4]. For the salt spray test, samples are exposed to the spray per 8 h under electrical load. After a rest period of 4 h, the samples are subjected to a simplified function test at

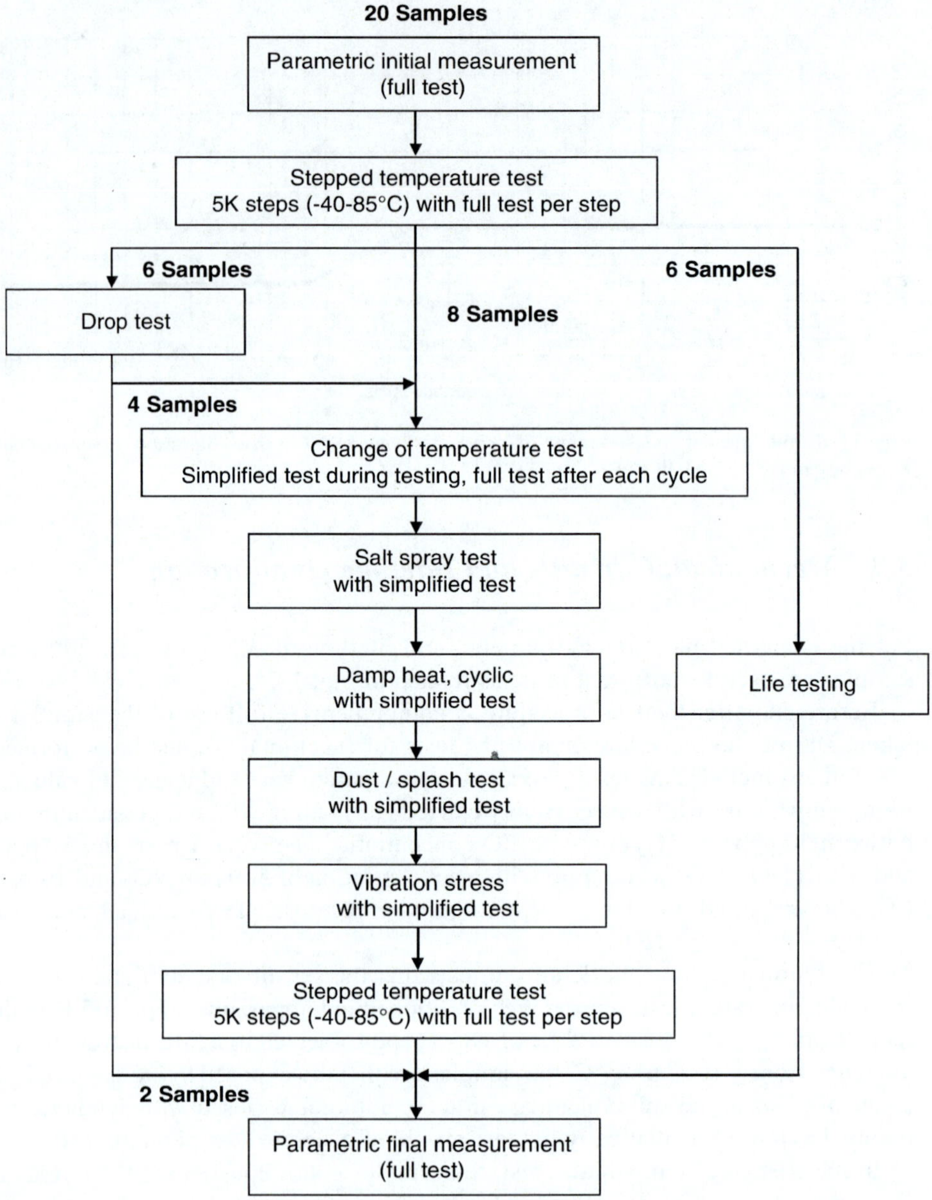

Fig. 12 Test procedure for mechanical, climatic, and lifetime classification of an automotive air quality gas sensor

$\vartheta = 50°C$. The test is carried out according to [6]. Figure 15 shows the test profile of the salt spray test. Eight cycles of the test profile are performed.

A cyclic damp heat test stresses the samples at high humidity ($H = 95\%$ r.h.) with temperature changes between 55°C and 25°C. The components are bedewed here.

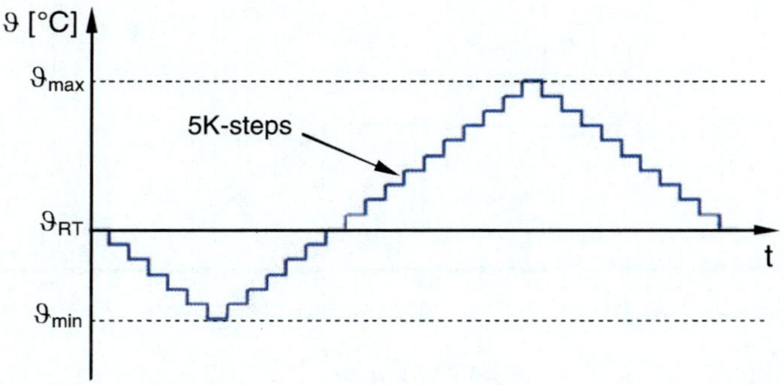

Fig. 13 Test profile for automotive stepped temperature test

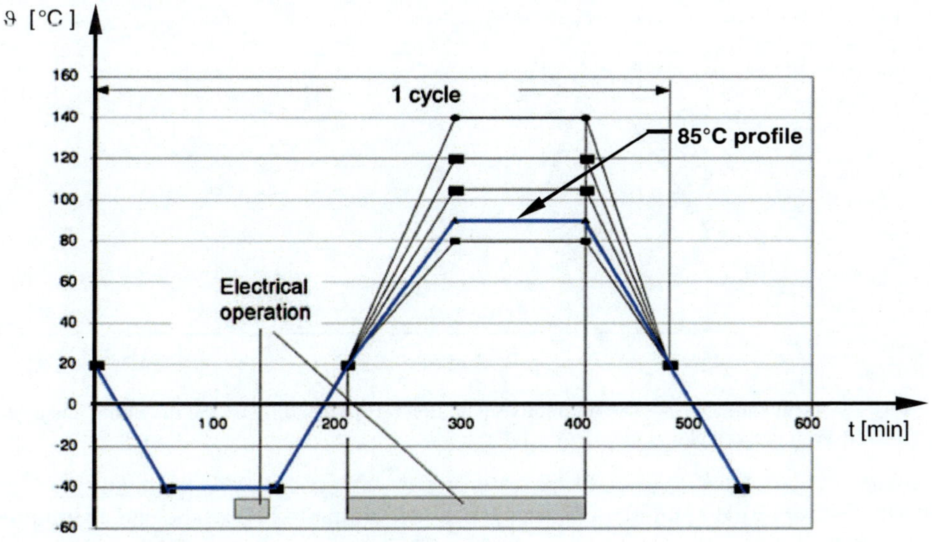

Fig. 14 Test profile for change of temperature test

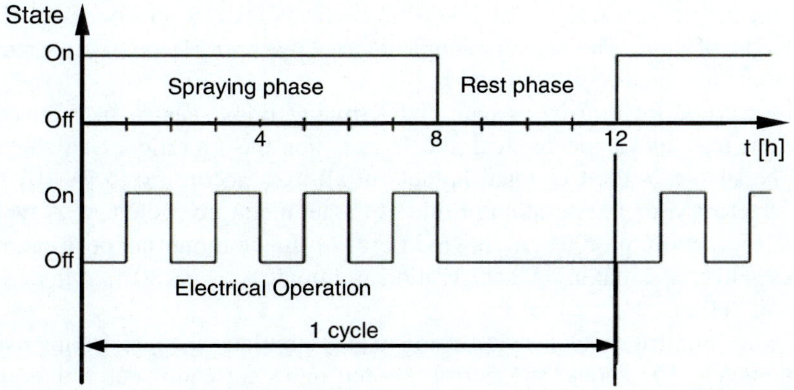

Fig. 15 Test profile for automotive salt spray test

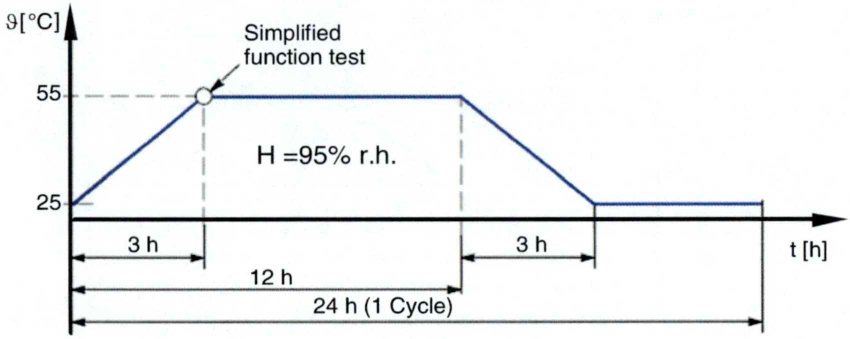

Fig. 16 Temperature profile for automotive cyclic damp heat test

Fig. 17 Typical test profile for automotive vibration test (depending on the mounting position of the sensor in the vehicle)

The test is carried out under electrical stress according to [7]. At the end a simplified function test at a temperature of $\vartheta = 55°C$ is carried out. Figure 16 shows the temperature profile of the cyclic damp heat test. Six cycles of the test are performed.

In the drop test, a sample falls twice in each spatial axis on a concrete slab from the height of 1 m. The sensors should then show no damage. The experimental conditions for this test are defined in [8].

The air quality sensor, as part of a growing body, has to be subjected to a vibration test and a mechanical shock test. For the vibration test, the samples must be swung in their original holders in all axes according to [9, 10] for 22 h with superimposed temperature profile of the temperature cycle test. A typical test profile for the vibration test is shown in Fig. 17. In the mounting position of the air quality sensor, the maximum acceleration amplitude a_{max} is 180 m/s^2 in a frequency range of 200 Hz $\leq f \leq$ 320 Hz.

The mechanical shock test according to [11] simulates the fast driving over curbs and potholes. The shocks are performed ten times for each spatial axis direction with acceleration of 500 m/s^2 over 12 ms. In vibration and shock test, the samples are electrically loaded and their values are evaluated. The lifetime test is to simulate a lifetime of 15 years. For this purpose, we propose a strain on the air quality sensors of 12,000 operating hours with 54,000 on/off cycles with a superimposed temperature cycle test.

3.4 Chemical Qualification

Through ingredient tests the chemical resistance of the sensor is tested against the following test media: water, snow/ice, acid rain, road salt, transmission fluid, motor oil, refrigerants (R134a, R152a), hydrogen, battery acid (sulfuric acid), coolant, premium fuel [12], diesel fuel [13], biodiesel [14], natural gas, brake fluid, glass cleaner, alcohol, ethanol, methanol, hydraulic oil, kerosene, petrol, preservative agents, engine cleaner to engine wash, soap, axle fluid, and starting means. The test samples are wetted at room temperature with the respective test medium and then aerated for 1 h. Oils, detergents, and preservative agents be ventilated at a temperature of $\vartheta = 70°C$. The test samples must function fully 2, 8, and 24 h after the test.

4 Implementation in the Automotive Air Conditioning System

4.1 Detection of the Air Quality in the Vehicle Exterior

The positioning of air quality sensors in the vehicle is usually near the fresh air supply of the air conditioning system or close-by the electrical fan in the engine area. A sufficient flow behavior of the sensor through the surrounding air that will be analyzed must be ensured. Figure 18 shows schematically the implementation context of an air quality sensor R_1 in a vehicle HVAC system.

The activation of recirculation control by the HVAC system can be carried out by involving additional vehicle-specific data such as outside temperature ϑ_A, vehicle speed v and closing frequency n of the recirculation flap. Figure 19 shows the principle for controlling the adjustment of the recirculation flap of a vehicle HVAC system using an air quality sensor R_1 in the outside section.

The air quality sensor R_1 shown in Fig. 19 is needed to capture the ambient air quality and to control the recirculation flap of the HVAC system. The automatic activation of the recirculation or the control of the flap actuator M_1 via an electronic control unit (ECU) is realized by analyzing the output signal of the air quality sensor R_1 and generating the control signal α_U for the flap actuator. In addition to the current AQLs, further information is considered by an evaluation algorithm to activate the recirculation operation. These include the outside temperature ϑ_A (measured by thermistor NTC_7) and additional vehicle-specific data such as vehicle speed v and closing frequency n of the recirculation flap.

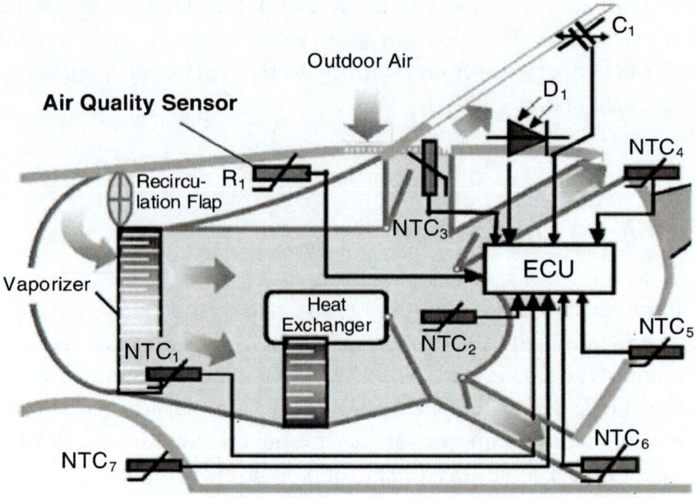

Fig. 18 Arrangement (schematic) of thermal comfort sensors in a vehicle HVAC system and their linking

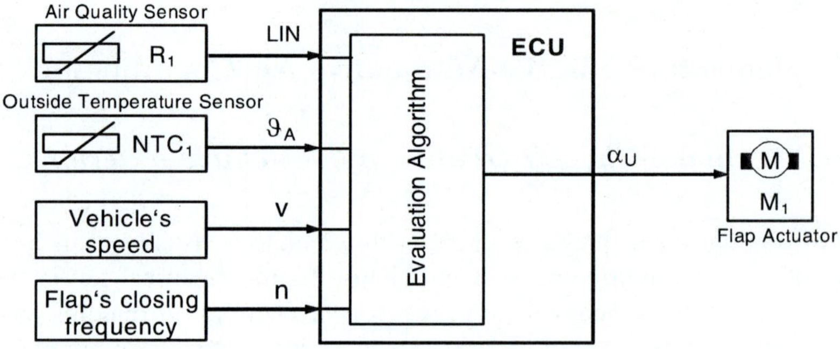

Fig. 19 Principle for controlling the recirculation flap of a vehicle air conditioning system using an outdoor air quality sensor in the vehicle and other vehicle-specific data

The main factors used currently for the detection of traffic smell situations are exhausts from gasoline and diesel vehicles. Representative substances in exhaust gases of gasoline vehicles are carbon monoxide (CO) and certain hydrocarbons (C_xH_y). For diesel vehicle exhaust emissions, nitrogen oxides (NO_x) are suitable as indicator gases, particularly nitrogen dioxide (NO_2). An exemplary characteristic is shown schematically in Fig. 20. The characteristic illustrates the CO concentration X_{CO} and NO_2 concentration X_{NO2} in the ambient air and the resulting switching behavior of the recirculation flap of a vehicle air conditioning system.

Figure 20 shows that overstepping a certain threshold concentration X_{CO} of CO and X_{NO2} of NO_2 leads to a corresponding closing response of the recirculation flap. During the time span $t_1 \leq t < t_2$, the closed flap state is caused by an increased NO_2 concentration. At $t = t_3$ an activation of the recirculation mode occurs by an

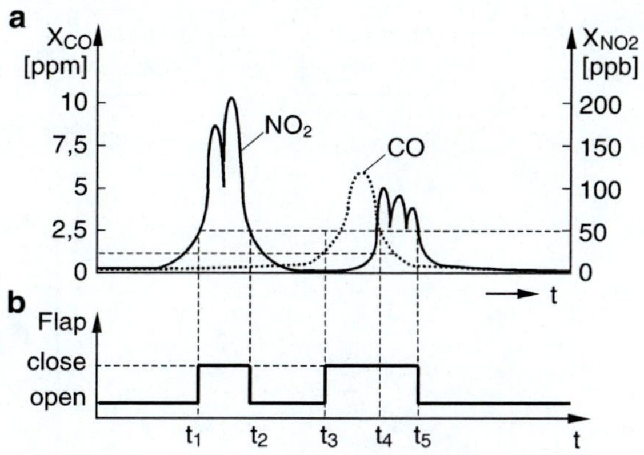

Fig. 20 (**a**) Schematically characteristic of traffic-specific concentration changes of CO and NO_2, (**b**) Concentration-dependent switching behavior of the recirculation flap

increased CO concentration in ambient air. The CO concentration will go below the threshold at $t = t_4$, but at the same time there is an increased NO_2 concentration, which results in retaining the recirculation flap in the closed state. This state is lifted only at $t = t_5$ when the NO_2 concentration is below the threshold concentration. Sensible detection limits for the corresponding traffic-specific smell situations are, for example, 50 ppb for NO_2 and 1 ppm for CO. Also C_xH_y with a detection limit of 1 ppm can be used as an indicator gas for gasoline exhaust. The detection limits given above refer to uncontaminated ambient air (zero air). The thresholds, on the other hand, reflect a change relative to background levels of concentration. Accordingly, these switching thresholds depend on the current background concentrations of the corresponding indicator gas in the ambient air. For the detection of traffic-specific gases such as CO, NO_2, and C_xH_y, metal oxide gas sensors with SnO_2 as working layer are particularly advantageous. This working layer has good properties with regard to reaction kinematics, robustness, and manufacturability.

The air quality sensor can be designed as a single-element sensor to detect one substance or as a sensor system to detect multiple substances. The sensor system can be constructed from sensor elements that are differently doped or varied in temperature, each optimized for the corresponding gas regarding selectivity and sensitivity. Broadband sensor elements can be used as well. They can detect multiple substances simultaneously.

4.2 *Detection of the Air Quality Inside the Vehicle*

In addition to the currently used air quality sensors in the vehicle exterior, an analysis of indoor air quality is technically also possible in the future. For this purpose, the acquisition of odor patterns by multigas sensors is considered.

Odors can be classified in terms of their effect by coupling several individual sensors to a sensor array and by using the sensor output signals to generate odor patterns. The required signal processing is based on the principle shown in Fig. 21.

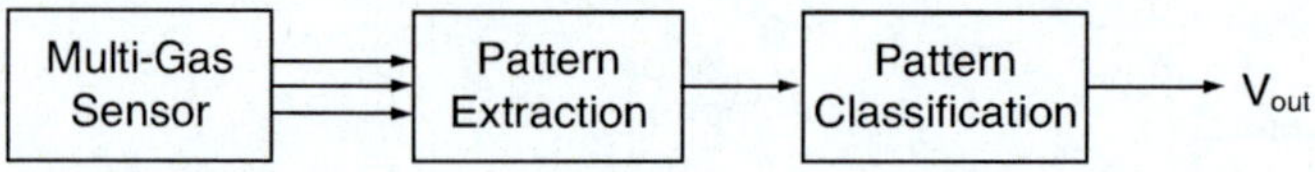

Fig. 21 Principle of signal evaluation of a multigas sensor

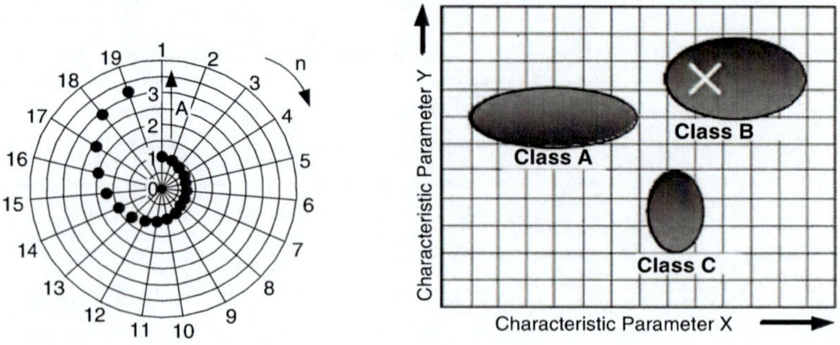

Fig. 22 Example of an odor pattern and the assignment of this pattern to a feature class (*n* number of sensor signals, *A* normalized sensor signal amplitude)

Fig. 23 Principle diagram for control of the recirculation flap of a vehicle HVAC system using air quality sensors in the vehicle exterior and inside the occupant compartment

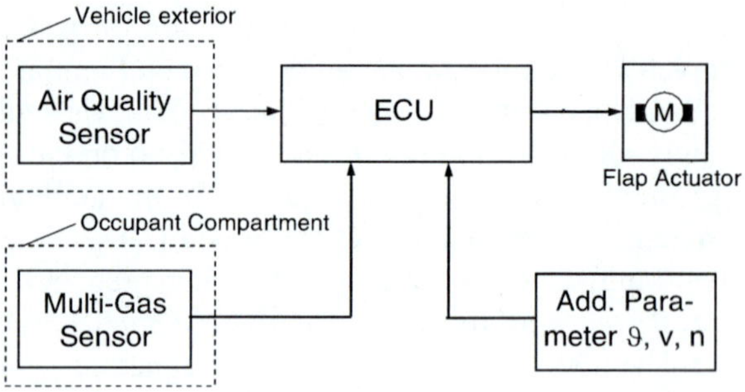

Via pattern extraction, an odor pattern which is dependent on substances and concentration is generated from the characteristics of the sensor output signals. This odor pattern is assigned to a defined class that represents a specific scenario of substantial characteristics. The classification can be carried out by the principal components analysis (PCA), in which the characteristics of the odor pattern are projected on weighted vectors and compared with a reference data set [15]. After allocation the odor pattern to a class, an output signal is generated representing an AQL associated with each class. Figure 22 shows an example of an odor pattern generated by several sensor signals, and the assignment of this pattern to a particular feature class.

The conditioning of the output signal and its further processing in the automatic climate control may be analogous to Sects. 2.2.2 and 2.2.3. Figure 23 shows a schematic circuit diagram for controlling the recirculation flap of a vehicle HVAC system using air quality sensors in the vehicle exterior and inside the occupant compartment.

5 Conclusion

In this article, the most important requirements for gas sensors using in air conditioning systems of vehicles were described. The required criteria for detecting air quality and the application to control the operation of the vehicle air recirculation control were presented. In particular, the design of sensors for the strict automotive requirements with regard to life time and environmental conditions, and integration into the vehicle electrical system were paramount.

References

1. Tille T, Schmitt-Landsiedel D (2004) Microelectronics. Semiconductor devices and their applications in electronic circuits. Springer, Berlin
2. Tille T, Kirchhoff R, Leinfelder R, Gaigl D, Wenkebach U, Kaufmann T, Wieszt H (2007) Multi-gas air quality sensor: a standard of the German automobile manufacturer. In: Hofhaus J (ed) Vehicle air-conditioning V. Expert, Renningen, pp 150–165. ISBN 978-3-8169-2766-2
3. LIN Specification Package, Revision 2.0, 2003
4. DIN EN 60529 (2000) Degrees of protection provided by enclosures (IP code)
5. BMW Standard N 113 99.0 (1997) Recycling of motor vehicles—Recycling optimized vehicle construction
6. DIN EN 60068-2-11 (1999) Environmental testing—Part 2: Tests, Test Ka: Salt mist
7. DIN EN 60068-2-30 (2005) Environmental testing—Part 2: Tests, Test Db: Damp heat, cyclic (12 + 12 hours)
8. DIN EN 60068-2-32 (2007) Environmental testing—Part 2: Tests, Test Ed: Free fall
9. DIN EN 60068-2-29 (1993) Environmental testing—Part 2: Tests, Test Eb and guidance: Bumping
10. DIN EN 60068-2-64 (1994) Environmental testing—Part 2: Test methods, Test Fh: Vibration, broadband noise (digital control) and guidance
11. DIN EN 60068-2-27 (1993) Environmental testing—Part 2: Tests, Test Ea and guidance: Shock (IEC 60068-2-27, 1987)
12. DIN EN 228 (2004) Automotive fuels—Unleaded petrol—Requirements and test methods, German version EN 228
13. DIN EN 590 (2004) Automotive fuels—Diesel—Requirements and test methods
14. DIN V 51 606 (1994) Liquid fuels—diesel fuel made from vegetable oil methyl ester (PME)—Minimum requirements
15. Blaschke M, Tille T, Robertson P, Mair S, Weimar U, Ulmer H (2006) MEMS gas sensor array for monitoring the perceived car cabin air quality. IEEE Sens J 6(5):1298–1308

Automotive Hydrogen Sensors: Current and Future Requirements

C. Kübel

Abstract Hydrogen concentration monitoring in ambient air and in exhaust gas of fuel cell vehicles is important for a safe vehicle operation. This can be done by using automotive hydrogen sensors. This chapter describes the current and future sensor requirements and their implication on manufacturing and calibration. Its objective is to provide a requirement-based guideline for future sensor development.

Keywords Fuel cell vehicle, Hydrogen sensor, Requirements, Safety

Contents

1 Introduction

At the end of 1997, GM and Opel started a global fuel cell vehicle development project. A major milestone was taken with the release of the "HydroGen3," a compact and powerful fuel cell vehicle based on the Opel Zafira. The next step to commercialization is currently being taken with the launch of more than 100 Chevrolet Equinox Fuel Cell, GM's fourth generation of hydrogen fuel cell vehicles. The Equinox is designed for operation from subfreezing to high-temperature environments.

C. Kübel (✉)
Adam Opel AG, GMAPCE, Peter-Sander-Str. 43, IPC MK-01, D-55252 Mainz-Kastel, Germany
e-mail: Christoph.Kuebel@de.opel.com

M. Fleischer and M. Lehmann (eds.), *Solid State Gas Sensors – Industrial Application*, Springer Series on Chemical Sensors and Biosensors (2012) 11: 35–38,
DOI 10.1007/5346_2011_11,
Published online: 10 December 2011

It is an important step on GM's pathway to an automotive-competitive fuel cell propulsion system.

2 Hydrogen Sensor Modules in Chevrolet Equinox Fuel Cell

Hydrogen concentration sensor modules are one major part of the vehicle's safety architecture. In case of a system leakage, they enable the safety system to react in an appropriate way. Depending on their mounting position in the vehicle, they can be grouped as follows.

2.1 Ambient Hydrogen Concentration Sensor Module (AHS)

Engine compartment
Passenger compartment
Tank compartment (under-floor)

2.2 Exhaust Hydrogen Concentration Sensor Module (EHS)

Fuel cell exhaust tail pipe

According to the Equinox Vehicle Technical Specification, the AHS and EHS specifications are very close to the *G*eneral *M*otors *W*orldwide (GMW) standards which are valid for GM's internal combustion engine vehicles. For details, please refer to the corresponding GM Worldwide standards such as GMW3172 (General Specification Electrical/Electronics) and others. These documents can be purchased at http://www.ihs.de (http://global.ihs.com).

The main properties of the current AHS and EHS are listed in Table 1.

The sensor module is designed in accordance with IEC 61508. To achieve the Safety Integrity Level 2 defined in the IEC 61508, the following measures are taken:

- Self-diagnosis routines detect internal hardware or software failures. They are executed at sensor module startup and at continuous operation.
- Self-diagnosis routines detect if the sensors accuracy is impaired. They are executed at sensor module startup and at continuous operation.

Even with the above-mentioned measures implemented, the high demands on accuracy and the conformance to SIL2 requires an additional measure during vehicle operation:

- The sensor modules have to be tested with calibration gas at regular intervals.

Table 1 Specification of hydrogen sensor modules in Equinox vehicles

Measurement range	0–4.4 Vol% H_2 in air
Accuracy at startup (t = 0–3 s)	±0.5 Vol%
Accuracy at run-time (t > 3 s)	±0.2 Vol%
Cross Sensitivity max. ±0.2 Vol%	Rel. humidity 0–100% and condensation
	CH_4: 0.1 Vol%
	C_8H_{16}: 0.1 Vol%
	CO: 1 Vol%
	CO_2: 5 Vol%
	NO_2: 20 ppm
	SO_2: 500 ppm
	H_2S: 50 ppm
	Cigarette smoke
	Combustion engine exhaust gas
Accuracy due to drift and aging	±0.2 Vol%
Accuracy when switched on in H_2	±0.5 Vol%
Response time (t_{90})	<3 s
Recovery time (t_{10})	<30 s
Power on time	<700 ms
Temperature range (ambient and gas)	−40 to 125°C
EMC	GMW standards, tested with and without hydrogen
Electrical interface	CAN bus
Sensor module weight	<0.1 kg
Sensor module size (L × W × H)	90 × 45 × 20 (mm)

Trying to summarize the experience we gained so far at vehicle operation, the specification has to be considered as conservative with respect to the hydrogen measurement properties. Concerning accuracy the modules are precision devices. This requires a complex module design and a great deal of calibration during the production process, which makes the sensor expensive.

3 Future Requirements

For GM's next generation of hydrogen-fueled vehicles, the fuel cell variant of the Chevrolet Volt, a new generation of hydrogen sensor module is required. It is intended to develop the specification in cooperation with other car manufacturers.

Measurement requirements will be reduced to an appropriate level, based on the results gained during the Equinox Project. Resolution for both, AHS and EHS, may be reduced to at least one alarm point. Accuracy requirements for both applications will decrease. Main challenges will be startup time, and for the EHS, the continuous or periodical exposure to moisture and condensed water combined with continuous or periodical exposure to hydrogen. "Automotive" requirements

(EMC, vibration, temperature shock, etc.) will no longer differ to ICE vehicle component requirements.

Consequentially the sensor modules have to be designed and produced in accordance with relevant automotive guidelines and have to meet the GMW standards. In addition, applicable laws, regulations, and industrial standards have to be fulfilled. An automotive derivate of the IEC 61508 is currently under discussion. It is called ISO WD (working draft) 26262 Road Vehicles Functional Safety. It will be available in August 2011. It is not planned to be mandatory for the fuel cell vehicles safety system or the hydrogen sensor module itself. Nevertheless, an error handling and error diagnostics system is required. In addition, the sensor module shall have an accuracy indication to determine whether the specified accuracy is impaired. The new generation of hydrogen sensor module shall be a reliable component that does not require to be periodically replaced, supervised, or calibrated.

During the design, production, and validation phases, the Global Advanced Product Quality Planning Process (APQP) has to be applied. Sensor calibration must also be included in this process. Important tools are Design- and Process-FMEA. Quality and control plans have to be developed and executed. The *A*nalysis/*D*evelopment/*V*alidation *P*lan & *R*eport (ADVP&R) process has to be implemented and executed.

The next generation sensor module has to show the potential to meet cost targets comparable to standard automotive sensor modules. To achieve this goal, a primary interest is the producibility of the sensing element:

- Is the production process suitable for mass production with high yield?
- Can the sensing elements be tested and calibrated on a substrate/wafer level?
- Are the electrical (bonding etc.) and mechanical connections to the module's electronics suitable for mass production?

A positive answer to these questions is mandatory to make the sensing element feasible for the next generation hydrogen sensor module and, later on, mass production.

Requirements for Fire Detectors

Hubert Scherzinger

Abstract In the field of fire detection, for more than 30 years the opto-electric principle was used for detecting a fire. In recent years, increasingly fire detectors are based on gas sensors. This chapter, based on some theoretical aspects, describes the different kinds of sensors used for fire detection. In the end, the role of standardization on different levels will be highlighted.

Keywords AUBE, Electrochemical cell, EN 54, Fire detector, Fire detection, Gas sensor, Standards, Standardization, Test fire, Types of fire detectors

Contents

H. Scherzinger (✉)
Shift-consult Hubert Scherzinger, Hessenweier 9, 79108 Freiburg, Germany
e-mail: hubert.scherzinger@shift-consult.de

M. Fleischer and M. Lehmann (eds.), *Solid State Gas Sensors – Industrial Application*, Springer Series on Chemical Sensors and Biosensors (2012) 11: 39–50,
DOI 10.1007/5346_2011_2,
Published online: 10 January 2012

1 Theory

One of the most important and noticed conferences in the field of Fire Detection and Fire Alarm Systems is the so-called "AUBE," which stands for "Automatic Fire Detection." The first document found dealing with gas sensor is a lecture from the 8th conference in 1982. Up to the 14th edition of the conference, an increasing number of lectures can be recognized. An overview of the quantity of lectures is shown in Table 1. The data are extracted from the proceedings list [1–7].

The very first lecture in 1982 [1] dealt with two main characteristics in relation to fire detectors, reliability and long-term stability, and is discussed in three sections. The first section reflects on how an ideal gas sensor has to look like for use in fire detection. The second section examines the kinds of gas sensors and which of them in theory are usable for fire detection. The third section discusses the advantages and disadvantages of these sensors. Also, there is a discussion on all kinds of sensors mentioned which are available but cannot be used as a fire detector. The reasons for that are commented.

In Pfister's opinion [8], fire alarms with gas sensors will be available in foreseeable time; however, he asks rightfully whether these fire alarms will offer the expected reliability with the detection of combustible gases as well as the necessary long-term stability.

In Pfister's conclusion, it is improbable that there will be a fire alarm with only combustible gas sensors. Rather it is his opinion that combustible gas sensors will be one part of sensing elements of multicriteria fire detectors. The other sensing elements will be the traditional ones, smoke and heat.

The two contributions during the AUBE 89 [2] dealt with both CO sensors and their ability to detect the test fires according to prEN 54-9. This standard describes the existing test fires used for the approval procedure for fire detectors. Other aspects mentioned are the power consumption of CO sensors and the maintenance and mass production of these sensors. Both lectures examine semiconductor sensors as well as electrochemical cells.

At AUBE 1995 one article only dealt with gas sensors [3]. Kohl and Kelleter dedicated their essay to the "question of the detection of smoldering fires by gas emission."

In the middle of the last decade of the twentieth century, both science and industry put their focus on the subject of gas sensors. This can be seen in the number of lectures at AUBE 99 [4]. In sum, 12 of the lecturers reported their experience with and knowledge on gas sensors. To go deeper, the subjects are

- Types of sensors
- Basics of combustible gases and aerosols and their contribution
- Use in multicriteria fire detectors
- Practical use and application

At AUBE'01, seven lectures were concerned with the topic of gas sensor system [5], where three lectures discussed the theoretical basics of three different types of

Table 1 Gas sensor in the aspect of the AUBE conference

Year	1982	1989	1995	1999	2001	2004	2009
No. of lectures	33	55	54	68	55	68	100
No. of lectures dealing with "gas sensor"	1	2	1	12	7	14	11
Percentage distribution (%)	3	4	3	17	13	22	11

sensors. One lecture was concerned generally with the combination of different sensors for fire detection, whereas two examined the requirements for combustible gas sensors used in fire detectors in private homes. The last contribution was application orientated and lighted up the special requirements in an Insulation Material Factory.

Like at the conference 2001, at the AUBE'04, the kinds of the contributions were different again [6]. On the one hand, additional types of gas sensors were examined and those already mentioned were examined more deeply. On the other hand, further possible operational areas and applications were described. Two aspects, which were added again, were the standardization and/or standardization of multicriterion alarm units as well as the examination of multiple sensor alarm units in the laboratory, also for the purpose of approval. The aspect of the standardization will be discussed later.

The last AUBE conference took place in 2009. At this conference, in addition to the already known sensors based on metal oxide semiconductors, new kinds of sensors and new methods were presented and discussed, such as acoustic method and infrared sensors to detect fires. Again, there were a number of contributions to the application of different sensors.

2 Applications

2.1 Introduction

Every year in Germany approximately 600 people die as a result of a fire in household. Another 6,000 are seriously injured and 60,000 are slightly injured [9]. Most of them die during night hours. The main reasons normally are not the fires and the heat itself but the poisoning with smoke. Just three breaths of the highly poisoning smoke can be fatal, the victims are unconscious in sleep and suffocate.

The majority of the fire detectors used to protect are based on proven technologies. These technologies are smoke detection using scattered or transmitted light, ionization, and heat detection. They have a high degree of reliability achieved, but further progress of these technologies is not predictable. There is a wide area of application for these detectors, for example industry plants, office buildings, private housing, hotels, and shopping malls.

In most countries, the use of fire detectors is regulated by government. The requirements for the products themselves are stated in different standards, such as ISO-standards 7240-x for a lot of countries outside Europe, or EN-standards EN54-x in Europe. Section 3 "Standardization" will give a more detailed overview to the field of standardization.

The basis for approval and certification in both series of standards is a summary of different test fires, which is shown in Table 2, where test fire 2 (TF2) up test fire 5 (TF5) are used to qualify a pure smoke detector. Multicriteria fire detectors are usually certified with additional test fires. The way and the implementation of these test fires are not uniform. There are differences between the ISO-standards and the EN-working papers in the quantity of substances used and in the qualification criteria.

The goal of using a variation of test fires is to offer a set of artificial stimuli to the detector under test to show its ability to detect a broad range of fire. For example, for a pure smoke detector the most difficult test fire to detect is TF5, because of its dark smoke. One result of this behavior is that most of the manufacturers adjust their product to fulfill the requirements of TF5, which results in a far too low alarm threshold level for TF2. This fact will be recalled later when discussing the gas sensors in combination with proven sensing technologies.

Each of the proven smoke and heat detecting technologies has itself some advantages and some disadvantages, which are discussed in the following section.

2.2 *Proven Technologies*

2.2.1 Ionization Sensors

The characteristic of an ionization smoke detector is to react very sensitive for visible and invisible smoke. Nevertheless, a big disadvantage is its sensitivity against air movement. This reduces the number of application where ionization detectors are applicable. With establishing optical smoke detectors, the acceptance of ionization detectors has dropped. This is to explain with strong national regulations for storage, transportation, and handling of the detectors on the one side. On the other side, it is because of the mindset of the people living and working in buildings supervised by ionization detectors. They do not want to stay all day in a radioactive environment, although the dose of material is quite lower than the allowed daily dose.

2.2.2 Optical Smoke Sensors

It is possible to build up an optical smoke sensor in two different ways. On the one hand, there are so-called scattered light sensors which depend on the scattering effect of light when colliding with smoke particles. On the other hand, the

Table 2 Characteristics of test fires [10]

Designation TF = Test fire	Types of fire	Development of heat	Upcurrent	Smoke	Aerosol spectrum	Visible portion
TF-1	Open cellulosic fire (wood)	Slow	Strong	Yes	Predominantly invisible	Dark
TF-2	Rapid smoldering Pyrolysis fire (wood)	Can be neglected	Weak	Yes	Predominantly visible	Light, high scattering
TF-3	Glowing smoldering fire (cotton)	Can be neglected	Very weak	Yes	Predominantly invisible	Light, high scattering
TF-4	Open plastics fire (Polyurethane)	Strong	Strong	Yes	Partially invisible	Very dark
TF-5	Liquid fire (n-heptane)	Strong	Strong	Yes	Predominantly invisible	Very dark
TF-6	Liquid fire (methylated spirits)	Strong	Strong	No	None	None
TF-7	Slow smoldering (Pyrolysis)	Can be neglected	Weak	Yes	Predominantly visible	Light, high scattering
TF-8	Low temperature black smoke	Can be neglected	Weak	Yes	Predominantly visible	Dark
TF-9	Slow smoldering	Weak	Weak	Yes (and carbon monoxide)	Predominantly visible	Light, high scattering

extinction of a transmitted light beam caused by fire aerosols is measured and judged. The disadvantages of this technology are low resistance against dazzling light or humidity. Another aspect is that this technology shows some problems in detecting dark smoke and some other disturbances. Some manufacturers have taken this disadvantage into account and applied more than one sensor to a single fire detectors. By using different scattering angles and/or light sources with different spectral colors as well as using special algorithms, it is possible to compensate these disadvantages.

2.2.3 Heat Sensors

For the use of heat sensors, we see two different types of algorithm. One possibility is to act with a fix or maximum alarm threshold level, whereas the other possibility is to calculate and judge the rate-of-rise to get an alarm signal. It is obvious that this sensor is not able to detect any smoke, and therefore it is applicable only when a fire is characterized by an open flame and not smoldering fires. These kinds of detectors are often used where disturbing environmental condition for the use of smoke detectors is usual, for example, catering areas with big amount of humidity, areas loaded by cigarette smoke, or automobile exhausts.

2.2.4 Interim Results

To sum it up, it can be said that there exist several disturbing factors in normal environment, which lead to a high rate of false alarms. That high rate of false alarms was one reason for the low acceptance of fire detectors in the past. A summary of the several disturbances is given in [11], such as cigarette smoke, natural and artificial fog, and heater fans. By developing multisensor detectors, which are more resistant to these disturbances, it was possible to reach a good acceptance in the meantime.

Despite all these efforts, these multisensor smoke detectors are still detectors not able to identify other criteria of a fire than smoke and heat.

2.2.5 Gas Detection

The most dangerous gases in the event of fire will be released in the earliest phase when a fire is developing. Conventional fire detection technologies and even some kinds of multisensor detectors are not able to detect these lowest concentrations of gases, as no smoke is released in that phase. This is a further big disadvantage of these traditional sensor technologies.

Having all the disadvantages of the proven technologies in mind, it is possible to discuss which gas sensors can be considered as a standalone solution and as an additional sensor in combination with a proven technology sensor.

2.2.6 Electrochemical Cell Based

An overview of existing sensor technologies including a list of the characteristics is given in [11]. In the field of fire detection, electrochemical CO gas sensor was used for a long period. This type of sensor shows also some disadvantages, for example, interdependencies with other gases than the monitored one, response to humidity variation, temperature influences, and limited lifetime. Even a lifetime-test due to maintenance reasons will shorten the overall lifetime of the sensing element. The existing standards take this behavior into account and state some additional requirements to show the resistance of a detector against these pollutants [10].

The question rises in which way CO-gas-sensors are able to support the detection of a fire to give a more reliable alarm signal? Linden [12, 13] shows the amount of CO, which is released during the test fires. From the results of Linden's examination, it can be clearly seen that not only the absolute value is of importance but the rate-of-rise of the signal is taken into account too. Especially the test fires with an open flame such as TF4 and TF5 do not produce a high amount of CO. Therefore, it is difficult to detect these fires by the means of a maximum CO-sensors-threshold only.

The combination of a CO-sensor and a heat sensor circumvents the weaknesses of a pure CO-sensor. By using this combination it is possible to reach a fast response for smoldering fires as well as for fires with open flame.

The next step in evolution is the combination of a smoke sensor, a heat sensor, and a CO-sensor. This combination of sensors allows detecting all the test fires as fast as possible by being resistant against pollution effects.

Most of the fire detector's manufacturers offer detectors with one or both combinations nowadays. On the other hand, the disadvantages of short lifetime of the CO sensor are still to be solved.

2.2.7 Semiconductor Based

Metal-oxide-Semiconductor-sensors are suitable for the use in fire detection by accepting the disadvantages of a high power consumption of the heating element and the dependencies of the detector to other than the gas to detect. Like a smoke detector, these types of semiconductor gas detectors are reacting to humidity and fluctuation of temperature.

Therefore, in the early 2000s, several projects were started dealing with gas sensors to reach the aims of reducing power consumption, reducing the interdependencies to environmental disturbances, or building up multisensor-arrays. An overview on public funded project in Germany can be found in [14].

One of the many number of projects was called KOBRA [15]. A new kind of gas sensor was developed in the early 2000s. The advantage of this new semiconductor sensor is the fact that no heating of the sensing element is necessary. Other advantages are an estimated lifetime of more than 10 years and the insensitivity against dust and humidity by combining different semiconductor technology in one single chip. The development is at an early stage. At the moment, we can see a sensor, which is able to detect the aerosols released during test fires and to suppress signals from disturbances. This sensor was presented to the public at the trade show "Security'08" in Essen in fall 2008 [16]. The development of this sensor is ongoing. The latest findings were shown at "Security'10" in late 2010 where no detectors based on that kind of sensor were exhibited. It seems that there is to do more research to provide a sensor which is ready for use in application.

2.3 Perspective to the Future

The situation in building automation is characterized by a couple of sensors measuring the same parameter for different applications such as

- Heat sensors for air condition and fire detection
- CO/CO_2 sensors for air quality and fire detection
- Presence of sensors for intrusion alarm system and for fire detection to influence the alarm threshold
- Time information to control various profiles within all the application mentioned

The potentials coming out of that situation by interconnecting the applications and therefore having one single detector information for all the applications belonging to a single room are evident, both the economical and the ecological ones [16, 17].

The challenge for use of combined sensors for different kind of building automation is to bring up theoretical models which shows in which way the parameters measured at the ceiling of a room have to be interpreted and which algorithm are necessary to extract reliable data for other building management applications.

3 Standardization

3.1 Introduction

The market of fire detection and fire alarm systems is a strong regulated one, by a multitude number of standards, regulations, and guidelines. This section highlights the influence and the importance of these standards and guidelines for this market.

From a European point of view the standardization of products of fire detection and alarm system is well structured and organized, whereas the regulations of planning, installing and maintenance differ from country to country. Additionally, in some countries national standards and regulation are given even for products not covered by international papers, e.g. fire brigade panel.

In the field of standardization we have to recognize that there are at least two organizations, which are influencing the worldwide market, the ISO—International Standardization Organization and CEN—Comité Européen de Normalisation, European Committee for Standardization.

These two organizations have agreed their relationship and work together in the so-called "Vienna agreement" [18]. This agreement states that as long as one party has started a work item on a standard the other one will not start a second work item with the same scope or focus. One goal of this agreement is to have the same set of requirements all over the world. This is an urgent need for manufacturers delivering all over the world to be able to offer one and the same product for all countries.

3.2 Standardization at ISO-Level

In 2001 the work item "Carbon Monoxide Fire Detectors Using electrochemical cells" was started. In the period of discussing this paper the title changed to "*Point-type fire detectors for detection of carbon monoxide*." In May 2003 the first draft of ISO_DIS_7240-6 was published for comments. Up to October 2003 the comments were collected and later discussed in the working group. In July 2004 the formal vote of the final version started which ends in September 2004. This version was accepted and published as ISO 7240-6:2004 [19].

Already in January 2005 a new work item proposal was started to work out a standard named: *ISO 7240-8: Carbon monoxide fire detectors using an electro-chemical cell in combination with a heat detector*. The first draft of this work item was published for voting in October 2005. Sums of comments were given to this draft and the discussion in the working group took up to 2007. Finally in May 2007 started the formal vote on this draft which was at least accepted in 2007-08-13.

Based on the results of these two standards in February 2006 the new work item proposal started: ISO 7240-27: Point-type fire detectors using a scattered-light, transmitted-light or ionization smoke sensor, an electrochemical-cell carbon-monoxide sensor and a heat sensor. After a working period of 16 months the draft paper was open to the publics' comments starting from 2007-05-25 up to 2007-10-25. The work on the comments is already finished. It is adopted and issued in 2009.

To sum it all up it can be said that within the ISO-papers carbon monoxide sensors based on electrochemical cells are allowed either as a standalone detector or in conjunction with other sensors.

3.3 Standardization at CEN-Level

Due to the lack of progress on ISO-level the responsible technical committee of CEN, TC 72, decided to bring up an own work item. Therefore working group twelve (WG 12) was established and got the task to bring up a standard [20]. WG 12 met for a first meeting dealing with point-type multisensor fire detectors on 2001-10-11. The very first scope as a basis for discussion was:

> *The European Standard EN 54-15 specifies requirements, test methods and performance criteria for point-type multisensor fire detectors for use in fire detection systems installed in buildings, incorporating in one mechanical enclosure sensors which shall detect more than one physical or chemical phenomena of a real fire.* The overall fire performance *shall be determined by utilizing a combination of the detected phenomena.*

Physical or chemical criteria of a real fire may be:

- *Aerosol, i.e. smoke*
- *Temperature*
- *Combustion gases*
- *And more*

The idea of this scope was to create a standard which does not take care of specific types of sensors. The goal was to write a standard to deal with the phenomenon of fire to detect instead of some kind of sensors [20].

To get a paper for formal vote the working group met for 14 two-day-meetings with intensive discussion. Finally in February 2006 the working group was able to release a draft version which was spread to the standardization bodies of the CEN-Members for comments.

At the end this proposal was rejected by board level of CEN TC 72 with the following arguments:

- Testing according to the proposal needs to build new test equipment
- No experience with new test equipment
- To high costs for approval due to additional tests

As the result of the following discussion at CEN TC 72 it was agreed to split and divide the scopes of EN 54-15 in different standards as follows:

- Part 26: Point fire detectors using carbon monoxide sensors
- Part 30: Multisensor fire detectors, Point fire detectors using carbon monoxide sensors with heat sensors
- Part 31: Multisensor fire detectors—Point detectors using a combination of smoke, heat and carbon monoxide sensors

Working group 20 took the task to write the standard EN 54-26. This first paper was published in April 2008. The period of giving comments was finished by 21 June 2008. Working group 12 got the challenge to write the other two papers starting by the paper for a standard for point-type detectors combining CO and heat.

Regardless of all these papers the initial goal to create a paper which is open to new technologies or even other technologies than carbon monoxide sensors for gas detection is still not reached. This discussion will be a task and a challenge for WG 12 for the future.

Table 3 gives an overview of the existing and anticipated standards of the series "EN 54-x: Fire detection and fire alarm systems" in summer 2010.

At least at that moment when semiconductor sensor-based fire detectors are ready for use in quantities and put on the market the standardization bodies

Table 3 Relevant detector product standards of EN54-series

Standard	Contents	Issue
EN 54-5	Heat detectors—Point detectors	2001-03
EN 54-7/A2	Smoke detectors—Point detectors using scattered light, transmitted light or ionization	2006-10
EN 54-15	Point detectors using a combination of detected fire phenomena	Rejected in 2006, to be discussed again
EN 54-26	Point fire detectors using carbon monoxide sensors	Draft 2008-4, adopted ~2012
EN 54-29	Multisensor fire detectors—Point detectors using a combination of smoke and heat sensors	Draft 2009-10, adopted ~2012
EN 54-30	Multisensor fire detectors—Point detectors using a combination of carbon monoxide and heat sensors	Draft 2009-3, adopted ~2012
EN 54-31	Multisensor fire detectors—Point detectors using a combination of smoke, carbon monoxide and optionally heat sensors	Still open

have to decide whether to write again a sum of standards, each dealing with one combination of sensors or to establish a general purpose standard dealing with fire phenomena instead of some kinds of sensors, and so coming back to the initial task of working group 12.

3.4 *Other Regulation Bodies*

In the early stage when new technologies are introduced to the applications and products the international bodies often do not have started a work item dealing with products based on that new technologies.

In this phase the national standardization bodies are allowed to start their own work item, which may be initiated by manufacturers, test laboratories, or the national standardization body itself.

Other interested parties in the field of fire detection and having comparable basics for testing and approval of products are the insurance companies. In Europe the insurance industry is organized in the Comité Européen des Assurances – CEA. This organization is also dealing and preparing guidelines for fire detectors, e.g. CEA 4021 [21]. Some of the test laboratories all over Europe belong more or less direct to the insurance companies and therefore are able to influence the preparation of regulations.

For the manufacturers of new technology products there is the possibility to sign a bilateral agreement with a test laboratory for testing and approval such a product. If a test laboratory notes that the inquires are coming from a number of manufacturers the test laboratory sometimes decides to write a paper as a guideline open to the market, e.g. VdS 2806 [21].

At that moment, when standardization bodies start their work on a specific work item these regulation papers are often the basis for the beginning discussion.

This clearly shows the usefulness and the value of the work and the results of these committees and demonstrates the effect on the following standardization activities which shall not be underestimated.

References

1. AUBE'82 Conference Proceedings. http://www.eusas.ch/publications/aube_82.html. Accessed 02 Sept 2008
2. AUBE'89 Conference Proceedings. http://www.eusas.ch/publications/aube_89.html Accessed 02 Sept 2008
3. AUBE'95 Conference Proceedings. http://www.eusas.ch/publications/aube95.html. Accessed 02 Sept 2008
4. AUBE'99 Conference Proceedings. http://www.eusas.ch/publications/aube_99.html. Accessed 02 Sept 2008
5. AUBE'01 Conference Proceedings. http://www.eusas.ch/publications/aube_01.html. Accessed 02 Sept 2008
6. AUBE'04 Conference Proceedings. http://www.eusas.ch/publications/aube_04.html. Accessed 02 Sept 2008
7. AUBE'09 Conference Proceedings. http://www.eusas.ch/publications/aube_09.html. Accessed 15 Nov 2010
8. Pfister G (1982) Detektion von Brandgasen mit Festkörpersensoren, Schwerpunkte der Forschungsaktivitäten (Detection of fire aerosols by solid state sensors, priorities in science activities) in AUBE 82 Conference Proceedings
9. http://www.rauchmelder-lebensretter.de/warum-rauchmelder.html. Accessed 03 Sept 2008
10. ISO/DIS 7240-9: 2003-10-07
11. Oppelt U (2001) In: Beall K, Grosshandler W, Luck H (eds) Proceedings of 12th international conference on automatic fire detection AUBE 01
12. Linden O (2003) Entwicklung von Prüfmethoden zur Sicherstellung der Schutzfunktion von Brandmeldern und Gassensoren. VdS Schadenverhütung, Köln. ISBN 3-936050-05-8
13. Linden O, Höleman H (1999) In: Luck H (ed) Proceedings of 11th international conference on automatic fire detection AUBE 99
14. http://www.mstonline.de/foerderung/projektliste, http://www.mstonline.de/foerderung/projektliste/printable_pdf?vb_nr=V2230. Accessed 03 Sept 2009
15. Kostengünstige Gassensoren für die Brand- und Gefahrdetektion—Kobra, www.mstonline.de/foerderung/projektliste/printable_pdf?vb_nr=V2230. Accessed 03 Sept 2008
16. n.n.; PROTECTOR 1-2/09, S. 18 f
17. Fleischer M, Katrin N (2008) When buildings come to life, pictures of the future 2008/02; http://www.siemens.com/innovation/pool/en/publikationen/publications_pof/pof_fall_2008/pof-2-2008-e-doppel.pdf. Accessed 03 Sept 2009
18. http://www.cen.eu/Boss/supporting/reference+documents/vienna+agreement/vienna+agreement.asp. Accessed 29 Sept 2008
19. http://www.iso.org/iso/iso_catalogue/catalogue_tc/catalogue_detail.htm?csnumber=38771. Accessed 29 Sept 2008
20. Brupbacher T (2004) Standardising multi-phenomena fire detectors—Status report from CEN TC72 WG12. In: Luck H, Laws P, Willms I (eds) Proceedings of 13th international conference on automatic fire detection AUBE'04, Duisburg, Germany, September 2004
21. Hesels T (2004) Multiple sensor and multiple criteria fire detectors—Current and Future test procedures in the VdS fire test laboratories. In: Luck H, Laws P, Willms I (eds) Proceedings of 13th international conference on automatic fire detection AUBE'04, Duisburg, Germany, Septmeber 2004

The Power of Nanomaterial Approaches in Gas Sensors

Camilla Baratto, Elisabetta Comini, Guido Faglia, and Giorgio Sberveglieri

Abstract The challenge of nanotechnology is to discover new effects on already known materials and to convert exciting new findings into advanced technologies that are useful for industrial applications.

In these years, researchers have achieved the ability to produce quasi-one-dimensional (Q1D) structures in a variety of morphologies such as nanowires, core shell nanowires, nanotubes, nanobelts, hierarchical structures, nanorods, nanorings. In particular, Q1D Metal OXides (MOX) are attracting an increasing interest in gas sensing application: nanosized dimension ensures high specific surface that leads to the enhancement of catalytic activity or surface adsorption. Moreover, single-crystalline structures with well-defined chemical composition and surface terminations are not prone to thermal instabilities suffered from MOX polycrystalline counterpart. All these peculiarities can help to fill the gap between research and industrial application needs, aiming at the development of a reliable, low cost gas sensor.

This chapter presents an up-to-date survey of the research on Q1D metal oxide materials for gas sensing application, addressing the preparation techniques of sensing nano-crystals in connection with their electrical and optical properties. The application as resistive, transistor-based or optical-based gas sensors will be treated.

Keywords Conductivity, Gas sensor, Growth mechanism, Metal oxide semiconductors, Nanostructures, Nanowires, Photoluminescence, Working principle

---

C. Baratto (✉), E. Comini, G. Faglia, and G. Sberveglieri
Sensor Laboratory, University of Brescia and CNR-IDASC, Via Valotti, 9, 25133 Brescia, Italy
e-mail: baratto@ing.unibs.it; comini@ing.unibs.it; gfaglia@ing.unibs.it; sbervegl@ing.unibs.it

M. Fleischer and M. Lehmann (eds.), *Solid State Gas Sensors – Industrial Application*, 53
Springer Series on Chemical Sensors and Biosensors (2012) 11: 53–78,
DOI 10.1007/5346_2011_3, © Springer-Verlag Berlin Heidelberg 2011,
Published online: 16 December 2011

Contents

Abbreviations

CTM	Charge transfer model
CW	Continuous wave
DMMP	Dimethyl methylphosphonate
MOX	Metal oxides
NW	Nanowire
PL	Photoluminescence
Q1D	Quasi-one-dimensional
RGTO	Rheotaxial growth and thermal oxidation
SCR	Space charge region
SLS	Solution–liquid–solid
SNT	Single nanowire transistor
TFT	Thin-film transistors
TRPL	Time resolved photoluminescence
VLS	Vapor–liquid–solid
VS	Vapor–solid

1 Introduction

Increasing attention on health hazard caused by pollution [1] or by terrorist attacks [2] stimulated the research on gas sensing for real-time environmental monitoring. Industrial requests are for a sensor that shows high sensitivity, high selectivity and good stability, possibly coupled with low cost and portability.

Among all possible technological approaches, conductometric gas sensors based on MOX semiconductors are the most promising to develop low cost and reliable sensors for the simplicity of the transduced physical quantity and the possibility of integration into Si technology. For conductometric gas sensor, it has been demonstrated that the sensitivity of these materials is increased as their crystallite size is decreased down to nanometer size [3], as a consequence of increased surface-to-volume ratio and of the depletion of carriers inside the grains that modifies the transport properties of these materials.

Indeed, polycrystalline materials showed temperature-induced grain coalescence [4] that affects sensor stability over time; to overcome this problem, single-crystalline nanosized MOX can be employed. The increasing scientific interest in Q1D systems such as nanowires and nanorods has stimulated their functional exploitation, and single-crystalline 1D nanostructures are nowadays emerging as building blocks for a new generation of electronic [5–7] and optoelectronic nanometer-scaled devices with superior performances [8, 9].

The fabrication techniques of homogeneous 1D nanostructures have pursued the control over shape, aspect-ratio, and the crystalline arrangement to a considerable degree, and the improvement of the synthesis methods [10] has recently achieved the fabrication of chemically nonhomogeneous 1D nanowires, heterogeneous structures, and eventually the direct integration of functional nanostructures into nano-devices.

After the first publication demonstrating the capability of NWs to detect gaseous species [11], huge research work was carried out as demonstrated by the high number of publications and conferences on this topic. The research is still debating on understanding sensing performances of MOX NW sensors: the challenge for the introduction of these materials into the real sensor market is mainly the nanowires integration with sensing platforms. This chapter starts introducing to the deposition techniques and growth mechanisms along with the sensing principles of MOX, and then is focused on the use of NWs of metal oxide for sensing application, as conductometric, transistor-based or optical-based sensor.

2 Deposition Techniques and Growth Mechanisms

Nanowires (NWs) can be prepared by bottom-up or top-down approaches. The latter consist in reducing the lateral dimensions of a film to the nanoscale using standard microfabrication techniques. The removal can be obtained, for instance, by electron beam, focused ion beam, X-ray lithography, nano-imprinting. Top-down approach uses the fully developed semiconductor industry technology and moreover the prepared NWs are on planar surface easing the manipulation and contacting issue. The drawbacks are a long preparation time and elevated costs. This technology can produce highly ordered NWs-based structures [12–15], but does not accomplish the requirements for industrialization of NWs-based devices.

On the contrary, bottom-up approach consists in assembly of building blocks at molecular level or in synthesis by vapor phase transport, solution, electrochemical- or template-based techniques. The better crystallinity and purity together with smaller diameter, the possibility to easily form heterojunctions, the lower production costs make it the most promising method for NWs fabrication. The main disadvantage is the difficulty in integration in planar structures for nano-devices development.

We can classify the deposition techniques considering the growth phase: vapor phase (Sect. 2.1) or liquid phase (Sect. 2.2)-based techniques. Furthermore, another classification can be made between catalyst-assisted and catalyst-free synthesis involving both liquid and vapor phase techniques. The growth mechanism, far from being completely understood, can be vapor–liquid–solid (VLS), solution–liquid–solid (SLS) or vapor–solid (VS).

2.1 Vapor Phase Growth

The deposition of semiconductor whiskers of micrometer dimensions from the vapor phase has been proposed almost 50 years ago. It consisted in sublimation of source material in controlled environment or by reduction of a metal halide precursor. Later on it has been extended for metal oxides NWs growth. The experimental setup consists in a tubular furnace for source material evaporation and the gradient between source and substrates allows condensation of 1D structures. An inert gas is used as carrier for vapor transport from the source to the substrates. The growth mechanisms can be VS (Vapor–Solid), VLS (Vapor–Liquid–solid).

Wagner and Ellis reported in 1964 VLS for the production of silicon whiskers [16]. It was named thanks to the phases involved, that is vapor phase of the precursor, liquid phase of the catalyst, and solid phase of the NWs as reported in Fig. 1. The liquid catalyst offers a preferential condensation site for the vapor, when the segregation limit is reached it condenses forming the NWs.

If a metal tip is found at the apex of the NW, the growth is commonly attributed to VLS mechanism, but of course this does not determine the catalyst phase during the growth.

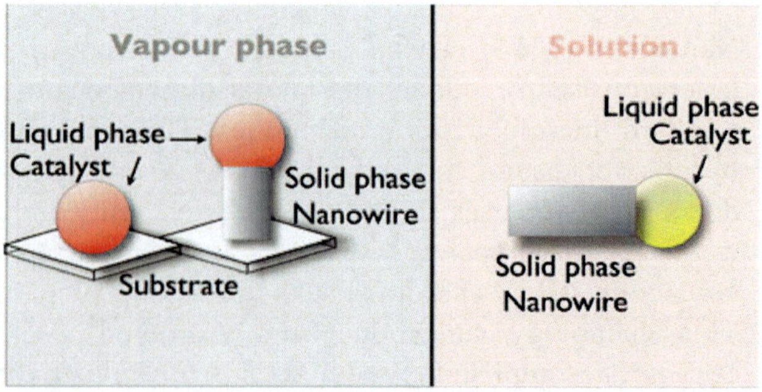

Fig. 1 Schematic representation of vapor–liquid–solid growth and solution–liquid–solid growth mechanism

The preliminary step for VLS growth is the preparation of the catalyst. Different methods can be used for its preparation, or it can be prepared by simple evaporation of a commercial colloidal solution onto the substrate. If the metal does not wet the substrate, there is the formation of clusters following Volmer–Weber growth, otherwise during the high temperature step for NWs growth process it will form clusters due to Ostwald ripening [17].

It has been reported an initial dead time before the NWs starts forming, when the growth is made by vapor phase processes [18]. This dead time is ascribed to the time necessary for the formation of nucleation sites. As an example, the catalyst in order to be active may require in situ reaction or modification. In VLS case, the catalyst droplet has to form a mixture with the condensed material, this can change the cluster dimensions and in turn the NWs diameters.

When the NW section is constant, the condensation and the incorporation should be occurring only at the interface between catalyst and NW, while in the opposite case during the growth there is evaporation, consumption, or incorporation of the catalyst that leads to conical shape NW. A big advantage of VLS growth is the capability to control NWs diameters playing with catalyst particles dimensions. Nevertheless, lot of factors must be taken into account to correlate them such as clusters curvature. The solubility depends on the radius curvature, so as cluster size is reduced the solubility increases and higher supersaturation has to be reached. Furthermore, the thermodynamic limit for the minimum radius of a metal liquid cluster at high temperature should be taken into account.

In the case of NWs growth from the vapor phase without a catalyst, the vapor–solid mechanism is attributed. The source material is vaporized and then it condenses on a substrate placed in a lower temperature region as in VLS, but in VS the initially condensed particles serve as seeds for NWs nucleation, facilitating the unidirectional growth. At the beginning it was ascribed to lattice defects, but the presence of defect-free NWs was reported. The adsorption on all the NWs faces followed by diffusion on the principal NW growth surface was proposed to explain a measured growth rate higher with respect to the condensation rate calculated from the vapor phase on the nucleation section.

2.2 Solution Phase Growth

NWs can be prepared also with different solution-based methodologies. The advantages are, for some of them, room temperature operation and the low costs of equipment compared to vapor phase growth. Among solution-based growths, there are template-assisted and template-free methods.

Concerning template-assisted growth, there is a physical confinement of the growth in one-dimensional nanostructures thanks to the template. The preparation of the template and the technique used for filling the template pores are crucial points. Furthermore, a critical issue is the template removal.

For template preparation, the most used technique is anodization, a well-established process for the preparation of hexagonal closed packed highly ordered alumina membranes and other porous templates. In the last decades, it was used for the preparation of porous films and afterward for the preparation of NWs and nanotubes. The pore diameter and pore-to-pore distance, which can be easily controlled by the anodization voltage, control the NWs section and NW–NW distance. Particular care must be taken for the template dissolution that can destroy NWs alignment.

Electrochemical growth [19] and sol gel method are the most used procedure for the deposition inside the template pores.

The main drawback is the low crystallinity of NWs prepared with most of the solution-based techniques. This can limit their potentialities for fundamental studies and applications.

The deposition of NWs can be made in liquid environment following different procedures that have been proposed in literature also without the use of a template for the growth confinement. In the last years, a great research effort has been put into this field, we can divide the most important methods into: surfactant-assisted, sonochemical, hydrothermal, and electrospinning methods.

Concerning template-free methods, the growth mechanism can be similar to VLS, that is solution–liquid–solid growth mechanism (Fig. 1). A metallic droplet serves as an ideal surface for precursor adsorption and decomposition, catalyzing the NWs growth. Catalyst must be carefully chosen controlling melting point, solvating ability and reactivity.

Surfactant method consists in the use of a surfactant for the promotion of an anisotropy in the growth necessary for NWs production. During the growth, there are three phases: oil, surfactant, and aqueous phase. The role of the surfactant is to confine the growth as in a microreactor. Particular care must be made in the selection of precursors, surfactant and reaction parameters such as temperature, pH, and reactants concentration. Most used surfactants are oleic acid, hexylphosphonic acid, tetradecylphosphonic acid, trioctylphosphine oxide, and trioctylphosphine. This method, such as most of the solution-based ones, is a trial-and-error procedure. Work has to be made for the selection of the right recipe, the appropriate capping agents and reaction environment [20–22].

Another solution-based method is the sonochemical one; in this case, crystal growth anisotropy is obtained thanks to ultrasonic wave that changes the reaction conditions by acoustic stirring. During sonication, there is the formation, growth and collapsing of bubbles and in such environment temperature greater than 5,000 K, pressure larger than 500 atm, and cooling rate higher than 1,010 K/s can be reached. Such extreme reaction conditions can allow the formation of 1D nanostructures [23, 24]. For example, magnetite (Fe_3O_4) nanorods were synthesized by ultrasonically irradiating aqueous iron acetate in the presence of beta-cyclodextrin, which serves as a size-stabilizer [25].

Hydrothermal growth is another well-known procedure for material synthesis. It has been proposed since the 1970s for the production of crystalline structures. The crystallization happens at high temperature in aqueous solutions at high vapor pressures. The crystal growth is performed in an autoclave (an apparatus consisting of a steel pressure vessel), in which an aqueous mixture of soluble metal salt (metal and/or metal–organic) of the precursor materials is supplied. A temperature gradient is kept at the opposite sides of the growth chamber. Temperature can range between 373 K and 573 K and pressure can exceed 1 atm. ZnO nanorods [26–29] and other oxides such as CuO [30], ceria [31] and titania [32] have been synthesized using wet-chemical hydrothermal approach.

Finally, electrospinning is another template-free synthesis method. Electrical field is used to produce anisotropic growth. A liquid jet is formed thanks to the high voltage. The electrostatic repulsions between the surface charges and the evaporation of solvent stretches continuously electrified jet giving rise to the wire morphology. The preparation with electrospinning results normally in the fabrication of uniform NWs with nanometer-scale diameters.

The first patent describing the operation of electrospinning was filed in 1934, when Formalas proposed an apparatus for producing polymer fibers by taking advantage of the electrostatic repulsions between surface charges [33].

Oxides such as TiO_2, SiO_2, V_2O_5, ZnO, Co_3O_4, Al_2O_3, CuO, NiO, MoO_3, Nb_2O_5, and $MgTiO_3$ have been fabricated as wire structures by this technique [34–42].

2.3 Devices Fabrication

For the fabrication of a conductometric sensing device, the transduction of the signal produced by the sensing material is important. NWs are the sensing material doing a receptor function, thanks to the reactions between the gas and their surface, but the choice of the transducer and operation mode can affect the device response. Having in mind practical applications a proper design and choice of the transducers, electrical contacting, bonding, and packaging should be made to accomplish each peculiar application requirements.

Silicon and alumina are the most used substrates: they can be in bulk or micromachined shape. If silicon is used as a substrate, an insulating layer must be foreseen. The transducers have to be equipped with at least two electrical contacts to measure NWs resistivity and a heater to maintain the NWs at the working temperature that is of the order of hundreds degrees centigrade.

In many cases the sensing device is realized using a bulk substrate with electrical contacts on top, and a heater on the backside. The working temperature of the metal oxide gas sensors is maintained applying a constant voltage to the heating meander.

The choice of the metal selected for electrical contacts is a crucial step. The contact has to be good: that is it must provide as low sheet resistance as possible in order to minimize the voltage drops along the interconnections and it should be ohmic. Platinum is one of the most used metallization since it has a good ohmic contact with most of the metal oxide, it does not oxidize at high temperatures, it has a low diffusivity, and it is resistive to corrosive gases. In the case of NWs, the contacting issue is much more difficult due to their peculiar morphology, they can be deposited after the NWs formation or NWs can be transferred to a transducer after removing from the growth substrate, dissolving in a solvent and using drop coating technique.

3 Electrical Properties

Deviation of stoichiometry of high gap MOX is responsible for their semiconducting properties [43]: cation vacancies are acceptors, producing holes and negative charged vacancies, while shallow states made up of oxygen vacancies are double n-type donors, the electrons on the adjacent cation are easily removed and donated to the conduction band [44].

The gas sensing properties of MOX arises from adsorption of molecules from the gas phase on reactive surface atoms, which lack binding partners. The first step of adsorption is physisorption, which is a slightly exothermic process characterized by high coverage at low temperature and a low coverage at high temperature. If the partial pressure is very low, the amount physisorbed is simply proportional to the partial pressure (Henry's Law). Physisorbed species can be chemisorbed (ionosorbed) when the adsorbate acts as a surface state capturing an electron or a hole [45].

As far as the relation between chemisorption and electrical properties is concerned in the simple Charge Transfer Model (CTM) [45, 46], the physi- and chemisorbed atoms and molecules are represented by surface-localized states in the semiconductor energy gap, whose occupation statistic is given by the same Fermi–Dirac distribution, physisorption corresponding to unoccupied and chemisorption to occupied states. The appearance of surface-localized acceptor states in n-type semiconductors induces therefore charge transfer between bulk and surface in order to establish thermal equilibrium between the two. The charge transfer results in a nonneutral region (with a nonzero electric field) in the semiconductor bulk, usually referred to as the surface space charge region (SCR).

Figure 2 reports the band showing the formation of an SCR in an n-type SC in presence of surface acceptor states, starting from an ideal surface before charge is transferred to the surface (a) and after the formation of the SCR (b); the potential of the electron in vacuum is used as a constant reference.

When neglecting the free carrier concentration in the SCR (depletion approximation), the potential, supposed null at surface, is parabolic and the height of the bulk surface potential barrier is

$$V_s = \frac{qN_d}{2\varepsilon} {x_0}^2 = \frac{qN_{ta}^{-2}}{2\varepsilon N_d},$$

where x_0 is the width of the depletion layer and N_d donor density (oxygen vacancies) and N_t^- the density of charged surface states with $N_d = qN_{ta}^-$.

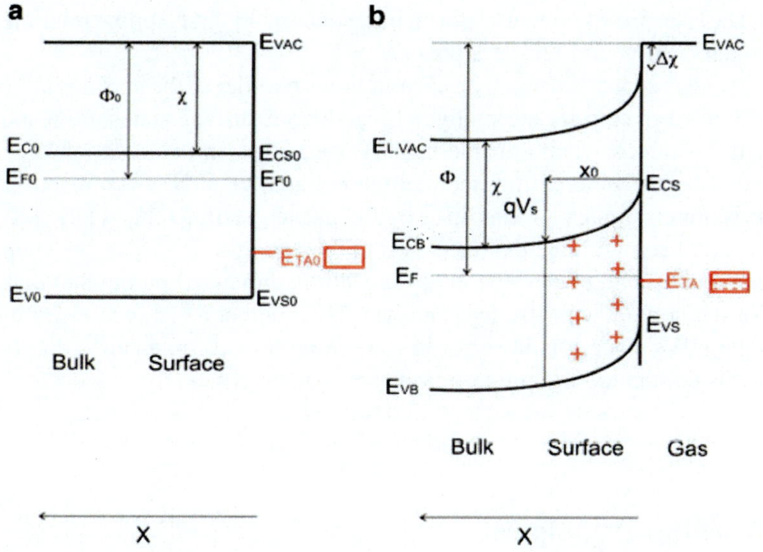

Fig. 2 Band model showing the formation of an SCR in an n-type SC in presence of surface acceptor states before (**a**) and after (**b**) adsorption. The potential of the electron in vacuum E_{vac} is used as a constant reference

For reasonable values, for example $V_s = 0.4\,\mathrm{V}, \varepsilon = 11.8\,\varepsilon_0$ and $N_d = 5 \times 10^{15}\,\mathrm{cm}^{-3}$, we obtain $x_0 = 300$ nm, a result that reveal the striking influence of surface states on the semiconductor bulk: states localized over no more than several monolayers affect the semiconductor electrically even thousand monolayers away from the surface.

In addition to surface states, another important phenomenon associated with a semiconductor surface is the surface dipole [47]. An adsorbate layer may result in a surface dipole, the magnitude of which depends on the ionicity of the adsorbate–substrate bond. The surface dipole manifests itself as a step in the electric potential at the surface because the potential changes abruptly over several monolayers. This is in contrast to the macroscopic dipole created by the surface states and surface SCR.

As for conduction, metal oxide gas sensors are generally operated in air in the temperature range between 500 and 800 K where conduction is electronic and oxygen vacancies are doubly ionized and fixed. When operated in the semiconducting temperature range, the overall resistance of the sensor element is determined by the charge transfer process produced by surface reactions and by the transport mechanism from one electrode to the other through the sensing layer. In polycrystalline gas sensors, a high resistance SCR is created around the surface of each grain and in particular at the intergrain contact: the carriers must overcome the energy barrier in order to cross from one grain to the neighbor.

The conductance of a polycrystalline gas sensor can therefore be expressed as

$$G = G_0 e^{\frac{-qV_s}{kT}} = G_0 e^{\frac{-q^2 N_{ta}^{-2}}{2\varepsilon N_d kT}},$$

where G_0 is a pre-exponential term independent in first approximation from the surface adsorption and temperature.

In Q1D gas sensors instead the current flows parallel to the surface. When the NW is fully depleted, carriers thermally activated from surface states are responsible for conduction. Indeed when considering NWs bundles, the conduction mechanism is dominated by the inherent intercrystalline boundaries at NWs connections—like in polycrystalline samples—rather than by the intracrystalline characteristics.

The metal semiconductor junction that forms at the interface between the layer and the contacts can play a role in gas detection, enhanced by the fact that the metal used for the contact acts also as a catalyst. The contact resistance is more important for single NWs since it is in series to the semiconductor resistance that for bundles where it is connected to a large number of resistances.

4 Working Principles

When a gaseous species interacts with the surface of a MOX, the adsorption isobar, that is the volume adsorbed as a function of temperature at a constant pressure, is characterized at low temperature by physisorption and at high temperature by equilibrium chemisorption – that decreases exponentially with temperature. In the intermediate region, irreversible chemisorption takes place because the desorption energy is greater than chemisorption one.

For a change of the electrical properties of semiconducting metal oxides to take place, it is essential the formation of charged chemisorbed species. Identification of optimal operating temperature for a MOX sensor is therefore a crucial step, since they must be obviously operated in a temperature range where reversible chemisorption takes place or alternative ways to induce desorption must be provided. The activation energies for adsorption and desorption can be supplied either thermally or by a nonequilibrium process such as illumination.

The process of gas detection is intimately related to the reactions between the species to be detected and ionosorbed surface oxygen. In the temperature range between 400 and 800 K oxygen ionosorbs over SnO_2 and other oxides in a molecular (O_2^-) and atomic form (O^-) [48]; when a reducing gas like CO comes into contact with the surface, it oxidizes to CO_2 by reacting with ionosorbed oxygen, releasing electrons from surface states to the conduction band. The overall effect at equilibrium is shrinking of the density of ionosorbed oxygen, that is occupied surface acceptor states and consequently the height of the surface barrier. Indeed surface reactions are still debated: CO sensing, for example, could take place through reaction with hydroxyl groups, producing atomic hydrogen that recombines with oxygen lattice and releases a free electron [49] and even by direct adsorption as CO^+ [50].

Direct adsorption is also proposed for the gaseous species– like strongly electronegative NO_2 – whose effect is to decrease sensor conductance. The occupation of surface states, which are much deeper in the bandgap than oxygen's, increases the surface potential and reduces the overall sensor conductance.

An important ubiquitous species that ionosorbs over MOX surfaces is water [45]. The chemisorption of water onto oxide from air can be very strong, forming a "hydroxylated surface," where the OH^- ion is bounded to the cation and the H^+ ion to the oxide anion. The overall effect of water vapor is to increase the surface conductance.

Addition of a small amount of noble metals over the MOX surface such as Au, Pd, Pt, and Ag can speed up surface reactions and improve selectivity toward target gas species. The activation of species involved in a surface reaction may be the dissociation of a molecule, the ionization of the species or some other intermediate reaction. From the energetic point of view, the effect of catalysis is to provide a more favorable reaction path. Catalyst should be dispersed as small crystallites over the surface of the oxide in order to be active near the grain boundaries where carrier transport takes place. Poisoning and mainly coalescence by surface migration of the metal particles at high temperature could deactivate the supported catalyst and should be carefully avoided.

As for signal transduction, the easiest measurable physical quantity that can be transduced is the sensor conductance in DC conditions. In laboratory tests, it is normally measured by a volt-amperometric technique at constant bias while in commercial gas sensors the layer is usually inserted inside a voltage divider.

The sensor response toward a target gas concentration for metal oxide semiconductor sensors is usually defined as the (relative) change of conductance. In case of an oxidizing target species, resistance increases following gas introduction and the sensor response is defined as the (relative) change of resistance.

Starting from the sensor response, it is possible to derive the sensor response curve, that is the representation of the steady state output as a function of the input concentration [51]. The sensor response curve is frequently called, erroneously, sensitivity curve. Instead, sensitivity is the derivative of the sensor response curve.

5 DC Conductometric Gas Sensors

The first use of NW bundle and conductometric gas sensor was demonstrated in 2001 [11] with SnO_2 nanobelts. Successively many oxides – synthesized in form of NWs, nanobelts, nanorods, and so on – were tested for gas sensing applications. The literature can be divided into two main categories: sensors where the sensing layer consists of a single NW (Sect. 5.1) and sensors where the sensing layer is made of bundles or mats of NWs (Sect. 5.2).

5.1 Single Nanowire Devices

As stated before (Sect. 3), when gas sensors are realized with single NWs, the current flows along the direction of the wire, parallel to the surface. Gas absorption creates an SCR that gives rise to a channel fully depleted of carriers if the diameter of the wire is comparable to L_D.

Even if in this case optimum results are expected, the technical difficulties in realizing reliable electrical contacts on individual nanostructure has restricted the number of works dealing with single NWs. To realize ohmic electrical contacts, a bunch of NWs dispersed in a solvent and placed on the substrate with prepatterned contacts by drop coating; a conveniently placed single NW is selected. Sometimes, a nanomanipulation step is required to align it between contacts. Then platinum or gold stripes are deposited by Focused Ion Beam (FIB) (Fig. 3) or electron-beam lithography (EBL) to realize contacts between predeposited contacts and NW or to improve existing electrical contacts.

Hernandez-Ramirez et al. [53] tested SnO_2 single NW at 570 K toward different combustion gases such as, CO, relative humidity, and NO_2, using AC Impedance Spectroscopy technique. They demonstrated that sensor response and sensitivity (slope of the response curve) were bigger when the NW radius is smaller than 25 nm (for bulk SnO_2 L_D is about 43 nm at 500 K [48]). More fundamental studies were carried out to determine the interaction of single SnO_2 NW with oxygen species [54]. Figure 4 reports the dependence of sensor response on the NW radius when atmosphere was changed from nitrogen to dry air. Again, diminishing the dimension of the NW, bigger response was observed. Long-term drifts of the resistance of NW after exposing it to changes of oxygen partial pressure in the air were observed in the resistance versus time plot. The authors proposed that oxygen diffusion into the bulk is responsible for this behavior.

MOX gas sensors usually work in the temperature range between 500 and 800 K to activate gas adsorption and desorption form the surface. An increase in working temperature usually means faster response and recovery times, as well as lower response. Another factor that influences the recovery time of Q1D NWs is the poisoning of the contact due to gas interaction. Yu et al. [52] evidenced that the deposition of a Pt extra contact, fixing single SnO_2 nanobelt over Pt contacts, is crucial to avoid contact poisoning. Their nanodevice was capable of sensing NO_2 down to 200 ppb at 473 K. The recovery of sensor signal is complete 3 min after

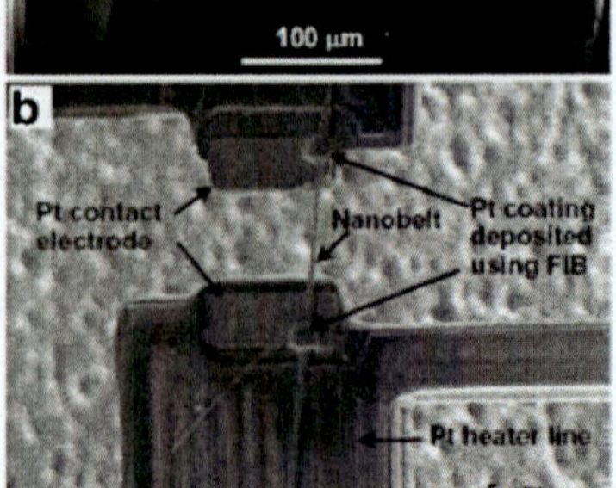

Fig. 3 SEM micrograph of the device with single NW realized by Yu et al. [51]. Pt contacts deposited by FIB are visible. Permission 2071440528477

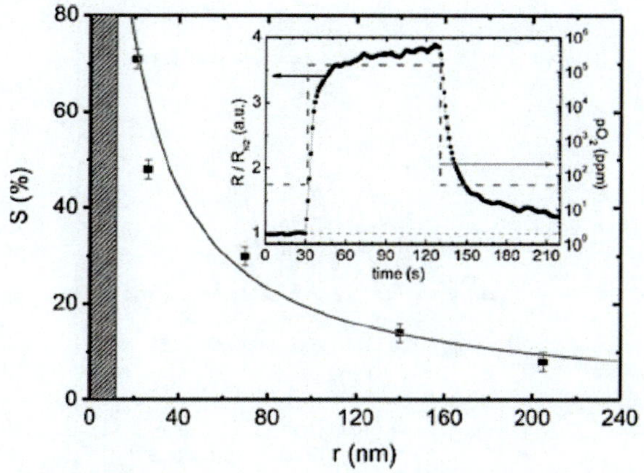

Fig. 4 Dependence of sensor response of single SnO_2 NW on the NW's radius when atmosphere was changed from nitrogen to dry air. The *inset* shows the response of an NW with radius 20 nm at 573 K [54]. Reprint permission 2071841099239

injection of fresh air, while the recovery time is much higher without the additional contact. Interestingly, the nanodevice was capable to detect dimethyl methylphosphonate (DMMP) a nerve agent simulant, important in the security field.

Functionalization of Q1D MOX is used to enhance sensing capabilities toward a specific gas, thus increasing selectivity (Sect. 4). Usually, the NW is coated with transition metal catalysts (such as Au, Pt, and Pd) or other metal oxide sensing materials (such as CuO, NiO, and ZnO).

Catalytic effect can be obtained changing surface band structure of metal oxide sensing materials by the formation of heterojunctions including n–n junctions and p–n junctions. For example, deposition of ZnO or NiO particles on SnO_2 NW was reported [55]. With ZnO, the effect was not so evident, with only a small improvement in the sensitivity and selectivity. Better results were obtained with NiO (p-type material) deposited by sputtering on SnO_2 NWs, with and enhancement of sensitivity to CO with no concomitant effect on the response to CH_4 (Fig. 5).

Kolmakov et al. [56] examined the influence of the surface sensitization of SnO_2 single NWs with catalyst particles of Ni/NiO and Pd. In this case, the metal accelerates the gas reaction on the metal oxide surface by catalytic action. Ni or Pd deposition improved response and kinetic of response of NWs to O_2, H_2, and CO. The conductance through the NWs in vacuum and in oxygen progressively dropped with the increase of the amount of catalyst on its surface as a consequence of formation of a catalyst-induced depleted region in the NW. There is an optimum amount of catalyst for which the effect is bigger, and a saturation effect for higher quantities.

Liao et al. [57] presented a work on ZnO used as single NW sensor; the diameter of the wire was quite big ($d = 400$ nm) and Pt contacts were grown by FIB. The single NW was implanted with He^+ at different doses in order to change the amount of defects in ZnO NW. At high doses, the material became amorphous with very low conductive electron. For optimal dose ZnO implanted NW showed higher H_2S sensitivity than unimplanted.

In order to decrease power consumption at high temperatures needed for sensor operation, Ryu et al. [58] realized single NW In_2O_3 device deposited over hotplate

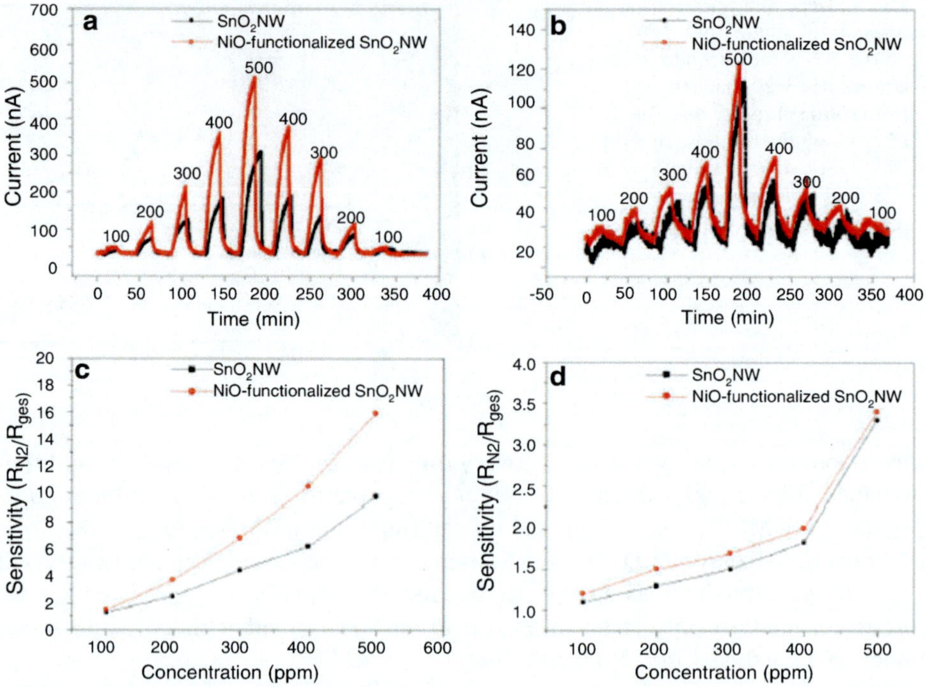

Fig. 5 Comparison of pure and NiO-functionalized SnO_2 single NW-based sensor to (**a** and **c**) CO and (**b** and **d**) CH_4. Reprint permission 2072001302317 [55]

for detection of H_2, ethanol, and CO down to concentration 1, 10, and 50 ppm, respectively. The device was capable to detect very low concentration of ethanol (1 ppm).

5.2 *Multiple Nanowire Devices*

The easiest way to realize a gas sensor with NWs is to place bundles or mats of NWs over a substrate. In this case, the sensing layer consists of a casual agglomerate of NWs placed between two contacts and the conduction mechanism is similar to polycrystalline samples (Sect. 3). Moreover, there is a natural distribution of the dimension of the wire between the bundles and so it is more difficult to obtain nonconducting channel like in the case of single NW: even if smaller wires switch to nonconducting behavior, the current can flow through bigger wires with a percolation path.

The work presented in literature regards the common metal oxides already used as gas sensors such as SnO_2, ZnO, In_2O_3, WO_3, and V_2O_5.

Bundles of SnO_2 nanobelts (200 nm large and 40 nm thick) were used for CO, NO_2, ethanol, and ozone sensing with good results [11, 59], as confirmed by the work of

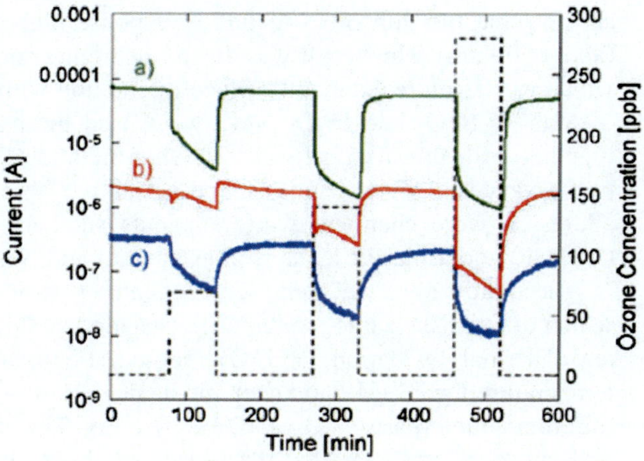

Fig. 6 Variation of current as a function of ozone concentration for (**a**) SnO_2 NWs operated at 400°C, (**b**) ZnO NWs operated at 350°C, and (**c**) In_2O_3 NWs operated at 400°C [59]. Permission 2071850973052

Ying et al. for ethanol sensing [60]. Fig. 6 reports the comparison between SnO_2, In_2O_3 and ZnO NWs for ozone sensing. As for thin film counterpart, the optimum sensing temperature for each type of analyte differs from one oxide to the other.

ZnO in form of bundles of nanorods with mean diameter between 10 and 30 nm were tested for ethanol and for H_2. It is interesting to note that ZnO prepared from different groups with different techniques [61, 62] gave similar results if one compares sensor response: the slight changes from one group to another in ethanol detection could be easily attributed to differences in surface to volume ratio between two types of sensor.

The effect of the intrinsic defect concentration on the ZnO surface has strong impact on the sensing properties of the sensors. Many authors prepared ZnO NWs in form of powders; afterward a bonding agent such as Polyvinyl alcohol solution was used to deposit a paste on Al_2O_3 tube. The bonding agent changed real surface coverage of the nanomaterials as well as the preparation conditions did, and can strongly affect sensing properties. Using NWs prepared in this way, a wide spectrum of gas was investigated, and the sensor exploited a nonselective behavior (also typical of metal oxide conventional sensor), showing response toward gases such as LPG, and 90# gasoline [63], ethanol and H_2S [64]. Test toward other gases such as HCHO, CO, NH_3, H_2, CH gave weaker, but not negligible, response.

Effect of catalysts deposition on ZnO NWs bundles was reported: Pd [65] or Pt [66] was deposited by sputtering on ZnO surface in order to enhance hydrogen detection properties. Better results were observed with Pd, since it gave higher response to a 50 times lower H_2 concentration.

In_2O_3 is widely used in gas sensor for detection of oxidizing gases, but is quite difficult to prepare in form of NWs due to its cubic cell. There are very recent works describing gas sensing application of In_2O_3 NWs [58, 67–69].

Again the control of lateral dimension of the wires was the key point for achievement of high sensitivities. Vomiero et al. [68] tested sensing capabilities of devices with different average size of the wires. The electrical conductance was lower for thinner wires (about 100 nm). The results on acetone and NO_2 sensing showed that the thin NWs feature a response higher than the one of thick wires (about 500 nm). This result was due to the higher surface-to-volume ratio of NWs, which causes more extended surface interaction with gas molecules.

Mats of SnO_2 and In_2O_3 NWs were used in an electronic nose configuration together with thin-film sensors [2]. This method allowed a comparison between performances of an emerging technology like NWs and the traditional ones. Gases addressed were chemical warfare agents such as dimethyl methylphosphonate (DMMP) and acetonitrile, as well as ethanol and NH_3.

The authors observed some sensing features common to NWs (SnO_2 and In_2O_3), which differ if the same oxides were prepared as thin films by Rheotaxial Growth and Thermal Oxidation (RGTO) method. For example, the optimum working temperature for DMMP and acetonitrile detection is higher for NWs than for their thin film counterparts (773 vs. 673 K, see Fig. 7).

Kaur et al. [69] reported the use of single-crystalline whiskers with very big diameter (100–300 μm) for H_2 and NO gas sensing at room temperature. Even if the dimension of the wire is quite huge, that means also high conductance of the wires (resistance was of the order of 10 Ohm), space charge induced by gas adsorption at the surface of whisker is sufficient to give a response to 200 ppb of H_2S at room temperature.

WO_3 is another well-known material for detection of oxidizing gases such as NO_2 and O_3; it was prepared in form of bundles of nanofibers by Polleux et al [70]. The width of a bundle was between 20 and 100 nm, while diameter of a single fiber was about 1 nm. The tungsten oxide NWs were nearly insensitive to CO in the concentration range between 1 and 10 ppm at all operating temperatures studied, as already reported for WO_3 thin-film gas sensor. High surface area and good crystallinity of WO_3 fibers were responsible for the high signal observed to low NO_2 concentrations (50 ppb).

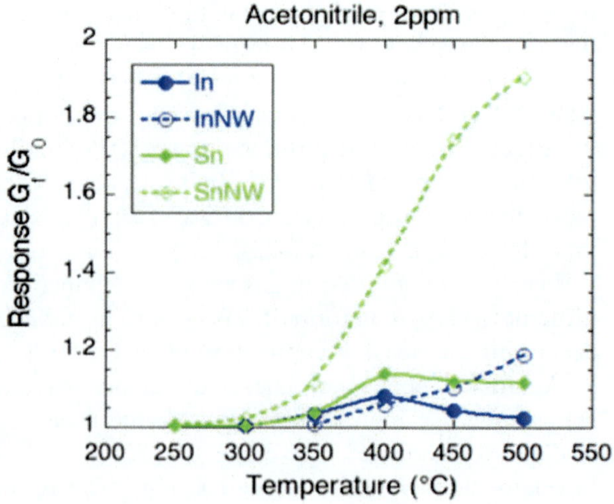

Fig. 7 Detection of acetonitrile with SnO_2 and In_2O_3 gas sensors as thin film and NWs. Reprinted with ©2008 IEEE permission from [2]

V_2O_5 nanobelts, with rectangular cross section 60–100 nm wide and 10–20 nm thick, were prepared by Liu et al. [71] by hydrothermal method. Sensors showed good response to ethanol, while response to other gases such as methanol, H_2S, H_2, NH_3 was very low and response to CO and NO_x was absent. Raible et al. [72] reported synthesis of V_2O_5 nanofibers with rectangular section (1.5–10 nm) by mean of a wet-chemical process that detected 10 ppm of 1-butylamine 10 ppm of ammonia. This result was explained considering high affinities to organic amines of V_2O_5 nanofibers.

6 SNT-Based Sensors

A Single Nanowire Transistor (SNT) is obtained by depositing a gate back contact over the Si substrate. Many independent studies showed how the additional gate voltage tool allows tuning the availability of electrons for surface adsorption and desorption reactions and the sensing and recovering performance of the devices.

By applying a gate potential, the well-known electroadsorbitive effect [73] can be exploited by modulating the position of the electrochemical potential and the availability of electrons for surface reactions. A majority carrier channel in accumulation mode is created upon application of a positive gate bias as in n-channel Thin-Film Transistors (TFT), see Fig. 8 [74, 75].

Li et al. [76] and Zhang et al. [76] first showed that at room temperature V_T of an In_2O_3 NWs-based SNT increases following the introduction of NO_2, as showed in Fig. 9. Since NO_2 ionosorbs over In_2O_3 and other metal oxides extracting electrons from the conduction band, as a consequence a greater threshold gate voltage V_T is required in order to build up the majority carrier accumulation conductive channel.

A different and still not fully understood behavior has been detected in presence of NH_3 a species, which is known to inject electrons in conduction band when adsorbed. V_T upon exposure to NH_3 has been found to decrease for lightly doped

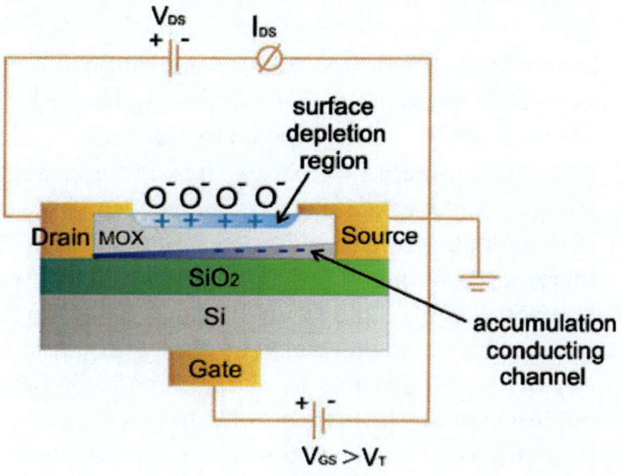

Fig. 8 Schematic cross section of a Single Nanowire n-type Transistor device in accumulation

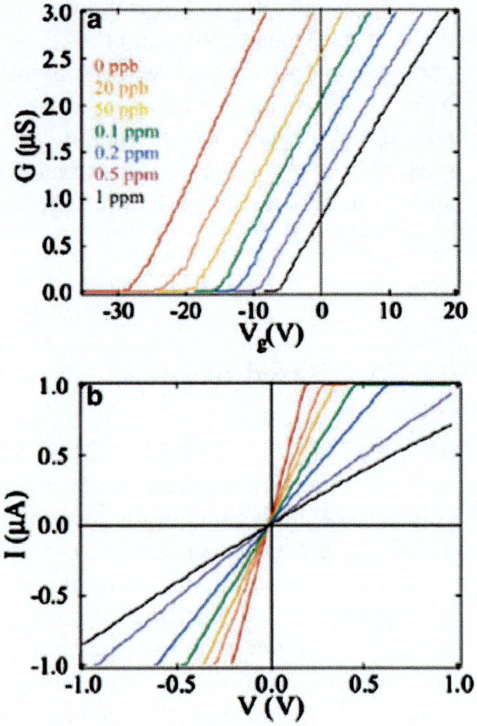

Fig. 9 (**a**) Conductivity of an In_2O_3 SNT as a function of VG at growing concentration (*left to right*) of NO_2, (**b**) I_{DS}–V_{DS} curves at $V_{GS} = 0$. Reprinted with permission 2523000823535 from [77]

In_2O_3 NWs, which surfaces were cleaned via UV illumination in a vacuum, and to increase for heavily doped NWs. Differently in ambient atmosphere V_T has been consistently observed to decrease upon NH_3 exposure, regardless of the doping concentration [76, 78].

Fan et al. [79] detected in ZnO-based SNTs that V_T changes from −7.9 V in pure Ar to −5.2 V in 5 ppm NO_2; moreover, they established that in the linear regime a linear relationship between conductance and NO_2 partial pressure exists. More important they showed that a strong negative $V_{GS} = -60$ V field is able to induce NO_2 electrodesorption at room temperature, an unseemly process because the thermal energy is usually lower than the activation energy for desorption, and to restore the ZnO SNT conductance, as shown in Fig. 10. A high gate voltage induces hole migration to the surface driven by the negative electric field and leads to recombination and discharge of NO_2^-. Besides, the negative gate-induced repulsive field stretches and weakens the bonding between NO_2 surface dipoles and adsorption sites.

Fan et al. [80] studied oxygen adsorption in ZnO-based SNT and found that V_T increases as a function of O_2 concentration in Ar, as expected since O_2 – like NO_2 – ionosorbs as O_2^- and O^- over metal oxide surfaces. The role of adsorbed oxygen was recently evidenced also by [81] who studied the electronic transport properties of ZnO NW transistors as a function of gate bias sweep rate under various oxygen environments. The current, threshold voltage, and carrier density were sensitively dependent on the gate bias sweep rate. Slower gate bias sweep rates result in greater

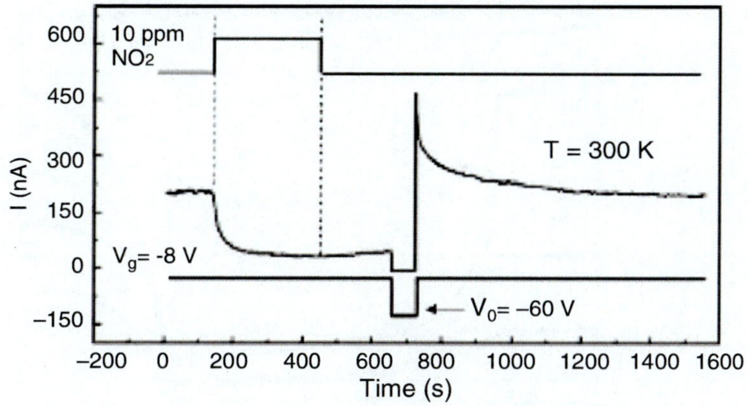

Fig. 10 A −60 V gate voltage pulse following the introduction of 10 ppm NO_2 in Ar induces ZnO SNT conductance recovery. Reprinted with permission 2524130451520 from [79]

oxygen adsorption, resulting in reduction in current and carrier density and shift of gate bias in the positive gate bias direction.

Zhang et al. [82] confirmed that the rates and extent of oxidation and reduction reactions taking place at the surface of an SnO_2 NW, configured as a field-effect transistor, can be modified by changing the electron density in the wire with a gate voltage. They carried out a detailed study of the effects of the gate voltage in a SnO_2-based SNT after O_2 introduction in N_2 and subsequent CO detection at 553 K.

As shown in Fig. 11, when operated in the active zone ($V_{GS} > V_T = -6.2$ V) the conductance decreases when O_2 is introduced, the amount of this reduction being lower at decreasing V_{GS}. After the introduction of CO, the conductance increases, eventually achieving a new steady state. Indeed, the observed conductance increase on admitting CO is not monotonic with V_{GS} but achieves a maximum value in the range $V_{GS} = -2$ to 0 V. Interestingly, even at $V_{GS} = -6$ V, when oxygen ionosorption no longer takes place, CO increases conductance.

Besides, [83] determined that the rate and extent of oxygen ionosorption and the resulting rate and extent of catalytic CO oxidation reaction on the NW's surface could be controlled and even entirely halted by applying a negative enough gate potential.

7 PL-Based Sensors

As seen before (Sects. 5 and 6), nanosized single crystalline MOX are very promising for gas sensing application, even if realizing stable and ohmic electrical contacts is still a challenge. So the discovery of transduction mechanism different than electrical one is very important. One possibility is to use the optical photoluminescence (PL) signal of ZnO and SnO_2 NWs. Faglia et al. [84] pointed out that SnO_2 NWs exhibit interesting gas-sensitive photoluminescence properties:

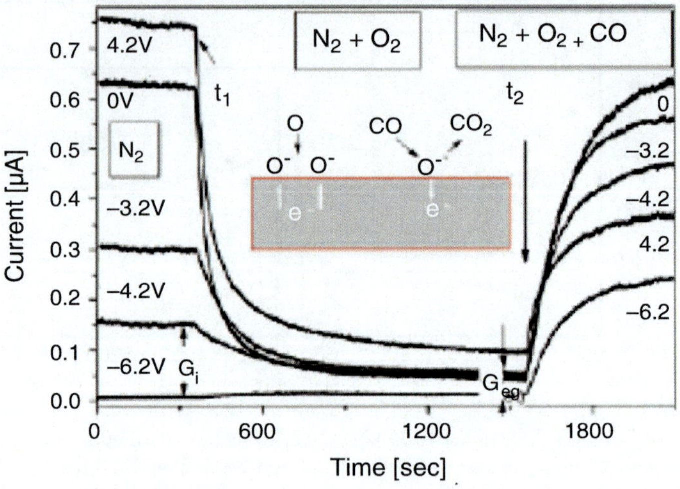

Fig. 11 Drain source current as a function of V_G of an SnO_2-based SNT following the addition of 10 sccm of O_2 to 100 sccm of flowing N_2 gas at time t_1 the addition of 5 sccm of CO at time t_2 ($V_{DS} = 0.2$ V, $T = 553$ K). Reprinted with permission 2524181449450 from [82]

exposure to low concentrations of nitrogen dioxide (few ppm) significantly quenched the visible PL emission of NWs in a reversible way. The influence of gas adsorption on the PL signal was studied for SnO_2 and for ZnO, by choosing different gases (ammonia, CO, ethanol, relative humidity) and by changing temperature of the sample. For both materials, the main result is a quenching with NO_2 at low concentrations (hundreds of ppb) that could be interesting for developing all-optical gas sensor for oxidizing gases using MOX NWs.

For the sake of clarity, the results on two oxides will be presented separately.

7.1 SnO_2

Nanostructured SnO_2 typically exhibits strong visible PL emission at room temperature peaked at photon energy at about 2 eV [84–90]. Due to wide bandgap energy of SnO_2 (3.6 eV), such emission band cannot be assigned to band-edge or to exciton recombination. Thus, it was attributed to transitions involving oxygen vacancy (OV) states or defective states within the bandgap.

Faglia et al. [84] studied behavior of PL spectrum of SnO_2 NWs, with NO_2 diluted in dry air (concentration down to 1 ppm). The sample was kept at 393 K. The observed reversible quenching was not associated with peak shift and the amount of quenching follows pulses of NO_2 diluted at constant relative humidity (0%, 70% and 30%). The measurements of PL where taken every 5 s, and the area under PL band was chosen as a feature. Humidity, CO and NH_3 did not change PL signal, thus giving a sort of selectivity toward NO_2 at 393K. The same gaseous species were active in affecting electrical transport [11].

NO_2 adsorbs over the surface and creates competitive nonradiative recombination paths. The adsorption reversibility is guaranteed, since UV excitation light gave NO_2 the needed energy to desorb from the surface and 90% recovery of the initial value is recorded after 600 s.

Lettieri et al. [91] investigated the same PL quenching as a function of temperature, demonstrating that bigger quenching is observed at room temperature. They studied in detail the mechanism of PL quenching by NO_2 adsorption with Continuous Wave (CW) and Time Resolved Photoluminescence (TRPL) [92]. Figure 12 reports typical PL spectra of SnO_2 NWs in air and after introduction of NO_2 in dry air. While the decrease in PL intensity is linearly proportional to surface density of adsorbed molecules (assuming a Langmuir-like adsorption of gas molecule on the semiconductor surface), the recombination rates are not significantly affected by interaction with NO_2. This suggests that NO_2 acts as static quenchers, determining a change in the number of states acting as radiative recombination centers.

7.2 ZnO

ZnO shows room temperature lasing activity and UV photoluminescence peak [93]: for these reasons, its optical features have been extensively characterized. For thin films and nanostructures (rods, tubes, and so on), PL spectrum is composed of UV

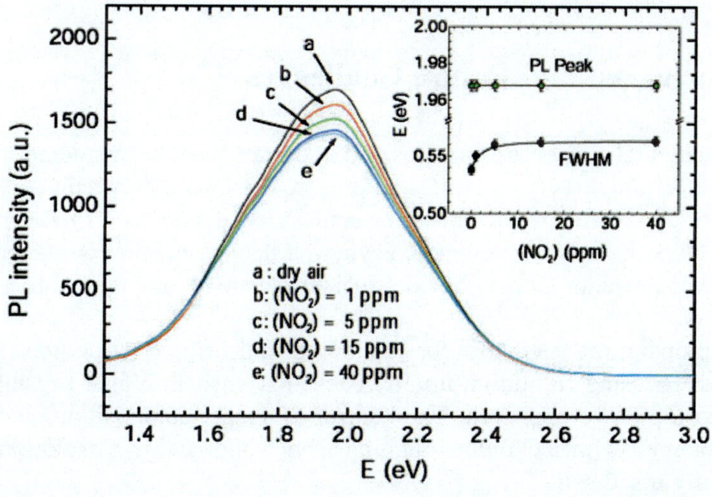

Fig. 12 CWPL spectra of SnO_2 NWs in air and at different NO_2 concentrations. NO_2 quenches the PL intensity without affecting peak position, as seen in the *inset*. Reprinted with permission from [92]

peak, attributed to free exciton emission at about 3.26 eV (380 nm), and broad band in the visible range depending strongly on the preparation conditions in the range from green to yellow [94, 95]. The visible emission has been assigned to the transition between the photoexcited holes and a single ionized oxygen vacancy [96], attributed to an antisite oxygen [96, 97] and donor–acceptor complexes [98, 99]. Surface states have also been identified as a possible cause of the visible emission in ZnO NWs [100].

As for SnO_2, the PL spectrum of ZnO is quenched by interaction with NO_2 in a reversible way at room temperature. UV and visible peaks are quenched in the same manner with no peak shift. PL signal should be related to radiative recombination of surface-localized electrons, and its quenching should be ascribed to charge capture from gas adsorbed onto the surface [91, 101].

The dynamic of quenching caused by introduction of low concentrations of NO_2 (2 ppm) at room temperature is fast with response time less than 30 s. The phenomenon was similar to that reported for SnO_2: the amount of PL quenching proportional to NO_2 concentration in nonlinear way. TRPL measurements are underway to determine whether the quenching is dynamic or static in this case.

Cross-interference of other gases such as ethanol, relative humidity, and CO in NO_2 detection was studied by Comini et al. [101]. No response was observed for CO, while a quenching was observed changing from dry air to 30% relative humidity and an increase was observed for 900 ppm of ethanol. The effect of humidity should be ascribed to room temperature operation of the sensor. Therefore, the optical sensor seemed quite insensitive to CO, even if there is still a dependence of optical signal from humidity and ethanol that must be taken into account.

8 Conclusions and Future Challenges

The fundamental properties of nanosized materials have been studied over the last years to determine novel properties arising from reduced dimensionality.

State-of-the-art findings in gas sensing application of innovative Q1D nanostructures of metal oxide have been presented, paying particular attention on the possibility to exploit the preparation of new devices such as electrical- and optical-based chemical gas sensors.

The preparation methods for obtaining anisotropic structures have been described, focusing to bottom-up approach that ensures higher crystallinity with respect to top-down approach. The control of shape, composition, and reproducibility obtained with techniques ranging from vapor phase to solution-based has been briefly discussed.

As for gas sensing application, MOX NWs approach ensures improved response to different gases as an effect of reduced dimensions and increasing of surface-to-volume ratio. Single NW devices are the most promising with respect to multiple NWs devices, even if the challenging contacts deposition remains an open issue.

To increase the selectivity of MOX gas sensors, functionalization of MOX surface has been described, with results comparable to conventional thin- or thick-film counterpart.

The stability of single crystalline devices is one of the main achievements that could lead to commercialization of MOX NWs gas sensors. Indeed, a report regarding stability tests of NWs-based devices over prolonged period has not been published yet. To ensure device stability, the contact between metal and NW should maintain its properties during high temperature operation.

Besides, the additional gate voltage, which is introduced by configuring a simple nanowire in a field-effect configuration, increases the sensing and recovery performances of the device. Optical transduction mechanism has been exploited as an alternative way to room temperature contact free detection of pollutant gases.

Acknowledgments The authors want to thank first of all the members of the SENSOR lab group in Brescia. This work was partially supported, within the EU FP6, by the ERANET project "NanoSci-ERA: NanoScience in the European Research Area" and by European Community's 7th Framework Programme, under the grant agreement n° 247768, and from the Russian Federation Government, under the State Contract 02.527.11.0008, within the collaborative Europe-Russia S3 project.

References

1. Hoeck G, Bruneckreef B, Fischer P, Van Wijnen J (2001) Epidemiology 12:355–357
2. Ponzoni A, Baratto C, Bianchi S, Comini E, Ferroni M, Pardo M, Vezzoli M, Vomiero A, Faglia G, Sberveglieri G (2008) IEEE Sens J 8:735
3. Yamazoe N (1991) Sens Actuators B 5–12
4. Guidi V, Carotta MC, Ferroni M, Martinelli G, Paglialonga L, Comini E, Sberveglieri G (1999) Sens Actuators B 57:197–200
5. Cui Y, Lieber CM (2006) Science 291:851
6. Cui Y, Duan X, Huang Y, Cui Y, Wang J, Lieber CM (2001) Nature 409:66, 291, 851
7. Huang Y, Duan X, Cui Y, Lieber CM (2002) Nano Lett 2:101
8. Lieber CM (2003) MRS Bull 28:486–491
9. Samuelson L (2003) Mater Today 6:22
10. Stellacci F (2006) Adv Funct Mater 16:15
11. Comini E, Faglia G, Sberveglieri G, Pan Z, Wang Z (2002) Appl Phys Lett 81:1869–1871
12. Haghiri-Gosnet AM, Vieu C, Simon G, Mejzas M, Carcenac F, Launois H et al (1999) J Phys 4:92–133
13. Marrian CRK, Tennant DM et al (2003) Nanofabrication. J Vac Sci Technol A21:S207–S215
14. Candeloro P, Comini E, Baratto C, Faglia G, Sberveglieri G, Kumar R, Carpentiero A, Di Fabrizio E (2005) J Vac Sci Technol B 23:2784–2788
15. Candeloro P, Carpentiero A, Cabrini S, Di Fabrizio E, Comini E, Baratto C, Faglia G, Sberveglieri G, Gerardino A (2005) Micro Eng 178:78–79
16. Wagner RS, Ellis WC (1964) Appl Phys Lett 4:89–90
17. Kolasinski KW (2008) Growth and etching of semiconductors. In: Hasselbrink E, Lundqvist I (eds) Handbook of surface science, vol. 3. Elsevier, Amsterdam
18. Kolasinski KW (2006) Catalytic growth of nanowires: vapor–liquid–solid, vapor–solid–solid, solution–liquid–solid and solid–liquid–solid growth. Curr Opin Solid State Mater Sci 10:182–191

19. Masuda H, Abe A, Nakao M, Yokoo A, Tamamura T, Nishio K (2003) Adv Mater 15:161–164
20. Xu CK, Xu GD, Liu YK, Wang GH (2002) Solid State Commun 122:175–179
21. Xu CK, Zhao XL, Liu S, Wang GH (2003) Solid State Commun 125:301–304
22. Xu CK, Xu GD, Wang GH (2003) J Mater Sci 38:779–782
23. Gao T, Li QH, Wang TH (2005) Chem Mater 17:887–892
24. Miao JJ, Wang H, Li YR, Zhu JM, Zhu JJ (2005) J Cryst Growth 281:525–529
25. Kumar RV, Koltypin Y, Xu XN, Yeshurun Y, Gedanken A, Felner I (2001) J Appl Phys 89:6324–6328
26. Liu B, Zeng HC (2003) Am Chem Soc 125:4430–4431
27. Wang JM, Gao L (2003) J Mater Chem 13:2551–2554
28. Guo M, Diao P, Cai SM (2005) J Solid State Chem 178:1864–1873
29. Sun Y, Ndifor-Angwafor NG, Riley DJ, Ashfold MNR (2006) Chem Phys Lett 431:352–357
30. Cao MH, Wang YH, Guo CX, Qi YJ, Hu CW, Wang EB (2004) J Nanosci Nanotechnol 4:824–828
31. Zhou KB, Wang X, Sun XM, Peng Q, Li YD (2005) J Catal 229:206–212
32. Yuan ZY, Su BL (2004) Colloids Surf A Physicochem Eng Aspects 241:173–183
33. Formalas A (1934) US patent 1 975 504
34. Dai H, Gong J, Kim H, Lee D (2002) Nanotechnology 13:674–677
35. Shao C, Kim HY, Gong J, Park SJ (2003) Mater Lett 57:1579–1584
36. Viswanathamurthi P, Bhattarai N, Kim HY, Lee DR (2003) Scripta Mater 49:577–581
37. Guan H, Shao C, Chen B, Gong J, Yang X (2003) Inorg Chem Commun 6:1409–1411
38. Yang X, Shao C, Guan H, Li X, Gong J (2004) Inorg Chem Commun 7:176–178
39. Guan H, Shao C, Wen S, Chen B, Gong J, Yang X (2003) Mater Chem Phys 6:1302–1303
40. Ding B, Kim H, Kim C, Khil M, Park S (2003) Nanotechnology 14:532–537
41. Viswanathamurthi P, Bhattarai N, Kim HY, Lee DR, Kim SR, Morris MA (2003) Chem Phys Lett 374:79–84
42. Dharmaraj N, Park HC, Lee BM, Viswanathamurthi P, Kim HY, Lee DR (2004) Inorg Chem Commun 7:431–433
43. Tsuda N, Nasu K, Fujimori A, Siratori K (2000) Electronic conduction in oxides, 2nd edn. Springer, Berlin
44. Samson S, Fonstad CG (1973) Defect structure and electronic donor. Levels in Stannic Oxide crystal. J Appl Phys 44:4618–4621
45. Morrison SR (1978) The chemical physics of surfaces. Plenum, New York
46. Madou MJ, Morrison SR (1989) Chemical sensing with solid state devices. Academic, San Diego
47. Kronik L, Shapira Y (1999) Surf Sci Rep 37:1–206
48. Barsan N, Weimar U (2001) J Electroceram 7:143
49. Barsan N, Weimar U (2003) J Phys Condens Matter 15:R813–R839
50. Hahn SH, Barsan N, Weimar U, Ejakov SG, Visser JH, Soltis RE (2003) Thin Solid Films 436:17–24
51. D'Amico A, Di Natale C (2001) IEEE Sens J 1:183–190
52. Yu C, Hao Q, Saha S, Shi L, Kong X, Wang ZL (2005) Appl Phys Lett 86:063101
53. Hernández-Ramírez F, Tarancón A, Romano-Rodríguez A, Casals O, Arbiol J, Morante JR (2007) Sens Actuators B 121:3–17
54. Hernández-Ramírez F, Prades D, Tarancón A, Barth S, Casals O, Jimenez-Diaz R, Pellicer E, Rodriguez J, Morante JR, Juli MA, Mathur S, Romano-Rodriguez A (2008) Adv Funct Mater 18:1
55. Kuang Q, Lao C-S, Li Z, Liu Y-Z, Xie Z-X, Zheng L-S, Wang ZL (2008) J Phys Chem 112:11539–11544
56. Kolmakov A (2006) Proc SPIE 6370:63700X1-8
57. Liao L, Lu HB, Li JC, Liu C, Fu DJ, Liu YL (2007) Appl Phys Lett 91:173110-1-3
58. Riu K, Zhang D, Zhou C (2008) Appl Phys Lett 92:093111

59. Sberveglieri G, Baratto C, Comini E, Faglia G, Ferroni M, Ponzoni A, Vomiero A (2007) Sens Actuators B 121:208
60. Ying Z, Wan Q, Song ZT, Feng SL (2004) Nanotechnology 15:1682–1684
61. Bie L-J, Yan X-N, Yin J, Duan Y-Q, Yuan Z-H (2007) Sens Actuators B 126:604–608
62. Wan Q, Li QH, Chen YJ, Wang TH, He XL, Lin CL (2004) Appl Phys Lett 84:3654–3656
63. Xu J, Chen Y, Li Y, Shen J (2005) J Mater Sci 40:2919–2921
64. Wang C, Chu X, Wu M (2006) Sens Actuators B 113:320–323
65. Wang HT, Kang BS, Ren F, Tien LC, Sadik PV, Norton DP, Pearton SJ, Lin J (2005) Appl Phys Lett 86:243503
66. Tien LC, Sadik PW, Norton DP, Voss LF, Pearton SJ, Wang HT, Kang BS, Ren F, Jun J, Lin J (2005) Appl Phys Lett 87:222106
67. Xianfeng C, Caihong W, Dongli J, Chenmou Z (2004) Chem Phys Lett 399:461–464
68. Vomiero A, Bianchi S, Comini E, Faglia G, Ferroni M, Sberveglieri G (2007) Cryst Growth Des 7:2500–2504
69. Kaur M, Jain N, Sharma K, Bhattacharya S, Roy M, Yagi AK, Gupta SK, Yakhmi JV (2008) Sens Actuators B 133:456–461
70. Polleux J, Gurlo A, Barsan N, Weimar U, Antonietti M, Niederberger M (2006) Angew Chem 45:261–265
71. Liu J, Wang X, Peng Q, Li Y (2005) Adv Mater 17:764–767
72. Raible I, Burghard M, Schlecht U, Yasuda A, Vossmeyer T (2005) Sens Actuators B 106:730–735
73. Bogner M, Fuchs A, Scharnagl K, Winter R, Doll T, Eisele I (1998) Appl Phys Lett 17:2524–2526
74. Dimitrakopoulos CD, Malenfant PRL (2002) Adv Mater 14:99–117
75. Mitzi DB, Kosbar LL, Murray CE, Copel M, Afzali A (2004) Nature 428:299–303
76. Li C, Zhang DH, Liu XL, Han S, Tang T, Han J, Zhou CW (2003) Appl Phys Lett 82:1613–1615
77. Zhang DH, Liu ZQ, Li C, Tang T, Liu XL, Han S, Lei B, Zhou CW (2004) Nano Lett 4:1919–1924
78. Zhang DJ, Li C, Liu XL, Han S, Tang T, Zhou CW (2003) Appl Phys Lett 83:1845–1847
79. Fan ZY, Lu JG (2005) Appl Phys Lett 86:123510
80. Fan ZY, Wang DW, Chang PC, Tseng WY, Lu JG (2004) Appl Phys Lett 85:5923–5925
81. Maeng J, Jo G, Kwon SS, Song S, Seo J, Kang SJ, Kim DY, Lee T (2008) Appl Phys Lett 92:233120
82. Zhang Y, Kolmakov A, Chretien S, Metiu H, Moskovits M (2004) Nano Lett 4:403–407
83. Zhang Y, Kolmakov A, Lilach Y, Moskovits M (2005) J Phys Chem B 109:1923–1929
84. Faglia G, Baratto C, Sberveglieri G, Zha M, Zappettini A (2005) Appl Phys Lett 86:011923
85. Hu J, Bando Y, Liu Q, Goldberg D (2003) Adv Funct Mater 13:493–496
86. He JH, Wu TH, Hsin CL, Li KM, Chen LJ, Chueh YL, Chou LJ, Wang ZL (2006) Small 2:116–120
87. Luo S, Chu PK, Liu W, Zhang M, Lin C (2006) Appl Phys Lett 88:183112
88. Luo SH, Fan JY, Liu WL, Zhang M, Song ZT, Liu CL, Wu XL, Chu PK (2006) Nanotechnology 17:1695–1699
89. Calestani D, Zha M, Zappettini A, Lazzarini L, Salviati G, Zanotti L, Sberveglieri G (2005) Mater Sci Eng C 25:625–630
90. Zhou JX, Zhang MS, Hong JM, Yin Z (2006) Solid State Commun 138:242–246
91. Lettieri S, Bismuto A, Maddalena P, Baratto C, Comini E, Faglia G, Sberveglieri G, Zanotti L (2006) J Non-Cryst Solids 352:1457–1460
92. Lettieri S, Setaro A, Baratto C, Comini E, Faglia G, Sberveglieri G, Maddalena P (2008) New J Phys 10:043013
93. Tak Y, Yong K, Park C (2005) J Electrochem Soc 152:794
94. Huang MH, Wu Y, Feick H, Tran N, Weber E, Yang P (2001) Adv Mater 13:113–116
95. Sun Y, Fuge GM, Fox NA, Riley DJ, Ashfold MNR (2005) Adv Mater 17:2477–2481

96. Vanhausden K, Warren W, Seager CH, Tallant DR, Voigt JA, Gnade BE (1996) J Appl Phys 79:7983
97. Hu JW, Bando Y (2003) Appl Phys Lett 82:1401
98. Lin B, Fu Z, Jia Y (2001) Appl Phys Lett 79:943
99. Studenikin SA, Cocivera M (2002) J Appl Phys 91:5060
100. Yao BD, Chan YF, Wang N (2002) Appl Phys Lett 81:757
101. Comini E, Baratto C, Faglia G, Ferroni M, Sberveglieri G (2007) J Phys D 240:7255–7259

Theory and Application of Suspended Gate FET Gas Sensors

C. Senft, P. Iskra, and I. Eisele

Abstract Looking at the literature about chemical sensors, it is evident that numerous sensor effects are reported but only a very few sensor concepts remain, which are suitable for industrial applications. This is in contrast to the measurement of physical parameters such as pressure, temperature, acceleration etc. There are several reasons for this discrepancy. First of all, for chemical reactions usually there exist much more cross correlations, which have to be considered in order to determine a correct concentration of chemical species, and secondly, environmental influences cannot be neglected at all. Therefore, it becomes extremely difficult to fulfill all the requirements, which have been listed in the introduction of this paper.

One interesting and successful gas sensor concept is based on the measurement of surface work function changes due to chemical reactions. Besides a Kelvin probe with a vibrating capacitance, field effect transistors (FET) are well suited to act as transducers for the determination of the corresponding potential variations. This chapter reviews the detection mechanisms, which lead to work function changes and gives an overview of the various transducer concepts. Beginning with the so-called Lundström FET, the historical development all the way to the hybrid mounted "floating gate FET (FG-FET)" is presented. This latest concept is extremely flexible and depending on the chemical-sensitive layer, it can be used for the detection of a large variety of gases.

As an example the development of a hydrogen sensor for future automotive application is presented in more detail. Using platinum as chemical-sensitive layer it is shown that under harsh environmental conditions, it is not sufficient to consider exclusively the reaction of hydrogen with the platinum, but rather, it is necessary to also take into account oxygen, which is always present in air. As a result, a characteristic catalytic ignition point, i.e., the adsorbed hydrogen atoms are consumed by a

C. Senft (✉), P. Iskra, and I. Eisele
Lehrstuhl für Mikrosystemtechnik, Institut für Physik, Universität der Bundeswehr München, 85577 Neubiberg, Germany
e-mail: csenft3@gmail.com

M. Fleischer and M. Lehmann (eds.), *Solid State Gas Sensors – Industrial Application*, Springer Series on Chemical Sensors and Biosensors (2012) 11: 79–112,
DOI 10.1007/5346_2011_12,
Published online: 10 March 2012

water-forming reaction, occurs at around 60°C. Only if the complete reaction scheme is considered, it is possible to understand the complex and partially intriguing transducer signals. The development of two-layer systems for the chemical-sensitive layer solves these problems by using a phase transition that comes along with the catalytic ignition and allows a stable sensor operation in the required temperature regime between −40°C and +120°C. Based on the theoretical modeling the temperature-dependent phase transition has been evaluated in order to measure hydrogen concentration up to 4% with high accuracy and to fulfill the automotive requirements.

Finally, in the last section, a new GasFET concept is presented, which extends the operating regime to temperatures as high as 400°C. In the future, this allows the incorporation of a variety of new gas-sensing materials, which need temperatures above 200°C in order to show a decent adsorption–desorption equilibrium.

Altogether, it has been proven that GasFETs are a very promising candidate for future industrial applications.

Keywords GasFET, High temperature FET, Hydrogen sensor, Vertical MOS-FET, Work function sensor, TPT-FET

Contents

Abbreviations

$\Delta\Phi$	Work function change
$\Delta\Phi_{Target}$	Target work function change
μ	Effective charge carrier mobility (channel region)
Φ	Work function
c	Gas concentration
C'	Total capacitance of air gap and passivation
C_A	Air gap capacitance
CC-FET	Capacitive-coupled field effect transistor
C_I	Gate oxide capacitance per unit area
CMOS	Complementary metal oxide semiconductor
C_{OX}	Oxide capacitance
C_{PASS}	Passivation capacitance
D	Diffusion coefficient/thermal diffusivity
e	Elementary charge
E_d	Desorption energy
E_r	Activation energy (reaction)
FET	Field effect transistor
FG-FET	Floating gate field effect transistor
HSG-FET	Hybrid suspended gate field effect transistor
HT-FG-FET	High temperature floating gate FET
I_{DS}	Source–drain current
I_{OFF}	OFF state current
I_{ON}	ON state current
k_a	Adsorption parameter
k_B	Boltzmann constant
k_d	Desorption coefficient
k_r	Reaction coefficient
L_{eff}	Effective channel length
MOS-FET	Metal oxide semiconductor field effect transistor
O_{ads}	Oxygen coverage
O_{max}	Oxygen saturation coverage
P(T)	Thermal dependent permeability
PMMA	Polymethylmethacrylat
p_x	Partial pressure
SEM	Scanning electron microscope
SG-FET	Suspended gate field effect transistor
SOI	Silicon on insulator
T	Temperature
TPT-FET	Temperature-controlled phase transition FET
T_R	Response Time
V_C	Counter voltage
V_{DS}	Source–drain voltage

V_G	Gate voltage
V_T	Threshold voltage
W_{eff}	Effective channel width
W_{Fermi}	Fermi energy level
W_{vac}	Vacuum energy level

1 Introduction

The major challenges for all types of solid-state gas sensors can be described by the following four "high" and three "low" criteria:

• High sensitivity	• Low power consumption
• High selectivity	• Low weight
• High stability	• Low price
• High signal/noise ratio	

Most of the existing sensor concept fulfill these requirements only partially or are suited just for a limited number of gas species. A typical example is the detection by heat conductivity, which can only be applied for hydrogen and helium [1, 2]. One promising gas-sensing concept with high flexibility regarding sensing effects and materials is the detection via field effect devices.

Suspended gate field effect transistor gas sensors (SG-FET) [3–5] belong to this group of sensors. In principle, they are capable of detecting a wide range of different gases by simply exchanging chemical-sensitive adsorption layers. As explained in more detail in the next section, this adsorption leads to a change in work function of the sensitive layer. The general idea of the suspended gate FET (SG-FET) is to measure these work function changes via a field effect transistor (FET). Simplified, the work function change acts as an artificial gate voltage, as displayed in Fig. 1. Overall, two different main device structures have been realized.

Historically, the first devices were based on metal oxide semiconductor field effect transistors (MOS-FETs). These GasFETs actually do not feature a suspended gate, but their operation method is similar. The gas sensitivity of these devices is

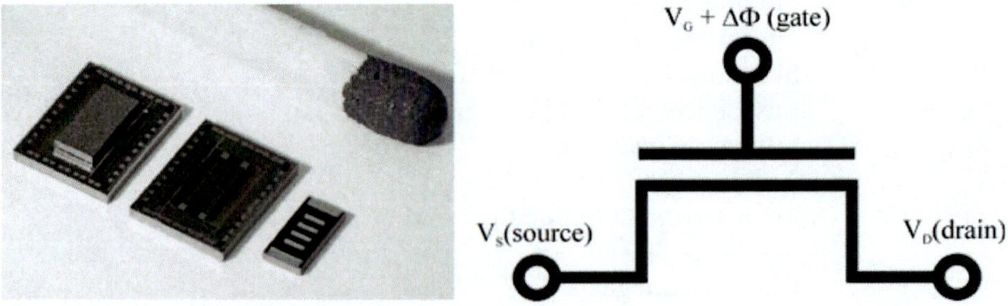

Fig. 1 FG-FET in CMOS technology [6] (*left*) and a schematic view of the device (*right*)

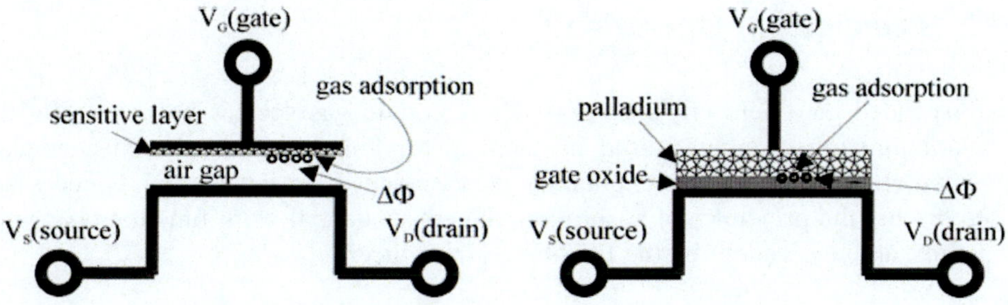

Fig. 2 Schematic view of an SG-FET (*left*) and a Lundström-FET (*right*)

provided by using a permeable palladium film as gate electrode, as can be seen in Fig. 2 (right).

As explained in Sect. 3, this so-called Lundström FET [7] is especially suitable for hydrogen detection. If this device is exposed to hydrogen, the molecules dissociate at the palladium surface and diffuse to the palladium–silicon oxide interface, where they are adsorbed. Therefore, a potential shift $\Delta\Phi$ is generated, which is added to a possible applied gate voltage. Consequently, the source–drain current due to the voltage drop $V_D - V_S$ is altered. This current change is used as indicator for the target gas concentration. To operate these devices, a metal gate is required, which is at least permeable for the target gas molecules. It is obvious that, this is a hard limitation to the number of possible gas-sensitive materials.

Later on, suspended gate field effect devices (SG-FETs) have been developed for gas detection. These devices have an air gap between the gate dielectric and the gate electrode, as displayed in Fig. 2 (left). Below the elevated gate electrode a chemical-sensitive film is deposited. The target gas molecules diffuse through the air gap and adsorb on the surface of the sensitive layer. As explained in Sect. 2, this induces a work function shift of the sensitive layer. The potential shift is added to a possible applied gate voltage V_G, and consequently, the source–drain current is altered. Similar to the Lundström device, this current change is used as indicator for the concentration of a target gas. In principle, these devices can be used to detect a wide range of gas species. To adopt a device to a certain target gas, one needs a matching sensitive film for the SG-FET or a matching permeable material for the Lundström device, respectively. Except for the chemically sensitive layer, the sensor device can be fabricated equally for all target gases, which is a unique feature for a gas sensor.

In this chapter we review theory and application of these two types of gas sensors. We explain the adsorption of gas molecules on a solid surface as well as the induced work function change (Sect. 2). Afterward, we provide an overview about several transducer concepts (Sect. 3) and their historical development [8]. In Sects. 4 and 5 as an example, the detection of hydrogen via these GasFETs is described. Furthermore, we will take a deeper look into several detection problems caused by the chemical behavior of the sensitive surface and the long way of development to solve these problems. Finally, in Sect. 6, we will give an outlook into the future of GasFETs, which are operable even at temperatures up to 400°C.

2 Principles of Operation

Suspended gate field effect sensors detect certain gas species by measuring the work function changes caused by their adsorption on a well-chosen chemical-sensitive layer. In order to explain the detailed sensing mechanism, it is necessary to discuss the principles of adsorption, the consequential work function change of solids, and the readout by the FET-based transducer.

2.1 Gas Adsorption on Solids

Under environmental conditions, all solid surfaces are covered by numerous molecule species like water, oxygen, and various types of hydrocarbons. The precise amount of adsorbents depends mainly on the surface chemistry and the available amount of the target gas molecules in the surrounding atmosphere. In general, we have to distinguish between two types of adsorption: physisorption and chemisorption.

In case of physisorption, which is shown in Fig. 3 (left), no chemical interaction takes place between the adsorbed gas molecules and the surface. This means the gas molecules are not chemically altered during the adsorption process, and as a consequence, physisorption is always reversible. The adsorbate is bound only via electrostatic Van der Waals forces. Therefore, the binding is very weak. In most cases, the binding energy is well below 100 meV, and consequently, these molecules can be easily desorbed thermally even at room temperature.

In contrast, chemisorbed molecules exhibit a much stronger binding to the surface. Incoming molecules are chemically altered during the adsorption. Figure 3 (right) displays the chemisorption in the special case of simple dissociative chemisorption. The adsorption is not governed by weak electrostatic forces, but rather by strong chemical bindings. Therefore, chemisorption is not always reversible.

Whether adsorption takes place mainly depends on the chemical relationship between the surface and the gas species. Table 1 shows some known combinations of adsorbing gases and reliable sensitive materials, which have already been deployed in an SG-FET.

Fig. 3 Gas adsorption on a solid: physisorption (*left*) and chemisorption (*right*)

Table 1 Known combinations of detectable gas species and associated sensitive material

Sensitive layer	Target gas
[9] Platinum	H_2
[10] Palladium	H_2
[11] Titanium nitride	NH_3
[12] Copper–phthalocyanine	NO, NO_2
[13, 14] Bariumtitanate, polysiloxanes	CO_2
[15] Tin dioxide (undoped a. Pt-doped)	CO, NO_2, O_3
[16] Silver oxide	H_2S
[17] Cobalt oxide	NO_2, NH_3
[17] Chromium titanium oxide	H_2S, NH_3, Cl_2
[15] Gold	Cl_2
[18] Potassium iodide	O_3
[19] Germanium	C_2H_5OH

$$\frac{d\Theta_X}{dt} = 2 p_X k_a (1-\Theta_X)^2 - 2 k_d \Theta_X^2$$

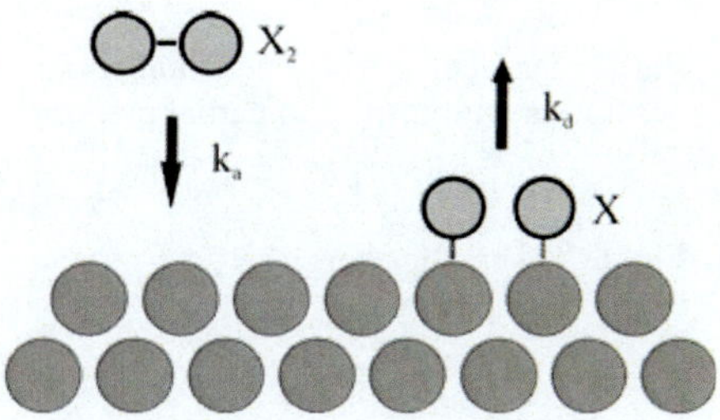

Fig. 4 Rate equation (*left*) and schematic view of dissociative adsorption and desorption (*right*)

The total amount of coverage depends on many parameters like partial pressure of the adsorbing gas, temperature, humidity, and disturbing adsorption of undesirable molecules. In case of dissociative chemisorption, the molecular coverage Θ_X is calculated from a simple rate equation, as displayed in Fig. 4.

p_x is the partial pressure; $k_{a/d}$ are rate constants for adsorption and desorption. Figure 5 displays the theoretical surface coverage at equilibrium state (steady state), i.e., $d\Theta/dt = 0$. A more detailed model of the adsorption behavior can be found [20]. The total coverage increases with increasing partial pressure of the adsorbing gas. As will be explained below, the continuous coverage leads to an analogous work function change of the surface.

2.2 *The Work Function*

The work function is defined as the minimum energy, which is necessary to extract an electron from a neutral solid. A simplified energy diagram is presented in Fig. 6. For a metal, the mathematical definition is $\Phi = W_{vac} - W_{Fermi} \cdot W_{vac}$ is the vacuum level energy; W_C is the conduction band, and W_{Fermi} is the Fermi energy. It should

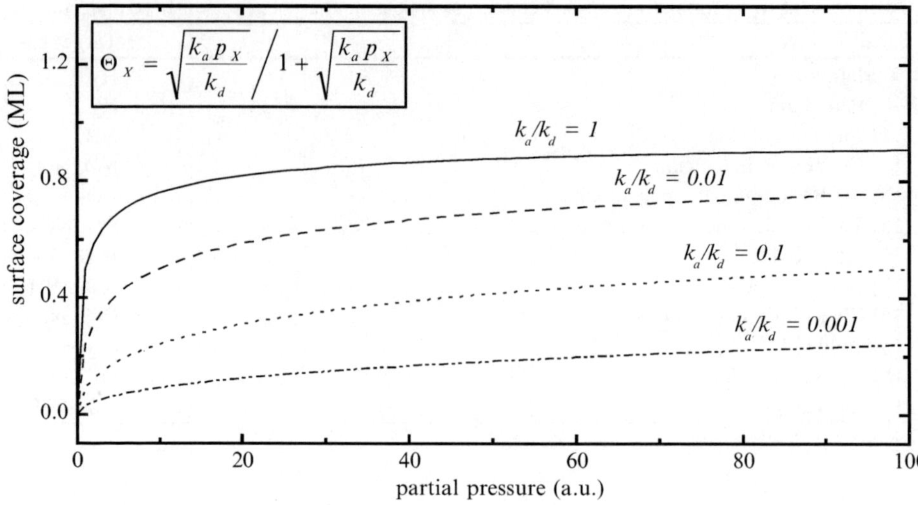

Fig. 5 Surface coverage equation (*inset*) and graphical solution of dissociative adsorption and desorption with respect to partial pressure

Fig. 6 Schematic energy diagram of a metal

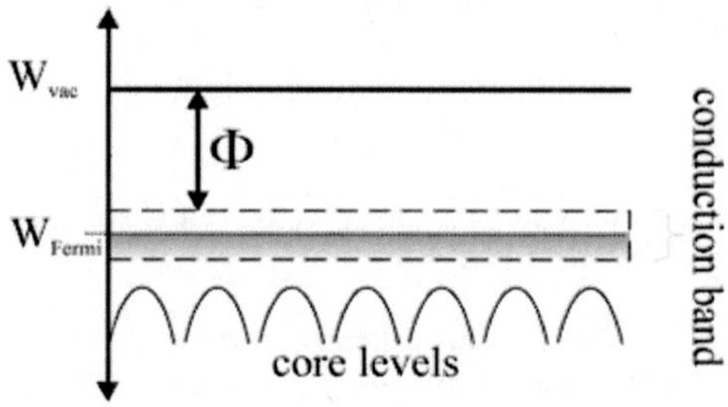

be noted that for semiconductors, the work function is mainly determined by the electron affinity, which is given by the energy difference between the vacuum energy and the corresponding band edges.

Since the vacuum energy is defined as zero energy level and the Fermi level is material dependent, the work function is a characteristic parameter for each solid. In addition, the surface orientation, reconstruction, and topology strongly influence the local Fermi energy and consequently the work function.

2.3 *Work Function Change Caused by Gas Adsorption*

If foreign atoms or molecules cover the surface of a solid, the work function is altered. In case of physisorption, a potential shift is provided by dipoles, which are formed by the adsorbed atoms or molecules on the surface. A much stronger

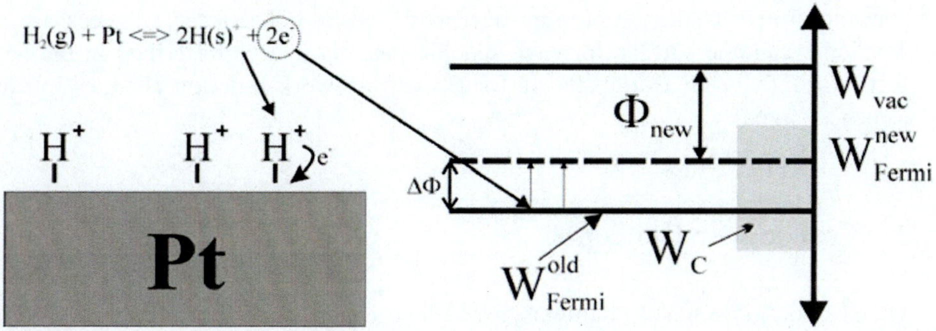

Fig. 7 Work function change caused by chemisorption of hydrogen on clean platinum

potential shift can be caused by chemisorbed species. To explain this behavior in more detail, we will discuss the chemisorption of hydrogen on polycrystalline platinum.

On platinum surfaces, the hydrogen molecules adsorb dissociatively [21], i.e., the molecules dissociate when they impinge at the surface and are bound as single atoms. These hydrogen atoms are strongly electropositive [21], and their electrons are partially transferred into the platinum. As displayed in Fig. 7, the additional electrons occupy former empty electronic states, and therefore, the Fermi level of the platinum is increased. Since the work function is defined as the energy difference between the Fermi energy and the constant vacuum energy level, the work function is decreased due to this adsorption. As has been shown by experiments [21] under UHV conditions, the adsorption of hydrogen decreases the work function of a platinum surface by 230 meV for Pt(111) and up to 300 meV for polycrystalline surfaces. The dependence of the potential shift with respect to coverage was quantified as: $\Delta\Phi = -0.23 \times (H_{ads}/H_{max})^{1.33}$ [eV],[1] where H_{ads} is the coverage in monolayers and H_{max} is the saturation coverage.

Another example of dissociative adsorption is the interaction of oxygen with a platinum surface. Similar to hydrogen, the oxygen is bound to platinum in the form of isolated single atoms. In contrast to hydrogen, the oxygen coverage is strongly electronegative, and therefore, bulk electrons are partly transferred to the adsorbed oxygen atoms. Consequently, the Fermi level of the platinum decreases and the work function increases. The dependence of this potential shift with respect to coverage was also measured in UHV experiments [22, 23] and can be written as $\Delta\Phi = 0.4 \times (O_{ads}/O_{max})$ [eV].

The given examples clearly show that measurements of work function changes can be used as indicator for the surface coverage of impinging species. Depending on surface properties, any gas adsorption leads to an individual work function change; this shift can be used as an indicator for gas species and their corresponding

[1] For Pt(111)

concentrations. Therefore, for a successful sensor concept, it is necessary to develop a suitable surface for each specific gas. As will be described in the next section, FET are ideal transducers for converting work function changes into an electrical signal.

3 Transducer Concepts

This section gives a short summary over the different transducer concepts of field effect-based gas sensors and their historical development. At first, the Lundström gas sensor is described, followed by the suspended gate devices and their evolution steps. The latest stage in development is represented by a combination of a Lundström FET and a SG-FET with a floating gate on a single sensor chip.

3.1 The Lundström FET

I. Lundström developed the first gas sensor based on a conventional FET in 1975 [7]. This device is shown in Fig. 8 (left).

In principle, it is a conventional MOS-FET with a palladium electrode above the SiO_2 gate dielectric. If such a device is exposed to hydrogen, the molecules are dissociated at the Pd surface, and subsequently, they diffuse in the form of single atoms through the Pd layer down to the Pd-SiO_2 interface as displayed in Fig. 8 (right). These hydrogen atoms occupy free sites on the buried Pd-SiO_2 interface, and local dipoles are generated. Consequently, the effective gate voltage is altered, and a possible source–drain current I_{DS} is changed. Simplified I_{DS} is given by:

$$I_{DS} = \mu \frac{W_{eff}}{L_{eff}} C_I \left(V_G + \frac{\Delta\Phi}{e} - V_T \right) V_{DS}. \quad (1)$$

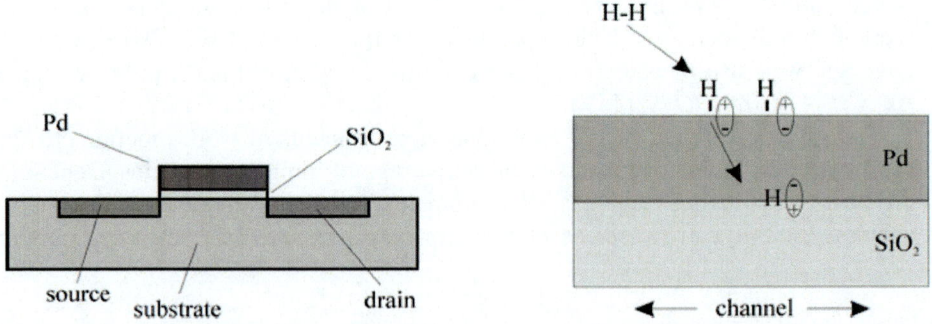

Fig. 8 Schematic view of the Lundström device (*left*) and interface adsorption of hydrogen (*right*)

Here, μ denotes the carrier mobility; W_{eff} and L_{eff} are the effective channel dimensions of the transistor; e is the elementary charge; C_I is the oxide capacitance with respect to the area; and V_G and V_{DS} are the applied gate and source–drain voltages. The Lundström FET can be operated like a conventional MOS-FET, and as the I-V characteristics of the device are known, the work function change induced by hydrogen exposure can be measured directly. Alternatively, one can keep the source–drain current by applying a counter voltage V_C to the gate. Operated in this mode, the work function change $\Delta\Phi$ simply corresponds to $-eV_C$. In principle a Lundström device can be used to detect various gas species, but one needs a permeable gate material for each species, which is not available for most target gases. In the case of hydrogen, the nonpermeability of Pd for other gas molecules is a big advantage, since it leads to a very low cross sensitivity to other gases [24]. In summary, the Lundström FET has been proven to be a reliable sensor for hydrogen concentrations up to 1.5%. Above this value, the occupation of interface sites and, as a consequence, the work function change as well as the sensor signal start to saturate [7].

3.2 *The Suspended Gate FET*

In order to abolish the disadvantages of Lundström-type gas sensors, which come along with the essential need of a permeable gate material, an air gap between the sensitive layer and the gate oxide was introduced. The first gas FET device with such an air gap was the monolithically integrated SG-FET. It based on a patent by Janata [4, 5], which describes the monolithic fabrication in silicon technology. A schematic cross section of this device can be seen in Fig. 9.

By etching a sacrificial layer (e.g., aluminum or silicon oxide) between the platinum gate electrode and the gate insulator, a bridge-like structure is monolithically formed. Underneath the bridge material platinum, which by itself is a sensitive

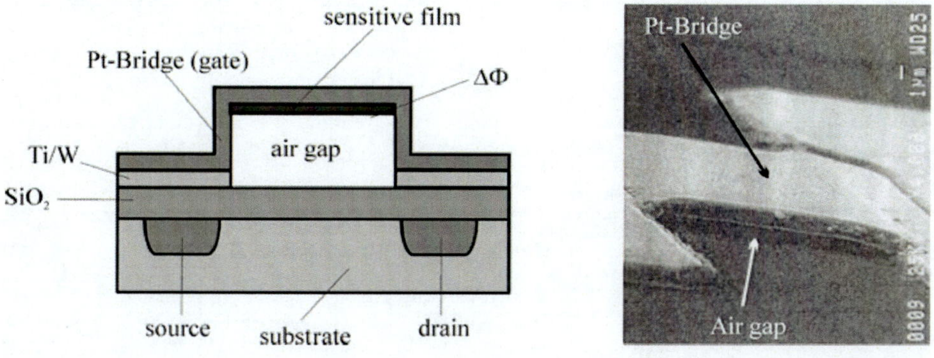

Fig. 9 Schematic view of a monolithic SG-FET (*left*) and an SEM image of the Pt-bridge (*gate*) [5] (*right*)

layer, various additional sensitive materials have been integrated by electrochemical deposition [25]. An SEM image of the platinum gate bridge is shown in Fig. 9 (right). The height of the air gap must be at least 1 μ, in order to keep gas diffusion to the sensitive layer fast enough. For the source–drain current I_{DS}, the effective capacitance C_{eff} with respect to area has to be considered according to (2):

$$I_{DS} = \mu \frac{W_{eff}}{L_{eff}} C_{eff} \left(V_G + \frac{\Delta\Phi}{e} - V_T \right) V_{DS} \quad \text{with} \quad C_{eff} = \frac{C_I C_A}{C_I + C_A}. \tag{2}$$

Here, C_I denotes the gate oxide capacitance, and C_A is the capacitance of the air gap and the sensitive layer. Because of the large air gap, C_A is very small in comparison to C_I, so $C_{eff} \approx C_A$. Consequently, the threshold voltage V_T is increased and can reach values up to 100 V. To avoid the application of such high voltages, which lead to isolation problems normally – ON transistors have to be used. In this case the device can be operated near ground potential for the gate electrode. Due to the small C_{eff}, the main disadvantage of this transistor type is a very low transconductance and consequently a low sensitivity to work function changes $\Delta\Phi$.

3.3 *The Hybrid Suspended Gate FET*

By far more flexible to the deposition of sensitive materials is the hybrid suspended gate FET (HSG-FET), which was developed by Eisele at al. [3] in 1992. In contrast to the monolithically SG-FET, a separated MOS-FET transducer device was combined with a hybrid mounted top electrode via flip chip bonding. A schematic layout (left) and an equivalent circuit diagram (right) is presented in Fig. 10.

The hybrid design opens up a wide spectrum of deposition methods for the sensitive materials since its preparation is no longer linked to the transducer fabrication. As the electrical design remains unchanged in comparison to the monolithic SG-FET, the source–drain current is given by (2) as well. HSG-FET also needs a normally – ON transistor, and consequently, the device has a low sensitivity to $\Delta\Phi$.

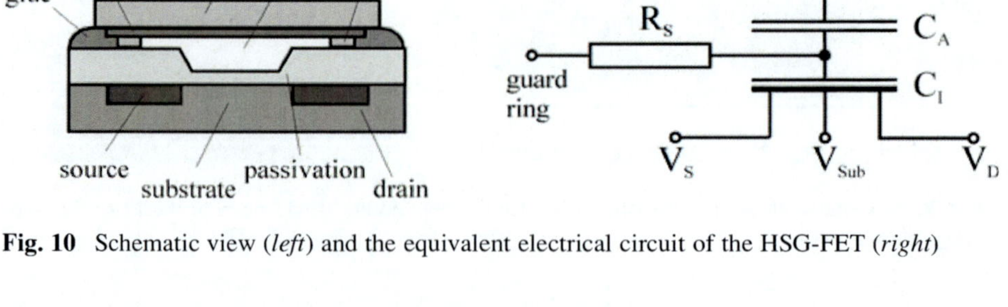

Fig. 10 Schematic view (*left*) and the equivalent electrical circuit of the HSG-FET (*right*)

3.4 The Floating Gate FET (FG-FET)

The next step in development was the capacitive-coupled FET (CC-FET). It was realized by Schipanski et al. [26, 27] and Jung [28] in the first place. This device combines the versatility of the HSG-FET with conventional MOS-FET technology. Based on this device, Burgmair et al. developed the floating gate FET (FG-FET) [29]. Until now, this device represents the latest stage of development. A schematic layout (left) and an equivalent circuit diagram (right) are presented in Fig. 11.

The fully integrated device consists of two parts: a capacitive voltage divider, which contains the chemical-sensitive layer, and a separated read out MOS-FET. The floating gate of this FET is electrically connected to the middle electrode of the capacitive voltage divider. The top electrode of the capacitive voltage divider is mounted again by hybrid technology, allowing the flexible deposition of sensitive materials.

The advantage of this concept is the fact that transistor and suspended gate can be operated independently. Due to the capacitive voltage divider, a change in work function $\Delta\Phi$ of the sensitive layer alters the potential of the buried floating gate electrode by ΔV_G, according to (3):

$$\Delta V_{\mathrm{G}} = +\frac{C\prime}{C\prime + C_{\mathrm{OX}}} \times \frac{\Delta\Phi}{e} \quad \text{with} \quad C\prime = \frac{C_{\mathrm{A}} C_{\mathrm{Pass}}}{C_{\mathrm{A}} + C_{\mathrm{Pass}}}, \tag{3}$$

where C_A is the air gap capacitance, C_{Pass} the capacitance of the floating gate passivation, and C_{OX} the capacitance between the floating gate and the bottom electrode.

Due to (3), the floating gate potential is lower than the work function change, which is not the case for the simple SG-FET. However, at the same time, the floating electrode is the gate of a separated conventional MOS-FET with an independent gate capacitance C_I. Since this capacitance exceeds by far all other capacitances of previous SG-FET devices in total, a significant amplification can be achieved. With respect to (3), the source–drain current can now be written as:

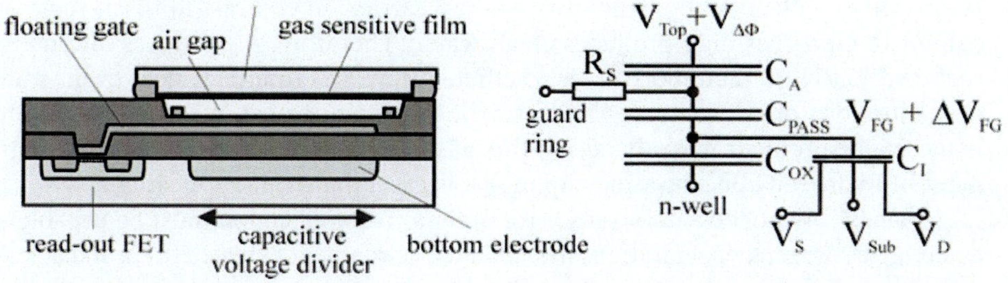

Fig. 11 Schematic view (*left*) and the equivalent electrical circuit of the FG-FET

$$I_{\mathrm{DS}} = \mu \frac{W_{\mathrm{eff}}}{L_{\mathrm{eff}}} C_{\mathrm{I}} \left(V_{\mathrm{ext}} + \frac{C'}{C' + C_{\mathrm{OX}}} \times \frac{\Delta\Phi}{e} - V_{\mathrm{T}} \right) V_{\mathrm{DS}} \quad \text{with}$$
$$V_{\mathrm{ext}} = \frac{C' V_{\mathrm{top}} + C_{\mathrm{OX}} V_{\mathrm{cap}}}{C' + C_{\mathrm{OX}}}. \tag{4}$$

The voltages V_{top} and V_{cap} can be applied to the top electrode and the capacitive well contact, respectively. They can be used to define the operating point of the device. Compared to the HSG-FET, this design offers a higher sensitivity to the target gas concentration. A more detailed description can be found in the literature [30].

To combine the advantages and disadvantages of the Lundström FET and the FG-FET, both transducers were combined on a single CMOS die in 2005 by Micronas GmbH. A detailed description of this concept will be discussed in Sect. 5.

4 The FET-Based Hydrogen Sensor

Work function-based field effect sensors are adaptable to various gas species. The applicability depends on the amplitude of the induced potential shift of the chosen sensitive layer. However, hydrogen detection via these devices was a central point of development from the very first time. In the beginning, this example was chosen because of the extraordinary large work function shift, induced by hydrogen adsorption on platinum [31] and palladium [10]. Today, the development of GasFETs for hydrogen detection is still an interesting subject, since this gas might be the energy carrier of the twenty-first century. This section gives an insight into the hydrogen-sensing mechanism of a platinum-based floating gate field effect transistor (FG-FET) [31] for automotive applications. The main focus is put on the sensing characteristics of the platinum layers and stability problems at operation temperatures above 60°C [32].

4.1 Requirements for Automotive Applications

Hydrogen is a clean and regenerative energy carrier. In contrast to fossil fuels, its combustion process only produces clean water. Therefore, its usage as alternative fuel especially in automotives is predicted for the future. Since even small concentrations of hydrogen (~4%) in surrounding air can cause explosions, safety is the major problem when handling this gas. Consequently, the precise measurement of hydrogen concentrations up to 4% is essential to all applications.

A reliable hydrogen-safety sensor for automotive application must be capable of detecting hydrogen concentrations in standard environment between 0.1% and at least 4% with an absolute accuracy of 0.2%. In order to prevent accidents, its reaction time has to be well below 3 s. The sensor also has to operate stable under a wide range of environmental conditions, e.g., humidity between 0% and 100% and surrounding temperatures between –40°C and 120°C must not disturb the precise detection. Besides the sensing specifications, there are high economical requirements, like low costs, low power consumption, and high integrability.

During the past years, many completely different sensor concepts were developed, like hydrogen detection via thermal conductivity [1, 2], electrochemical cells [33] or metal oxide sensors [34]. But up to now, none of these concepts has yield in a commercially available hydrogen sensor, which is suitable in mass products, especially not in fuel cell cars. The major problem is to fulfill all requirements, listed above at the same time.

After nearly 30 years of development, the actual SG-FET devices are now close to satisfying all physical and economical requirements. In this section, we provide an insight to the detection problems, as well as to the development of their solutions.

4.2 The FG-FET Hydrogen Sensor

In order to detect hydrogen via a work function change, platinum [31] and palladium [10] are used as sensitive layers in the FG-FET because of their strong catalytic behavior. On their surfaces, the hydrogen molecules adsorb dissociatively [21], which leads to a partial electron transfer into the metal. In consequence, the Fermi level increases, and the work function decreases. As can be seen in Fig. 12, the potential shift of these metals can reach values up to 800 meV, so it is quite easy to detect hydrogen via this work function change. However, as reported in the literature [21], the work function change of a clean platinum surface, generated by the adsorption of hydrogen in UHV, saturates at about only 300 meV. The difference can be explained if we consider an induced change in oxygen coverage as well. Under atmospherical conditions, a platinum surface is always covered by a high amount of environmental oxygen [35]. This coverage comes along with a work function, increased by up to 400 meV. If hydrogen adsorbs at an oxygen-precovered

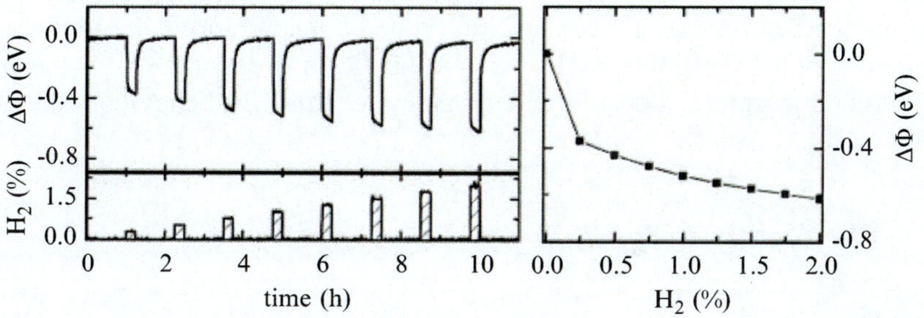

Fig. 12 Work function change of a platinum surface, if exposed to various hydrogen concentrations in synthetic air with 30% relative humidity

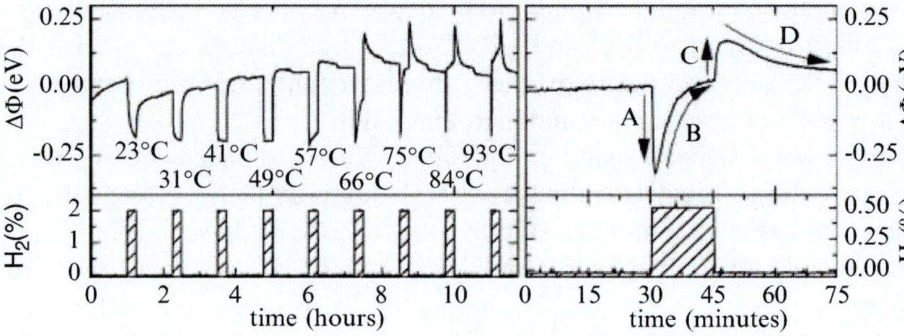

Fig. 13 Work function change of a platinum surface exposed to hydrogen in synthetic air with 30% relative humidity at different temperatures (*left*) and the drift behavior (*right*)

platinum surface, the oxygen coverage is lowered due to the catalytic surface reaction between the two species [36]. In combination, the adsorption of hydrogen and the reduction of the initial oxygen coverage lead to the high-observed work function changes [32].

As displayed in Fig. 12, the work function of polycrystalline platinum decreases nearly logarithmical with rising hydrogen concentrations [31] and starts to saturate at about 2%. Therefore, it is very difficult to differ between 2.4% and 2.5% [24].

Besides the very low upper detection limit, the platinum-based FG-FET faces another vital problem. The logarithmic work function change of a platinum surface is stable only below a certain temperature limit. As shown in Fig. 13 (left), hydrogen exposure above temperatures of about 60°C lead to instable work function changes.

Above this critical temperature, we observed only a short negative work function change (*A*) when exposed to hydrogen. After this initial decrease, the signal starts to drift in the opposite direction (*B*). Right after turning the hydrogen off, a small but fast increase in work function is observed (*C*), followed by a slow decrease (*D*), which shifts the work function back to its initial value. This whole drift behavior is accelerating with further rising of the temperature. As described in the beginning of this section, most applications and especially all automotive applications require the stable operation of hydrogen sensors up to temperatures of at least 120°C. Since the absolute value of the change in work function is used as sensor signal, this back drift prevents the determination of precise hydrogen concentrations. In consequence, a platinum-based FG-FET is useless above 60°C. In order to expand the possible operating temperature, a deeper understanding of this drift behavior is necessary.

4.3 The Catalytic Ignition

As mentioned before, the work function change of a platinum surface under hydrogen exposure in standard environment is determined by its actual surface coverage. Adsorbed hydrogen decreases the work function while oxygen coverage increases it.

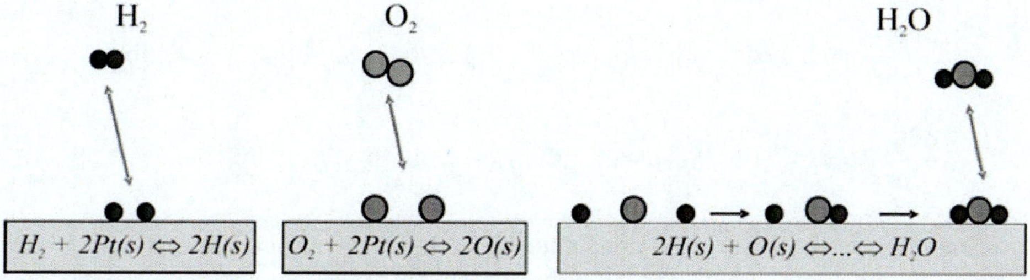

Fig. 14 Reaction steps of the catalytic hydrogen oxidation on a platinum surface

To understand the drift effects described in the previous section, a detailed knowledge of the covering adsorbents and their reaction kinetics is required.

In the gas phase, hydrogen and oxygen can react to water in a detonation reaction above temperatures of about 560°C.[2] Due to the strong catalytic properties of platinum on the metal surface, this reaction can take place even at room temperature since the reactants are adsorbed in an activated state. A schematic of the water-forming surface reaction steps are displayed in Fig. 14.

The single reaction paths are as followed:

$$\text{Hydrogen adsorption}: \ H_2(g) + 2Pt(s) \Leftrightarrow 2H(s) \tag{5a}$$

$$\text{Oxygen adsorption}: \ O_2(g) + 2Pt(s) \Leftrightarrow 2O(s) \tag{5b}$$

$$\text{OH} - \text{creation}: \ H(s) + O(s) \Leftrightarrow OH(s) \tag{5c}$$

$$\text{Water creation A}: \ OH(s) + H(s) \Leftrightarrow H_2O(s) \tag{5d}$$

$$\text{Water creationB}: \ OH(s) + OH(s) \Leftrightarrow H_2O(s) + O(s) \tag{5e}$$

$$\text{Water desorption}: \ H_2O(s) \Leftrightarrow H_2O + Pt(s) \tag{5f}$$

The overall reaction follows the Langmuir–Hinshelwood mechanism [37]. Assuming the mean-field approximation, where the reactants are randomized across the surface, this system can be described by four coupled ordinary, but highly nonlinear, differential equations, which determine the absolute coverage change $d\Phi_x/dt$ of the involved species in the time domain:

$$\frac{d\Phi_H}{dt} = k^a_{H_2}(\Phi)P_{H_2} - k^d_{H_2}\Phi^2_H - k_{r_1}\Phi_H\Phi_O - k_{r_2}\Phi_H\Phi_{OH} \tag{6a}$$

$$\frac{d\Phi_O}{dt} = k^a_{O_2}(\Phi)P_{O_2} - k^d_{O_2}\Phi^2_H - k_{r_1}\Phi_H\Phi_O + k_{r_3}\Phi^2_{OH} \tag{6b}$$

[2] Material safety data sheet, Linde AG, Germany

$$\frac{d\Phi_{OH}}{dt} = k_{OH}^{a}(\Phi)P_{OH} - k_{OH}^{d}\Phi_{OH} + k_{r_1}\Phi_{H}\Phi_{O} - k_{r_2}\Phi_{H}\Phi_{OH} - k_{r_3}\Phi_{OH}^{2} \quad (6c)$$

$$\frac{d\Phi_{H_2O}}{dt} = k_{H_2O}^{a}(\Phi)P_{H_2O} - k_{H_2O}^{d}\Phi_{H_2O} + k_{r_2}\Phi_{H}\Phi_{OH} + k_{r_3}\Phi_{OH}^{2}. \quad (6d)$$

Here, k_a is the adsorption constants; k_d is the desorption constants, and k_r is the reaction constants, which can be written as:

$$\text{Adsorption}: \; k_{x}^{a}(\Phi) = (1-\Phi)^{y} \times s \times \sigma \quad (7a)$$

$$\text{Reaction}: \; k_r = \nu_r \times \exp(E_r/k_B T) \quad (7b)$$

$$\text{Desorption}: \; k_{x}^{d}(\Phi) = \nu_d \times \exp(E_d(\Phi)/k_B T). \quad (7c)$$

Here, s is the sticking coefficient; σ is the adsorption area; E_r and E_d are the activation energies, and Φ is the total surface coverage. Exact kinetic parameters can be found in literature [37]. The work function change of a platinum surface, which is exposed to hydrogen depends on the change of the surface coverage (oxygen and hydrogen) and can be calculated by (8). This equation is described in more detail in Sect. 2.3:

$$\Delta\Phi = -0.23\left(\frac{H_{ads}}{H_{ML}}\right)^{1.33} + 0.4\left(\frac{O_{ads}}{O_{ML}}\right)[\text{eV}]. \quad (8)$$

Overall, these reactions consume hydrogen and oxygen from the gas phase and produce water, which desorbs from the surface. Since the reaction can be very fast, the mass transport of the reactants to the surface has to be considered. Assuming a simple diffusion behavior, it is described by Fick's Second Law:

$$\frac{\partial c}{\partial t} = \frac{\partial c}{\partial x}\left(D\frac{\partial c}{\partial x}\right). \quad (9)$$

Here, c is the concentration and D is the diffusion coefficient. Furthermore, because of the strong exothermal character of the reaction, the heat balance of the surface is not negligible. The surface is heated by the reaction and is cooled down by the surrounding gas phase and the substrate. The thermal diffusion can also be described by Fick's Second Law, if concentration c is replaced by temperature and D by thermal diffusivity.

The coupling of the surface rate equations and the diffusion equations results in a numeric model which explains the work function drift of the platinum surface very well. A more detailed explanation of this numeric model can be found in literature [32]. Figure 15 (left) shows the results of such a simulation. A hydrogen pulse of 2% in ambient air at 25°C was simulated (*g*), and the surface coverages (*c–f*) as well

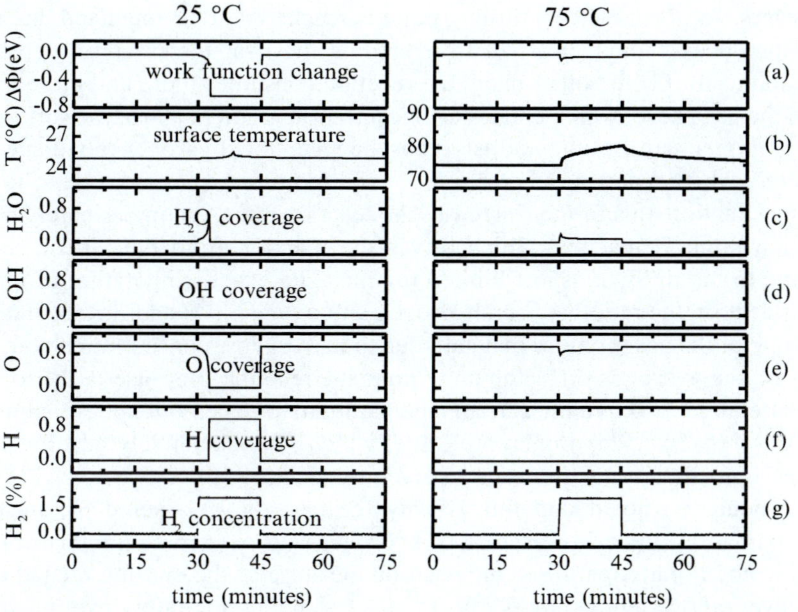

Fig. 15 Numerical simulation of the hydrogen adsorption on Pt and the corresponding work function change at a temperature of 25°C (*left*) and 75°C (*right*)

as the work function change (*a*) are calculated. The presented simulation results show a good agreement with our experimental data. Nevertheless, these simulations do only allow qualitative statements and predictions since they were carried out only in one dimension, and therefore, some parameters had to be fitted with respect to our experiments. The hydrogen is inserted after 30 min of simulation time and is turned off after 60 min (*a*). Without hydrogen, the simulation predicts a high coverage of oxygen. Shortly after the start of the hydrogen exposure, the surface is nearly fully covered by hydrogen (*f*). The preexisting oxygen coverage is consumed by reaction equation (5c/5d) and finally desorbed as water. The resulting free sites at the surface become occupied by hydrogen very fast. After reaching the maximum hydrogen coverage a steady state is reached, and until the hydrogen is turned off, no further alteration is observed. Moreover, as consequence of (8), the work function change (*a*) is stable during the whole hydrogen exposure. As can be observed in Fig. 12, this is in good agreement with our experimental results. Our calculations show that this behavior is similar up to a surface temperature of about 60°C. Because of the high hydrogen coverage, oxygen is prevented from adsorbing, following (6b/7a). This leads to a very low reaction rate, so the generation of heat as well as the mass transport through the gas phase are negligible. In other words, the high hydrogen coverage inhibits its own oxidation.

This situation changes at higher temperatures. By increasing the temperature of the platinum surface, adsorbed hydrogen starts to desorb. Due to this, free sites on the surface are created. More oxygen can reach the surface and so the reaction rate increases. At about 60°C, a turning point is reached. On the one hand, the reaction now produces enough heat (*b*) to overcome the heat losses and accelerates the reaction itself. On the other hand, the reaction consumes more and more hydrogen from the gas phase, which creates a concentration gradient above the surface. Thus, the hydrogen adsorption and, as consequence, the hydrogen coverage is also lowered, which leads to the creation of even more free sites for oxygen to adsorb. In that situation, this further increases the reaction rate. At approximately 60°C, the ignition point is reached, which drives the reaction from the kinetic-controlled regime to the mass transport-limited regime. The exact temperature of ignition is dependent on the ratio between hydrogen and oxygen [37] and the absolute partial pressure of the gas mixture. In detail, due to its very low concentration compared to oxygen, the hydrogen diffusion now limits the reaction. Because of the low partial pressure directly above the surface, the platinum surface no longer remains hydrogen covered. Every hydrogen molecule, which is dissociatively adsorbed, is consumed immediately by the reaction given by (5c) or (5d).

It should be noted that this is only the case if we assume the mean- field approximation. In reality, islands of hydrogen and oxygen could exist for a short time, without participating in the reaction, because of the missing related reactant. However, above temperatures of 60°C in this model, the only covering species if platinum is exposed to hydrogen in synthetic air are oxygen (*e*) and the final reaction product H_2O (*c*). The intermediate reaction product OH is also consumed rapidly through reaction steps (5d) and (5e). Consequently, no relevant coverage (*d*) has been observed during our calculations. In this temperature regime, considering (8), the work function change is provoked only by the lower oxygen coverage during the hydrogen exposure. The main issue now is that the oxygen coverage, unlike the hydrogen coverage at lower temperatures, is not stable in this situation. Figure 15 (left) shows that the initial reduction in oxygen coverage (*e*), which is triggered by the start of the hydrogen pulse, decreases during the exposure and so does the work function change, according to (8).

With respect to our calculations, the reason for that can be found in the diffusion and heat balance of the platinum surface. Because of the high reaction rate in this situation, diffusion and temperature gradients are generated during the presence of hydrogen. This leads to a higher temperature than the initial sensor temperature and a lower hydrogen partial pressure directly above the surface. Therefore, the oxygen coverage could rise up again despite the presence of hydrogen in the gas phase. Considering (8), the work function change drifts back. In fact, if the temperature is high enough, the work function can reach the same value as before turning on the hydrogen. Figure 13 confirms this situation in practice. As drawn in Fig. 15, our model does not explain the signal drift beyond the original baseline as displayed in Fig. 13. This might be a consequence of some other gas species like NO or CO. However, the exact hydrogen concentration in the ambient air is no longer detectable by our sensor since its output signal is directly proportional to the work function change of the platinum surface. In fact, given the right temperature, the sensor cannot even define whether it is exposed to hydrogen or not.

Summarized, the simple platinum-based FG-FET is not capable of satisfying the high requirements for automotive applications. The major shortcomings are the stability at high operational temperatures and the low upper detection limit. In the next section, we present an exceptional solution for these problems.

5 The Temperature-Controlled Phase Transition FET (TPT-FET)

In the past, many efforts were made [10, 32, 38] to deal with the inadequacies of hydrogen detection via the SG-FETs [4, 5] (particularly FG-FETs [9]), which are mentioned in the previous section. The common approach of all modifications is to stabilize the induced work function shift of platinum, due to the suppression of the catalytic hydrogen oxidation. In this section, we present a new approach that bypasses the stability problems above the catalytic ignition point [32, 37] as well as low saturation limit. In deep contrast to all previous sensitive layers utilized in the FG-FET, we developed a heterogeneous PMMA – platinum dual layer, which allows precise concentration measurements by determining the temperature-dependent phase transition at a buried platinum surface. We named this new sensor device: "temperature-controlled phase transition FET" (TPT-FET) [39]. In combination with the Lundström device [7] on the same CMOS chip [24], it yields a very reliable hydrogen sensor, which satisfies the tight specifications for automotive applications. Especially, the detection of hydrogen concentrations up to 4% is remarkable.

5.1 Fabrication and Embedding of the Sensitive Layer

Despite the completely different hydrogen-sensing mechanism, compared to the conventional operating FG-FET [9, 32], the TPT-FET is still based on the FG-FET transducer. The only physical difference consists in the design of the sensitive film. The new sensitive film is a heterogeneous dual layer, composed by a thin polymer film on top of a platinum surface. The following section describes the preparation of this dual layer.

A heavily boron-doped silicon (100) wafer was used as substrate for the hydrogen-sensitive layer. On top of 10-nm sputtered titanium, which is necessary for good adhesion, 50 nm of platinum was deposited by DC sputtering. However, the thickness of these layers does not influence their sensitivity significantly. In the next step, 2.3 μm of the PMMA dissolved in chlorobenzene were deposited by spin coating. It follows an annealing step at 180°C for 3 h to remove the remaining solvents. In Fig. 16 (left), a schematic layer stack of this system is presented.

Afterward, this layer stack is mounted as top electrode on the FG-FET transducer via flip chip bonding (see Sect. 3).

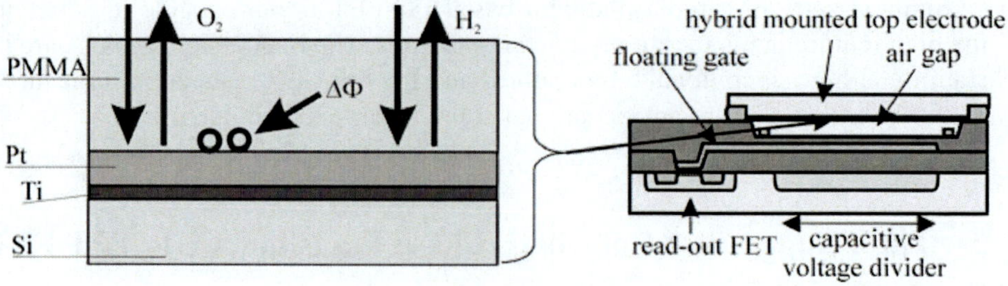

Fig. 16 Schematic view of the heterogeneous PMMA-platinum dual layer (*left*) and the finished FG-FET (*right*)

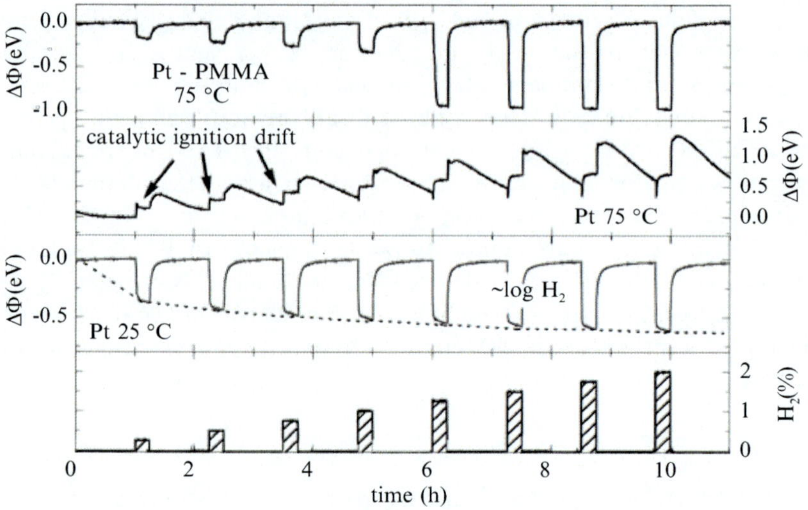

Fig. 17 Work function change of the PMMA-Pt system at 75°C (*top*) if exposed to various hydrogen concentrations in comparison with pure Pt at 75°C (*mid*) and 25°C (*bottom*)

5.2 Results and Discussion

If a PMMA-platinum layer is exposed to hydrogen, the work function change shows a behavior which is completely different from a pure platinum layer. These differences can be observed in Fig. 17. During hydrogen exposure, pure Pt (dashed line) shows a significant back drift at temperatures above the critical ignition point of 60°C, i.e., the sensor cannot be used for industrial application. This negative behavior disappears if platinum is covered by a thin layer of PMMA as is depicted by the dotted line.

Below the ignition point (60°C), pure Pt shows no back drift (full line), but a logarithmic response to increasing hydrogen concentrations limits the applicability above 2%. In contrast, the work function change of the PMMA-Pt layer (dotted line) increases linearly with rising hydrogen concentration up to a certain critical

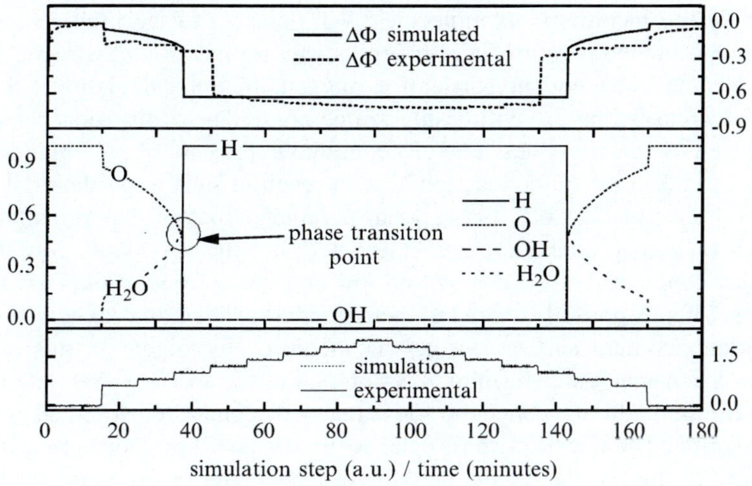

Fig. 18 Numeric simulation of the work function change (*top*) in comparison with the experimental results and the calculated surface average coverage (*mid*), if exposed to various hydrogen concentrations (*bottom*) at 75°C.

value. Beyond that limit, the work function shift jumps to a maximum of about −1 eV. Further rising of the hydrogen concentration does not result in a higher work function change at all. This mechanism is fully reversible (Fig. 18) when decreasing the hydrogen concentration again. There is no intermediate state, and we observed a sharp switching between these two states. Consequently, each exposed hydrogen concentration leads either to the proportional state or to the saturated state.

5.3 *Understanding the Phase Transition*

Since PMMA is a nonconducting polymer, the work function change has to be provided solely by the buried platinum surface as shown in Fig. 16. Similar to an open platinum surface, a potential shift can be induced either by a change of the hydrogen or the oxygen coverage. As described in Sect. 2, the final work function change is given by:

$$\Delta\Phi = -0.23\left(\frac{H_{\text{ads}}}{H_{\text{ML}}}\right)^{1.33} + 0.4\left(\frac{O_{\text{ads}}}{O_{\text{ML}}}\right)[\text{eV}]. \tag{10}$$

The coverage is altered by adsorption, desorption, and the reaction between both species. In the previous section, we described the surface reaction kinetics as well as their coupling to the hydrogen-containing gas phase and the heat balance following the Langmuir–Hinshelwood reaction mechanism. To understand the actual sensitive dual layer, we adapted our previous reaction model by considering the additional polymer layer. Numerically, we described the PMMA layer as a simple diffusion barrier for the incoming gas molecules. With respect to their molecular size, we chose a low and temperature independent barrier for incoming hydrogen and a high and temperature-dependent barrier for oxygen. In our calculations, the oxygen barrier decreases linearly with rising temperature. These diffusion barriers result in relative low reaction rates. Therefore, unlike in the case of pure platinum, we do not have to take care of the reaction heat. A result of such a simulation is shown in Fig. 18. Using the PMMA barrier as fit parameter for our experimental data, we obtain a very good qualitative agreement with our measurements. Simulating low hydrogen concentrations, we observed low and nearly proportional work function changes. This response is related to the reduction of the initial oxygen coverage of the buried platinum surface due to the incoming hydrogen. In this regime, we monitor a constant water-forming reaction on the surface.

If a certain hydrogen concentration is reached, a phase transition takes place, the buried surface becomes fully hydrogen covered, and all previous oxygen coverage is removed. In this state, the high hydrogen coverage inhibits other oxygen molecules from being adsorbed because there are no free adsorption sites left on the platinum surface. Therefore, we observe no further reaction for this surface state. The large step-like change of the work function is provided by the complete removal of the native oxygen and the nearly complete hydrogen coverage. If now the hydrogen concentration in the gas phase is lowered again, more and more free sites are generated, and therefore, some oxygen adsorbs. Due to the water-forming reaction, this in turn causes even more free sites. If the critical concentration is reached again, the surface coverage jumps back to the reactive O-dominated regime. It is remarkable that we observe neither an intermediated state nor any sign of hysteresis during our experiments.

5.4 Temperature Dependence of the Phase Transition

In Fig. 19 (left), the work function change is plotted for two different temperatures. The switching point is located at higher hydrogen concentrations if the temperature is increased from room temperature to 75°C. From our experiments, this is at least valid for temperatures up to 180°C. The logarithmic plot in Fig. 19 (right) presents the correlation between surface temperature and the hydrogen concentration necessary to reach the saturated state. All combinations of hydrogen concentrations and temperatures, that are located above the black line, lead to a nonreactive and fully hydrogen-covered platinum surface and therefore to a work function change of about -1 eV. Combinations located below this line keep the surface in the reactive regime covered by a decreased amount of oxygen (see Fig. 18). For temperatures above 60°C, we observed an exponential dependence. It should be noted that independent of temperature hydrogen concentrations below 1% always result in an oxygen-covered surface.

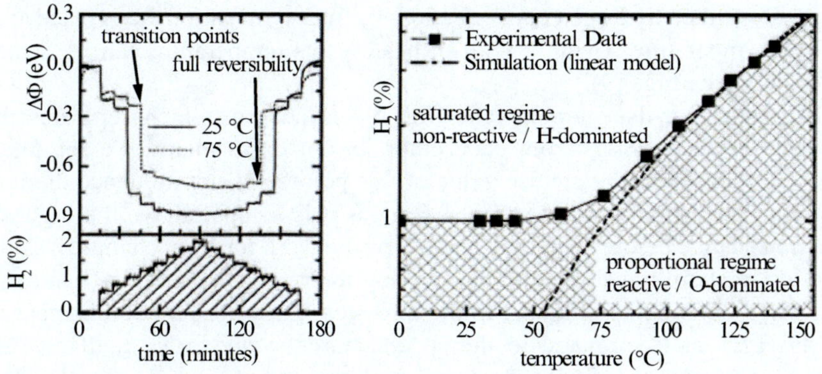

Fig. 19 Phase transition point with respect to hydrogen concentration and sensor temperature (*right*) and an example measurement (*left*) at constant temperature (illustration: [41] © [2008] IEEE)

5.5 *Modeling of the Temperature Dependence*

The temperature dependence of the phase transition point is a consequence of two different effects. First, there is the oxygen diffusion through the PMMA, which is increasing with temperature. Therefore, at higher temperatures, the effective equilibrium between hydrogen and oxygen at the buried platinum surface is shifted toward larger oxygen concentrations. On the other hand, due to its lower desorption energy [37], the simple and direct hydrogen desorption increases much stronger than the oxygen desorption with rising temperature. The combination of both effects leads to the temperature dependence of the critical hydrogen amount, which is necessary to reach the saturated state. To fit our experimental data, we have calculated the necessary permeabilities P_{O_2} and P_{H_2} as a function of temperature T (in K):

$$P_{O_2}(T) = 0.162 + 0.0018\,(T - 273).P_{H_2} = 0.92 \tag{11}$$

Using this normalized permeability functions for the incoming gas molecules, we obtain a temperature dependence as shown in Fig. 19 (right) (dashed line), which fits our experimental data very well at higher temperatures. In this regime, the temperature dependence is exponential.

5.6 *The Temperature-Controlled Mode*

As the transition between the two surface states is fast and fully reversible, we can use the temperature dependence of the phase transition point to measure exact hydrogen concentrations with the FG-FET. Due to the large work function difference, the FG-FET can accurately detect the phase transition of the PMMA-Pt dual layer. For a fixed hydrogen concentration, the phase transition takes place at a well-defined temperature. Using a built-in heater, this temperature can be determined accurately (see Fig. 19).

In practice, a fixed target value for $\Delta\Phi$ is chosen, which separates the two different surface phases from each other. In our experiments, we set $\Delta\Phi_{\text{target}}$ to -0.4 eV. However, the precise value of this point does not influence the temperature-controlled mode significantly as long as it is located above the highest work function change before switching and below the work function change in saturation. Due to the sharp transition and the missing intermediate state, one can set values between -300 and -700 meV. In the next step, the work function is measured by the FG-FET and compared to the specific target value $\Delta\Phi_{\text{target}}$. If $\Delta\Phi$ exceeds $\Delta\Phi_{\text{target}}$, the system is hydrogen dominated, and therefore, the controller heats up the sensor until the phase transition point is reached and $\Delta\Phi_{\text{target}} = \Delta\Phi$. If the observed work function change drops below $\Delta\Phi_{\text{target}}$, the PMMA-Pt layer is in the oxygen-dominated regime and the temperature controller cools down again. Operating in this mode, the sensor temperature is adjusted to the point of phase transition and represents the specific hydrogen concentration. A measurement example is shown in Fig. 20.

As displayed, each tested hydrogen concentration leads to a well-defined sensor temperature characteristic for the hydrogen concentration. This operational mode allows detecting high hydrogen concentrations up to at least 4%, which is a unique feature on the micro-sensor market. However, most important is the precise and stable hydrogen detection in the former impossible temperature regime above 60°C. An entirely new operating mode has been demonstrated, which uses a well-defined and temperature-dependent phase transition to measure exact hydrogen concentrations. Due to this fact, our sensor does not suffer from the stability or cross sensitivity problems of former sensing concepts, which simply analyzes the absolute work function change.

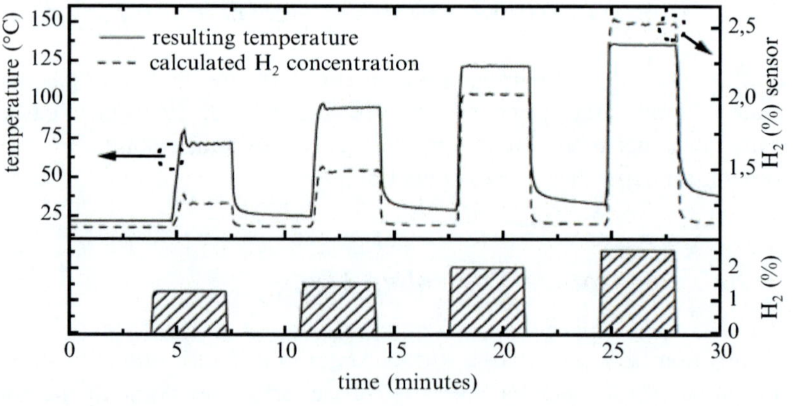

Fig. 20 Resulting sensor temperature and calculated H2 value, if operated in temperature-controlled mode (illustration: [41] © [2008] IEEE)

Concentrations below 1% are not detectable in this mode since no phase transition takes place in that concentration domain. To measure low concentrations, we operate the TPT-FET in the standard mode [9, 32], where the work function change is directly used as sensor signal, or we can utilize an additional LundströmFET [7], which is available on the same CMOS chip and very reliable at low hydrogen concentrations [24].

5.7 *Response Time*

As mentioned in Sect. 4.2, the response time is a critical parameter for hydrogen sensors in automotive applications. In this section, we will give a short overview about the response times of our sensor system and possible enhancements.

To dctcrmine response times, we equipped a TPT-FET with a 2.3 μm PMMA film on pure platinum. Different hydrogen concentrations were realized by mixing a fixed test gas concentration 3% of hydrogen in synthetic air with pure synthetic air due to a set of conventional mass flow controllers. Consequently, the maximum available hydrogen concentration was 3%. During exposure, the work function change $\Delta\Phi$, the signals from the mass flow controllers, as well as the regulated sensor temperature were recorded. The results of such an experiment with a hydrogen concentration of 2.5% at room temperature are shown in Fig. 21.

The hydrogen pulse starts at zero seconds on the time axis. After 1 s, the work function change $\Delta\Phi$ starts to increase. This delay is caused by the experimental setup with a finite gas flow time from the mass flow controllers to the sensor.

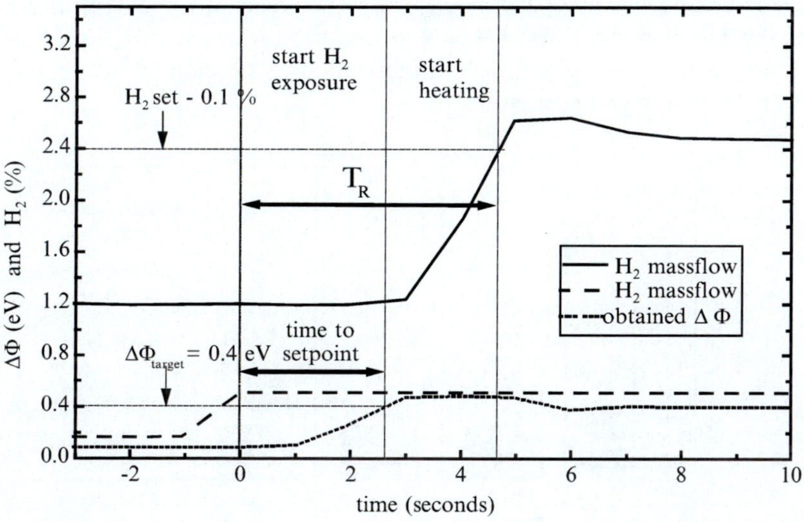

Fig. 21 Hydrogen response time measurements of the TPT-FET (PMMA 2.3 μm)

We estimate that it takes at least 1 s until hydrogen reaches the sensor and at least another second until the device is exposed to the full hydrogen concentration. However, about 2.5 s after starting the hydrogen pulse, the work function change reaches the critical value of 0.4 eV. At this time, the built-in heater is activated, and consequently, the temperature of the sensor increases. After 2 s of rapid heating, the phase transition point is passed, and therefore, $\Delta\Phi$ starts to decrease. At this point, the thermal controller decreases the heating power until a stable $\Delta\Phi = \Delta\Phi_{\text{target}}$ is achieved. The temperature now remains constant, and as described before, this temperature is a sharp indicator for the hydrogen concentration in air. We define the final response time T_R as the time period between the start of the hydrogen pulse at the mass flow controllers and the time where the sensors temperature has reached a value that is related to the exposed hydrogen concentration minus 0.1%. The resulting times T_R are summarized in Fig. 22.

For the 2.3 μm version, we observed relatively large reaction times T_R ranging from 7 to 27 s. Because diffusion mainly limits the response time, we also investigated thinner PMMA layers with 0.3 μm. As can be seen from Fig. 22, a reduction of T_R can be indeed observed. It is imported to note, that the presented data are not corrected with respect to the gas flow delay which was estimated to about 2 s. Therefore, the real response of the sensor is faster.

Even shorter response times are observed if the sensor temperature is increased to 75°C before hydrogen exposure. In this case, the response times T_R are reduced by 2 s.

As described in the next section, the TPT-FET can be combined with a Lundström FET on the same CMOS chip. In order to decrease response times, the sensor can be already heated, while the Lundström device is detecting low hydrogen concentrations. This increases the diffusion through the PMMA, and therefore, the rise time for $\Delta\Phi$ is reduced. On the other hand the effective heating time is decreased since less energy is necessary to reach the final temperature. Both effects lead to response times below 4 s.

To further lower these values, a minimization of the thermal heat capacitances, i.e., wafer thinning, is necessary.

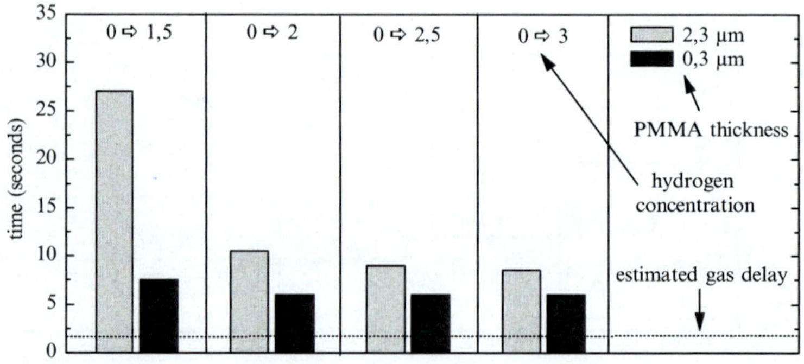

Fig. 22 Experimental (uncorrected) response times TR of a TPT-FET with 2.3-μm (*light gray*) and 0.3-μm (*dark gray*) PMMA on pure platinum recorded at room temperature. The estimated gas delay of the experimental setup is indicated

5.8 Hydrogen Dual Sensor: Lundström and FG-FET on One Chip

Since the Lundström sensor is a stable and precise hydrogen detector from 50 ppm up to 1% [7, 24], whereas the FG-FET-based TPT-FET is reliable for hydrogen concentrations between 1% and at least 4%, these devices are ideal candidates for a combination on one single hydrogen sensor chip.

This idea has already been realized in a research project of the Micronas GmbH in Freiburg, Germany. A Lundström FET and an FG-FET with Pt-suspended gate have been integrated on a single chip in CMOS technology. The whole transducer is fabricated in 0.5 μm CMOS technology. A schematic cross section (left) and a photographic top view (right) of this double sensor are presented in Fig. 23.

A detailed description of the manufacturing process and the device characteristics can be found in [24]. In an advanced version, the integrated FG-FET was modified to act as a TPT-FET. Right before the hybrid mounting of the top electrode, which contains the sensitive platinum layer, the PMMA film is deposited as described at the beginning of this section. The resulting device is a hydrogen sensor, which is capable of detecting precise hydrogen concentrations between 50 ppm and 4%. Furthermore, this sensor concept satisfies the tight economical and technical requirements for automotive applications, which were noted in Sect. 4.2.

The operational mode for the device is as follows: if no hydrogen is present, only the Lundström FET is active, and no or only a slight heating is required. If the device detects a certain hydrogen concentration, e.g., 0.7%, the TPT-FET is activated automatically and heated up to about 60°C in order to keep the reaction times small. When the hydrogen concentration reaches values above 1%, the TPT-FET starts the temperature-controlled mode, as explained in Sect. 5.7. Between 1% and 1.8% of hydrogen, both devices provide precise information about the concentration. Above 1.8%, the Lundström FET saturates, and the TPT-FET is the primary operating sensor. This procedure is controlled via an external micro-controller.

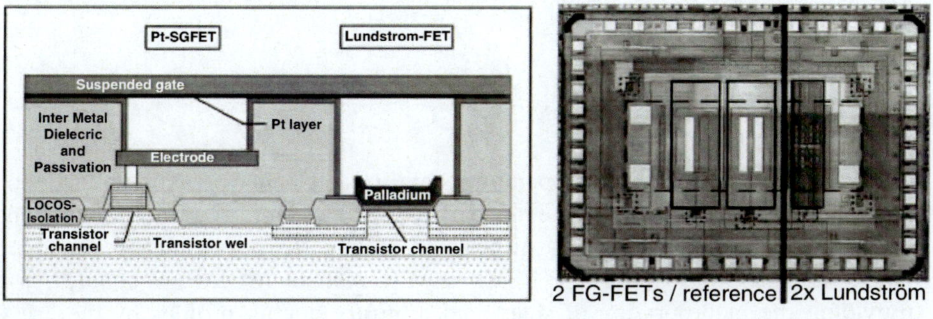

Fig. 23 Schematic cross section (*left*) and microscopic top view (*right*) of the dual sensor device (illustration: [24])

6 GasFET Concept for High Temperature Operation

The existing FG-FET concept described in Sect. 3 is operating in a temperature range between −40°C and 200°C. Higher temperatures lead to significant leakage currents in the transducer electronics and consequently to device breakdown. Because increasing temperature changes the adsorption and desorption behavior, many sensitive materials like metal oxides are only sensitive above 200°C [41, 42]. Hence, a new FG-FET concept for high temperature operation has been developed. The key benefits of the HT-FG-FET are achieved by using SOI substrates and a vertical MOS-FET with high doping concentrations in the channel region. Both improvements lead to an extension of the temperature range up to 400°C.

6.1 Requirements for High Temperature Operation of Silicon-Based MOS-FETs

The carrier concentration of doped silicon is constant over a wide temperature range. However, above a critical temperature, the intrinsic concentration exceeds the dopant concentration, i.e., the semiconductor becomes intrinsic [42, 43]. Increasing the doping concentration increases the critical temperature and consequently provides a higher operating temperature.

Another aspect to be considered is the fact that for highly doped pn-junctions, the space charge region becomes very small, allowing a transition of electrons by *Zener* tunneling [42] and thus a shortage occurs. Consequently, the aim is a trade-off between both effects.

6.2 Device Design of the FG-FET for High Temperature Operation (HT-FG-FET)

The new HT-FG-FET [44] concept for operation at high temperatures goes along with two essential improvements.

6.2.1 Highly Doped Vertical MOS-FET

The trade-off between high temperature stability and Zener breakdown leads to a target doping concentration in the range of 10^{18} to 10^{19} cm^{-3}. Using a conventional lateral MOS-FET concept, it is difficult to realize sharp and highly doped pn-junctions. Therefore, the HT-FG-FET uses a vertical MOS-FET concept, [45] providing the incorporation of sharp and arbitrary doping profiles by the use of epitaxy. Another important advantage of the vertical MOS-FET concept is the capability to precisely define the channel length by the thickness of the deposited layer rather than being limited by lithography. This allows an inexpensive scaling of the channel length down to 200 nm.

Therefore, the HT-FG-FET benefits from a large W_{eff}/L_{eff} ratio, and the sensitivity S is given by (see Sect. 3.4):

$$S = \frac{\partial I_{DS}}{\partial \Delta\Phi} = \mu \frac{W_{eff}}{L_{eff}} C_I \frac{C'}{C' + C_{OX}} V_{DS}. \tag{12}$$

The standard concept has been fabricated with a W_{eff}/L_{eff} ratio of 3.4 [46], whereas the vertical MOS-FET with a channel length of 200 nm and width of 110 μm provides a ratio of 550.

6.2.2 Fabrication on SOI Substrates

Thc standard FG-FET depicted in Fig. 11 uses an implanted well as part of the capacitive voltage divider. The drawbacks of this concept are a determined polarity and an unavoidable current of the reverse-biased pn-junction, which deteriorates the sensitivity.

The FG-FET for high temperature operation is fabricated on SOI substrates. The capacitive voltage divider and the MOS-FET are separated, assuring a complete isolation. Figure 24 shows a schematic cross section of the new HT-FG-FET concept.

6.3 *High Temperature Measurements*

Due to the floating gate of the sensor, it is difficult to achieve an exact characterization of the readout MOS-FET and its thermal stability. Therefore, each chip

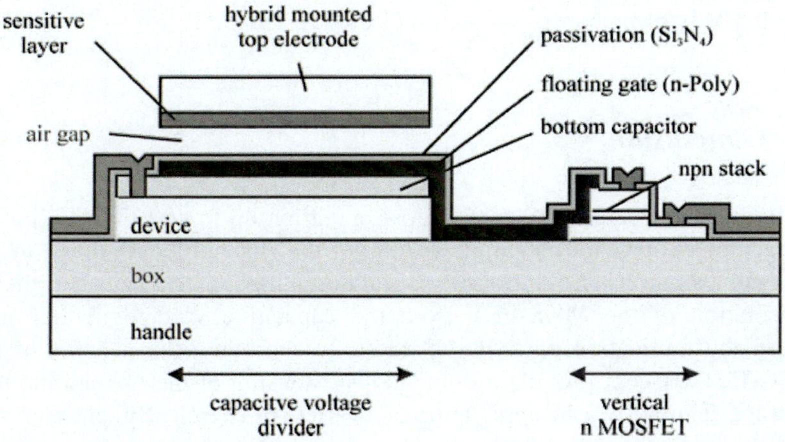

Fig. 24 Schematic cross section of the HT-FG-FET concept

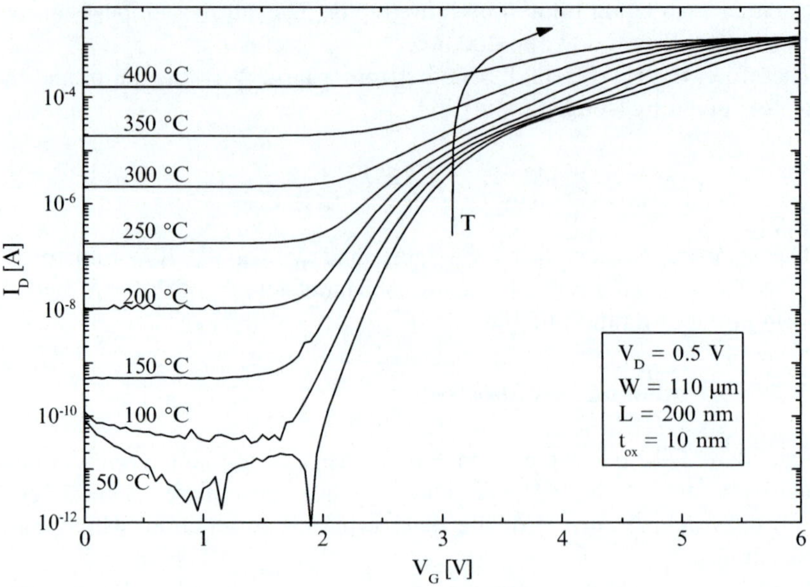

Fig. 25 Temperature dependence of the input characteristics of the MOS-FET

includes two reference transistors with the possibility to contact the gate directly. The results of the reference devices can be used to characterize the HT-FG-FET.

Figure 25 shows input characteristics of the reference MOS-FET with respect to temperatures up to 400°C.

The input characteristic for the lowest temperature shows a threshold voltage of 2.4 V and an on current I_{ON} of above 1 mA. As expected, increasing the temperature leads to a higher OFF-state leakage current. Nevertheless, the input characteristics show a gate-controlled drain current even at a temperature as high as 400°C. The I_{ON}/I_{OFF} ratio decreases to one decade, but the slope of 0.6 mA/V for a drain voltage $V_D = 0.5$ V is sufficient for operation of the HT-FG-FET.

6.4 Conclusion

Highly doped vertical MOS-FETs allow operation up to 400°C. Additionally, the vertical concept provides a simple scaling of the channel length down to 200 nm, leading to a large W_{eff}/L_{eff} ratio in (12) and thus a higher sensitivity. Furthermore, a full isolation of the MOS-FET from the capacitive voltage divider has been obtained by fabrication on SOI substrates. In summary, the benefits of the new HT-FG-FET concept provide a tremendous extension of the operational temperature range. This allows the application of a variety of chemically sensitive materials for work function measurements, which so far could not be used due to their limited sensitivity at low temperature.

References

1. Pollack-Diener D, Obermeier G (1993) Heat-conduction microsensor based on silicon technology for the analysis of two- and three-component gas mixtures. Sens Act B13: 345–347
2. Simon I, Arndt M (2002) Thermal and gas-sensing properties of a micromachined thermal conductivity sensor for the detection of hydrogen in automotive applications. Sens Act A97–A98:104–108
3. Doll T, Flietner B, Eisele I (1992) German Patent DE 4239319
4. Janata J (1983) US Patent 4,411,741
5. Lorenz H, Perschke M, Riess H, Janata J, Eisele I (1990) New suspended gate FET technology for physical deposition of chemically sensitive layers. Sens Act A23:1023–1026
6. Micronas GmbH, Freiburg, Germany
7. Lundström I, Shivaramann MS, Svensson C, Lundkvist L (1975) A hydrogen-sensitive MOSfield-effect transistor. Appl Phys Lett 26(2):55–57
8. Senft C, Iskra P, Eisele I, Hansch W (2011) Work function based gas sensors: Schottky and FET based devices. In: Ghenadii Korotcenkov (ed) Chemical sensors, vol 6. Momentum Press, New Jersey. Print ISBN: 978-1-60650-239-6
9. Eisele I, Doll T, Burgmair M (2001) Low power gas detection with FET sensors. Sens Act B78:19–25
10. Scharnagl K, Eriksson M, Karthigeyan A, Burgmair M, Zimmer M, Eisele I (2001) Hydrogen detection at high concentrations with stabilised palladium. Sens Act B78:138–143
11. Ostrick B, Pohle R, Fleischer M, Meixner H (2000) TiN in work function type sensors: a stable ammonia sensitive material for room temperature operation with low humidity cross sensitivity. Sens Act B68(1–3):234–239
12. Fleischer M, Simon E, Rumpel E, Ulmer H, Harbeck M, Wandel M, Fietzek C, Weimar U, Meixner H (2002) Detection of volatile compounds correlated to human diseases through breath analysis with chemical sensors. Sens Act B83(1–3):245–249
13. Simon E, Lampe U, Pohle R, Fleischer M, Meixner H, Frerichs H.-P, Lehmann M (2003) Novel carbon dioxide gas sensor based on field effect Transistors. In: Proceedings of the. Eurosensors XVII, Portugal
14. Stegmeier S, Fleischer M, Tawil A, Hauptmann P, Endres H-E (2011) Sensing of CO_2 at room temperature using work function readout of (hetero-)polysiloxanes sensing layers. Sens Act B 154(2):206–212
15. Sulima T, Knittel T, Freitag G, Widanarto W, Eisele I (2005) A GasFET for chlorine detection. IEEE Sens 3
16. Eisele I, Knittel T (2006) Work function based field effect devices for gas sensing, in Encyclopedia of sensors. In: Grimes CA, Dickey EC, Pishko MV (eds) American Scientific Publ, vol 10. 473. ISBN 1-58883-066-7
17. Burgmair M (2003) Thesis, Universität der Bundeswehr München. ISBN 3-89820-621-1
18. Knittel T (2005) Thesis, Universität der Bundeswehr München
19. Wöllenstein J, Böttner H, Jaegle M, Becker WJ, Wagner E (2000) Material properties and the influence of metallic catalysts at the surface of highly dense SnO_2 films. Sens Act B70 (1–3):196–202
20. Zhdanov VP, Kasemo B (1994) Kinetic phase transition in simple reactions on solid surfaces. Surf Sci Rep 20:111–189
21. Christmann K, Ertl G, Pignet T (1976) Adsorption of hydrogen on a Pt(111) surface. Surf Sci 54:365–392
22. Derry GN, Ross PN (1985) A work function change study of oxygen adsorption on Pt(111) and Pt(100). J Chem Phys 82:2772
23. Eisert E, Elg AP, Rosen A (1995) Adsorption of oxygen and hydrogen on Pt(111) studied with second-harmonic generation"'. Appl Phys A60:209–215

24. Wilbertz Ch, Frerichs H-P, Freund I, Lehmann M (2005) Suspended-Gate- and Lundstrom-FET integrated on a CMOS-chip. Sens Act A123:2–6
25. Josowicz M, Janata J (1988) Suspended field effect transistor. In: Seiyama T (ed) Chemical sensor technology. Elsevier, Amsterdam
26. Gergintschew Z, Kornetzky P, Schipanski D (1996) The capacitively controlled field effect transistor (CC-FET) as a new low power gas sensor. Sens Act B36:285–289
27. Schipansky D, German Patent DE 4333875 C2
28. Chung JS (1996) IMCS, Gaitersburg, Proc
29. Burgmair M, Frerichs H-P, Zimmer M, Lehmann M, Eisele I (2003) Field effect transducers for work function gas measurements device improvements and comparison of performance. Sens Act B95:183
30. Freitag G (2005) Thesis, Universität der Bundeswehr München
31. Eisele I, Doll T, Burgmair M (2001) Low power gas promising with FET sensors. Sens Act B 78:19–25
32. Senft C, Galonska T, Widanarto W, Frerichs H-P, Wilbertz Ch, Eisele I (2007) Stability of FET-based hydrogen sensors at high temperatures. IEEE Sens 189–192
33. Escube Space Sensor Systems (2002) Technologien – H2-Sensor. www.escube.de. Accessed 10 Dec 2002
34. Malyshev V-V, Pislyakov A-V (2003) Dynamic properties and sensitivity of semiconductor metal-oxide thick-film sensors to various gases in air gaseous medium. Sens Act B96 (1–2):413–434
35. Gland J-L, Sexton B-A, Fischer G-B (1980) Oxygen interactions with the Pt(111) surface. Surf Sci 95(2–3):587–602
36. Völkening S, Bedürftig K, Jacobi K, Wintterlin J, Ertl G (1999) Dual-path mechanism for catalytic oxidation of hydrogen on platinum surfaces. Phys Rev Lett 83(13):2672–2675
37. Rinnemo M, Deutschmann O, Behrendt F, Kasemo B (1997) Experimental and numerical investigation of the catalytic ignition of mixtures of hydrogen and oxygen on platinum. Combust Flame 111:312–326
38. Galonska T, Senft C, Widanarto W, Senftleben O, Frerichs H-P, Wilbertz Ch, Eisele I (2007) Cross sensitivity and stability of FET – based hydrogen sensors. IEEE Sens 1036–1039
39. Senft C, Wilbertz Ch, Frerichs H-P, Iskra P, Eisele I (2008) Temperature controlled phase transition as a detection principle for Gas-FETs (TPT-FET). IEEE Sens (in press)
40. Widanarto W (2007) Thesis, Universität der Bundeswehr München. ISBN -1 978-3-86727-193-6
41. Inoue H, Andersson M, Yuasa M, Kida T, Lloyd Spetz A, Shimanoe K (2011) CO_2 sensor combining an MISiC capacitor and a binary carbonate. Electrochem Solid-State Lett 14:J4–J7
42. Grove AS (1967) Physics and technology of semiconductor devices. Wiley, New York
43. Shoucair FS (1989) Scaling, subthreshold, and leakage current matching characteristics in high-temperature (25°C-250°C) VLSI CMOS devices. IEEE Trans Comp Hybrids Manufacturing Technol 12(4):780–788
44. Freitag G (2005) Thesis, Universität der Bundeswehr München. ISBN 978-3-86537-448-6
45. Moers J (2007) Turning the world vertical: MOSFETs with current flow perpendicular to the wafer surface. Appl Phys A 87:531–537
46. Oprea A, Frerichs H-P, Wilbertz C, Lehmann M, Weimar U (2007) Hybrid gas sensor platform based on capacitive coupled field effect transistors: Ammonia and nitrogen dioxide detection. Sens Act B 127:161–167

Chromium Titanium Oxide-Based Ammonia Sensors

K. Schmitt, C. Peter, and J. Wöllenstein

Abstract Chromium titanium oxide ($Cr_{2-x}Ti_xO_{3+z}$, CTO) is a solid solution with the corundum crystal structure of the pure chromium oxide if x is in the range of 0.01–0.45. When heated to temperatures above 300°C, CTO shows a very strong and fast resistivity response to the presence of ammonia in air. The conductivity of CTO is primarily determined by chromium imperfections. In gas measurements, CTO shows a p-type semiconductor behavior. At even higher temperatures (>400°C), CTO is an excellent material for ammonia (NH_3) detection with a reduced cross sensitivity to humidity. This has been the key to the successful development of ammonia sensors based on CTO.

We investigated CTO as a sensitive material for NH_3 sensors operating at room and slightly elevated temperatures. It is based on the change of work function of $Cr_{1.8}Ti_{0.2}O_3$ upon gas exposure. CTO exhibits fast response and relaxation, no baseline drift induced by exposure and little influence of changing ambient humidity. The cross sensitivity to other gases is low, in particular to NO_2.

Keywords Ammonia, Chromium titanium oxide, Gas sensor, Inkjet printing, Work function

---

K. Schmitt and C. Peter
Fraunhofer Institute for Physical Measurement, Heidenhofstr. 8, 79110 Freiburg, Germany

J. Wöllenstein (✉)
Fraunhofer Institute for Physical Measurement, Heidenhofstr. 8, 79110 Freiburg, Germany

Department of Microsystems Engineering, University of Freiburg, Georges-Köhler-Allee 102, 79110 Freiburg, Germany
e-mail: Juergen.Woellenstein@ipm.fraunhofer.de

M. Fleischer and M. Lehmann (eds.), *Solid State Gas Sensors – Industrial Application*, Springer Series on Chemical Sensors and Biosensors (2012) 11: 113–136,
DOI 10.1007/5346_2011_8,
Published online: 4 January 2012

Contents

1 Introduction

Ammonia (NH_3) is a colorless gas with a characteristic pungent odor. Ammonia contributes significantly to the nutritional needs of terrestrial organisms by serving as a precursor to food and fertilizers. Ammonia, either directly or indirectly, is also a building block for the synthesis of many pharmaceuticals [1]. Although in wide use, ammonia is both caustic and hazardous.

The human respiratory system can detect levels at around 10 ppm. Levels of 50–100 ppm affect performance, particularly daily gain which may be reduced by up to 10% during prolonged periods of exposure. At levels of 50 ppm and above, the clearance of bacteria from the lungs is also impaired, leading to a higher proneness to respiratory disease. Clinical signs include increased coughing and respiratory rates, irritation of the mucosa lining the respiratory tract and an increased incidence of pneumonia.

For ammonia detection, devices based on solid-state semiconducting materials have become indispensable due to the wealth of possible applications and their easy fabrication and small size. The most prominent semiconductor-based sensors used for ammonia detection are electrochemical cells and sensors with SnO_2 or CTO as sensitive layer. Some applications also use optical and polymer-based sensors for ammonia.

1.1 Piggery

Ammonia is the most common poison in the pig's environment. Generally, ammonia levels are less than 5 ppm in well run pig houses. If the pigs are exposed to high ammonia concentrations over a long time, they get restless, uncomfortable and may show increased levels of vice, such as tail biting, ear biting, and flank chewing [2]. Diagnosis is based on the assessment of air quality by the farmer and observed effects on the pig.

The air quality in piggeries is periodically controlled. In case of high ammonia concentrations, where levels are as high as 30–50 ppm, ventilation must be improved. For economic reasons, the ventilation has to be regulated adequately.

1.2 Leakage Detecting in Refrigeration Systems

Common food industries using refrigeration systems include: meat, poultry, and fish processing facilities, dairy, and ice cream plants, wineries and breweries, fruit and vegetable juice, soft drink processing facilities, cold storage warehouses, other food processing facilities, and petrochemical facilities. However, because of the worldwide prohibition of FRIGEN (chlorodifluoromethane, $CHClF_2$) in the last years the use of NH_3 as a refrigerant substance rises rapidly. However, the favorite FRIGEN substitute ammonia at concentrations above 50 ppm is considered a high health hazard. Even at much lower concentrations (<20 ppm), NH_3 acts as a poison for stored food in refrigerator systems. This is the reason why the European Union decreed an order that the leakage monitoring of ammonia in such systems for safely and healthy. But up to now no suitable long-term stable low-priced ammonia sensors are available on the market.

1.3 Selective Catalytic Reduction

One application for ammonia sensing is the so-called "selective catalytic reduction" (SCR) in most big diesel consuming engines. In this case, ammonia is used to reduce toxic nitrous gases, which are generated in the combustion process. The reduction of NO_x compounds using ammonia to harmless nitrogen (N_2) and water takes place while the gases are passed through the catalyst chamber. Before entering the catalyst chamber, ammonia is injected and mixed with the exhaust gas. Next to ammonia, also other catalysts are applicable. By using urea, carbon dioxide is another reaction product. Due to the actual European emission standards Euro IV and Euro V, all heavy diesel engines have to be equipped with an SCR-system [3]. The most common industrial catalyst for trucks is a 32.5% solution of water and urea with the trade name AdBlue™.

2 State of the Art

2.1 Electrochemical Cells for Ammonia Detection

The interaction of ammonia with a metal layer can be used, for example, in combination with conducting solid-state materials or liquids—this is known as electrochemical cell. The operation mechanism of electrochemical sensors is quite complex; the simplest analog is a battery, yet in the case of batteries all reaction partners are already present while in the electrochemical cell the target gas is required to permeate through a diffusion barrier to the working electrode inside the cell to produce a signal (Fig. 1).

The catalytic reaction at the working electrode causes a change in the electrode potential, thereby causing a current flow due to a potential difference between the working electrode and the counter electrode. This current is generally several orders of magnitude smaller than the current delivered by a battery. The sensing principle of electrochemical cells is basically determined by a chemical reaction involving the exchange of electrons between the reaction partners (redox reaction). The oxidation- and reduction process must be spatially separated, which is ensured by the two electrodes (half-cells). The electrons responsible for the signal current are present in the circuit outside the cell. The electrical signal is proportional to the concentration of the target gas because with the reaction of a target gas molecule at least one electron is released.

Electrochemical sensors usually work according to the amperometric principle, i.e. the electrical signal is detected as a current (at constant electrode voltages). In literature, an electrochemical cell for ammonia based on a Cu electrode with an Ag/AgCl counter electrode is described with a linear response up to 10% (v/v) [4].

Electrochemical cells are commercially available for many different gases. A detailed description on the electrochemical cells sold by Dräger is given in [5]. Their detection

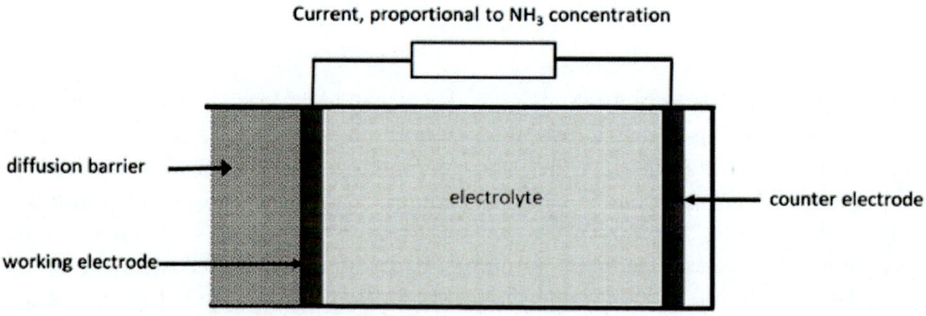

Fig. 1 Principle of an electrochemical cell used for the detection of gases: the target gas, in this case ammonia, permeates through the diffusion barrier to the working electrode and causes a potential difference between working electrode and counter electrode. This potential difference can be detected by a current flow, proportional to the ammonia concentration

limit is usually in the low ppm range. A disadvantage of the electrochemical cells is their limited lifetime since the electrolyte is used up, requiring a renewal of the cell. A more detailed description on electrochemical cells involving polymer electrolytes is given in Sect. 2.4.

2.2 Other Solid-State-Based Ammonia Sensors

The most frequently cited and often found solid-state-based sensors are metal oxide gas sensors. Materials used for these sensors include SnO_2, MoO_3, WO_3, V_2O_5, and CTO. The sensing principle of metal oxides, especially CTO, is described in greater detail in Sect. 3. Many companies all around the world, yet in particular in Japan, sell metal oxide sensors for ammonia detection (http://www.figaro.com, http://www.umweltsensortechnik.de). Yet, metal oxide sensors are not selective to one particular gas, so that a clear identification of ammonia in unknown environments is not possible, and their response times are very high (several minutes up to an hour). Several methods have been applied to make metal oxide sensors more selective, for example periodic temperature variations, since the sensitivity to different gases changes with temperature, or principle component analysis or artificial neural networks to evaluate the sensor signals, for example depending on temperature or sensing material. Catalysts combined with the sensing material can render them also more selective. Detection limits of such sensors are around 1 ppm.

Apart from these metal oxide sensors, other approaches can be found in literature. A sensor reported to be particularly suitable for ammonia detection is based on the resistometric measurement using thick-film copper sulfide (Cu_xS). The sensors showed very fast response times (100 ms) and work in a concentration range between 1 ppm and 1,000 ppm [6]. Another approach is proposed in [7]. They studied tellurium films for their ammonia detection properties using a resistometric setup. The response was investigated for ammonia concentrations up to 150 ppm, yet no cross sensitivities were studied. The same is true for the approach to use porous silicon carbide (SiC) for ammonia detection [8]. The detection limit is given with 0.5 ppm.

Another material class suitable for selective ammonia detection is zeolites, that is microporous aluminosilicates. Over 40 different zeolites are occurring naturally, but they can also be very easily synthesized. They consist of a scaffold of AlO_4^-- and SiO_4-tetrahedrons with various cations, for example Na^+, K^+, Ca^{2+}, Mg^{2+} accommodated. Zeolites were observed to show selective adsorption/desorption properties of gaseous compounds resulting in a change of the electrical properties. This effect can be used to fabricate selective gas sensors for the desired target gas. Zeolite-based sensors, in particular for automotive applications, are proposed in [9, 10]. The selected zeolite composition (H-ZSM-5) is reported to be very selective and sensitive to ammonia when operated at 800°C. The sensor is fabricated in thick-film, i.e. is very inexpensive and reliable. The detection limit is around 5 ppm. These sensors are commercially available from Synkera (http://www.synkera.com).

2.3 Polymer-Based Ammonia Sensors

Apart from solid-state materials, several polymers are suitable for ammonia detection. The most prominent are polyaniline (Fig. 2a) and polypyrrole (Fig. 2b). Good overviews on electroconducting polymers are given in [11, 12].

There exist two sensing mechanisms involving polypyrrole: the first is making use of the irreversible reaction of the polymer with ammonia, the second uses the conductive (reversible) properties of the polymer for the creation of amperometric or resistometric sensors. The irreversible reaction is illustrated in Fig. 2c. The presence of ammonia causes a mass increase in the polymer film, which can be detected, for example, by the frequency change in a resonator [13]. The reversible reaction is due to the reduction of the oxidized form of polypyrrole through ammonia, leading to a change in conductivity. The reaction mechanism is proposed as follows [14] (Table 1).

Lähdesmäki et al. [14, 15] investigated the amperometric detection principle using polypyrrole in detail. They found limits of detection in the low ppm

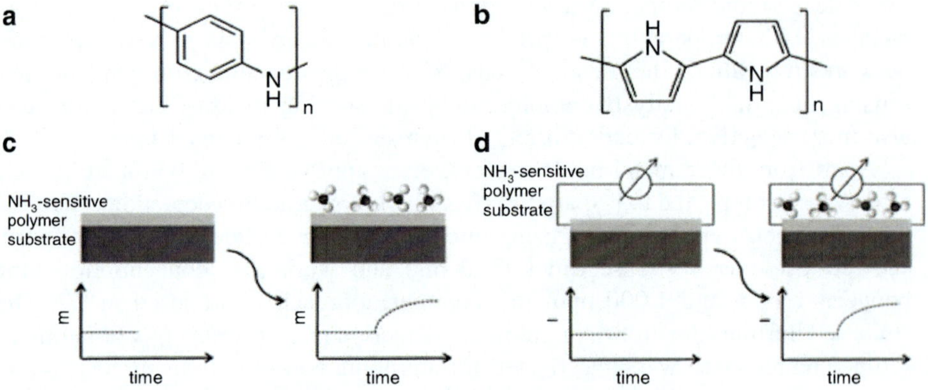

Fig. 2 Polymeric materials for ammonia detection and their detection mechanisms: (**a**) chemical structure of polyaniline, (**b**) chemical structure of polypyrrole, (**c**) detection of ammonia using the polymeric material as mass-sensitive device: the presence of ammonia leads to an increase in mass in the polymer film, (**d**) amperometric detection of ammonia using polymer films: the reaction of ammonia with the polymer leads to an increase in the current flow

Table 1 Reversible reaction of ammonia with polypyrrole (OxP: oxidation product of ammonia, PPy: polypyrrole)

$NH_3 \leftrightarrow OxP + e^-$	Oxidation of NH_3
$PPy^+ (Cl^-) + e^- \leftrightarrow PPy^0 + Cl^-$	Reversible reduction of PPy
$PPy^+ (Cl^-) + NH_3 \leftrightarrow PPy^0 + OxP + Cl^-$	Overall chemical reduction of PPy by NH_3
$PPy^0 + Cl^- \leftrightarrow PPy^+ (Cl^-) + e^-$	Electrochemical reoxidation of PPy
$NH_3 \leftrightarrow OxP + e^-$	Overall reaction of ammonia detection

range, and identified possible interference sources. Yet, the polypyrrole gradually decreases when applying higher concentrations of ammonia, probably due to ammonia oxidation compounds sticking to the electrode surface and partly an irreversible electrochemical degradation of polypyrrole.

Polyaniline generally shows a significantly higher sensitivity to ammonia at shorter response times. Furthermore, it is chemically more stable and therefore very interesting. The reversible reaction mechanism of polyaniline with ammonia is as follows:

$$PAH^{+} + NH_3 \leftrightarrow PA + NH_4^{+} \tag{1}$$

If ammonia is present, the equilibrium position lies far to the right because the ammonia molecules react with the polyaniline to the energetically more favorable ammonium, NH_4^+. Upon decreasing ammonia concentration in the ambient, the reaction goes toward the left due to the decomposition of ammonium into volatile ammonia, and the original protonated form of the polyaniline is restored. Quite a large number of ammonia sensors based on different variations of polyaniline films can be found in literature: Hirata and Sun [16] report on a polyaniline-based sensor where the change in resistivity is taken as sensor signal. The resistivity increases upon reaction with ammonia and is linear in the range from 5 to 20 ppm ammonia. A similar polyaniline sensor was developed by Kukla et al. [17] and tested for ammonia concentrations from 1 to 2,000 ppm. This sensor is reported to have a high chemical stability in the presence of oxidizing reagents, and can be easily regenerated by heating. A polyaniline sensor doped with acrylic acid is shown in [18]. Ammonia was applied in a concentration range from 1 to 600 ppm, yet the sensor response was linear only up to 58 ppm, and they observed a decrease in resistivity for higher ammonia concentrations, which is contrary to the effect reported by other groups. The same amperometric sensor principle was applied from Osakai et al. [19], but with a polyvinylchloride-nitrobenzene as ammonia-sensitive polymer.

The principle of ammonia detection using electroconducting polymers has been commercialized in the last decade (but is not on the market any more) by the Finnish company Vaisala (http://www.vaisala.fi). The absorption of ammonia in the polymer changes the capacitance of the sensor. Since the sensor has a cross sensitivity to water, the effect of humidity interference has been minimized with a temperature-controlled measurement cycle. By this, the ammonia and the water is alternately absorbed and desorbed.

2.4 *Optical and Colorimetric Ammonia Sensors*

For the optical detection of ammonia, two principles are used in general: the optical absorption of ammonia in the infrared spectral range for either nondispersive infrared sensors or spectroscopy setups. The second possibility is to use optical methods to detect color changes in pH indicator materials (colorimetric detection).

A very prominent example for the colorimetric detection of ammonia is the commercially available Dräger tubes. If broken at one end, the color change of the material inside can be easily read from a scale on the tube. A further example for colorimetric ammonia detection is pH indicator paper, which colorizes according to the ammonia concentration. The use of pH indicators has already been integrated in a variety of optical setups to read out the coloration change, from simple transmission configuration [20] to planar waveguide [21, 22] or fiber-optic setups [23, 24]. The transmission configuration involves the pH indicator bromophenol blue in a silicone matrix. The color of the indicator changes reversibly from yellow to blue with increasing concentration of ammonia, while the transmittance is determined at a given wavelength. With this setup, a detection limit in the lower ppb range could be achieved; however, the response times are quite long and the lifetime is limited to a few days.

Klein et al. and Yimit et al. coated planar waveguides with immobilized indicator dyes and used the sensor element for ammonia detection. With optical waveguides, the evanescent field can be used to detect the color change, which is much more sensitive than in direct illumination, so that detection limits of around 1 ppt can be reached. The response times of such sensors are around 10 s. A variation of the waveguide configuration is the implementation of optical fibers instead of planar waveguides, which increases the sensitive surface. Yet with these sensors the limit of detection was not below ppb, and the reaction of the sensitive layers with ammonia does strongly depend on temperature variations in the ambient.

3 Chromium Titanium Oxide-Based Resistive Sensors

3.1 Introduction

Chromium titanium oxide ($Cr_{2-x}Ti_xO_{3+z}$, CTO) is an excellent sensitive material for resistive semiconductor gas sensors because of both the gas-sensing behavior and the chemical stability. It exhibits fast response and relaxation upon exposure to ammonia and a low baseline drift. The cross sensitivity to other gases is low. One drawback of most metal oxide gas sensors is their high cross sensitivity to water vapor. CTO has an improved performance with respect to humidity changes. The effect of humidity on CTO on both baseline and sensitivity is much less in comparison with other metal oxides. This has been the key to the successful commercial development of thick-film sensors based on CTO, for detection of carbon monoxide and ammonia in the air [25]. CTO was first reported by Moseley and Williams [26].

After a few years of material development at University College London, Capteur Sensors and later City Technology provided CTO sensors at the market. After the take-over of City Technology by Honeywell, the sensors were withdrawn from sale. However, there is still a demand for CTO sensors.

3.2 Chromium Titanium Oxide

The mixed crystal CTO has no natural existence, but can be synthesized by so-called diffuse lattice substitution. It is synthesized of its single pure components TiO_2 and Cr_2O_3 by a firing process [26]. The resulting chemical structure is: $Cr_{2-x}Ti_xO_{3+z}$. If x is in the range of 0.01–0.45, the material retains a solid solution with the corundum crystal structure of the pure chromium oxide [27]. During this sintering process, the titanium atoms move to chromium lattice sites. In literature, the temperature of calcination is reported to be above 1,000°C up to 1,400°C [28–30]. The temperature is limited by the changing temperature of the lattice structure. One preferred proof of a successful material preparation is X-Ray Diffraction (XRD). Through a complete implementation of all TiO_2 no peaks due to its rutile structure are visible in the XRD measurement. The theta/2theta plot of CTO has to be identical to that of pure Cr_2O_3 [28–30].

After sintering, the conductivity of titanium-substituted chromium oxide ($Cr_{2-x}Ti_xO_{3+z}$) is primarily determined by chromium imperfections. The material shows, compared to most other semiconducting materials, a p-type conductivity and resistivity in an easily measurable range.

The implementation of titanium decreases in general the conductivity of the semiconductor compared to pure Cr_2O_3 [31]. The final conductivity is depending on the amount of substituted titanium.

CTO has a very high chemical stability. The melting point of crystalline Cr_2O_3 is above 2,000°C [32]. For CTO, a similar value is expected. CTO has a green color. The band gap of $Cr_{1.8}Ti_{0.2}O_{3+z}$ is 1.5 eV (Niemeyer Private Communications) (Table 2).

CTO shows a remarkable good gas sensitive behavior to nearly all oxidizing and reducing gases. The reasons therefore are segregation effects, as well as an enrichment of titanium ions on the surface [28, 33, 34]. The number of the surface near titanium ions is in good accordance with the number of the adsorption centers for O_2 on the surface.

Table 2 Basic properties of CTO

Property	Value
Melting point	>2,000°C
Band gap	1.5 eV
Activation energy	0.6–0.7 eV

3.3 Fabrication Technologies

Commonly, metal oxide gas sensors can be categorized into "thick-film" or "thin-film" devices. "Thick-film" devices are fabricated by screen printing, inkjet printing, and dip- or spray coating of a powder in a suitable vehicle, which is subsequently fired to produce a porous polycrystalline layer. "Thin-film" devices, conversely, are typically prepared by a sputtering technique so as to produce, ideally, a thin solid film of the sensing oxide. The main difference in behavior of the two film types derives from the fact that gas must diffuse into a porous "thick-film" to produce a response, whereas an ideal thin-film device is impermeable to gas and only gives a response from the surface. In practice, this generalization is an oversimplification, and thin film sputtered devices often have a polycrystalline porous structure.

3.3.1 Screen Printed CTO sensors

Screen printing is one possibility of thick-film deposition of different materials. Originally, these materials have to be powders, which can be immobilized into a screen printing paste. The paste normally consists of stabilizers, tensides, preservatives, and humidity storage. One problem of the paste generation is the preparation of a metal oxide powders with defined chemical structures. These can, for example, be prepared with sol-gel techniques [28]. The so prepared powder can be mixed with several stabilizing chemicals by a ball-milling process [29]. The paste is deposited onto the substrate using a mesh. Therefore, the mesh is positioned on top of the substrate. The paste is given all-over the mesh. By pressing down and moving of a scraper, the paste gets extrude through the mesh and is well defined deposited on the sensing field. This technique allows the deposition of very thick and porous films ($>10\ \mu m$). The main disadvantage is a low structural resolution.

3.3.2 Inkjet Printed CTO sensors

In case of a ceramic CTO ink, it has to be synthesized from the components TiO_2 and Cr_2O_3 by a firing process [28]. The basic materials are TiO_2 and Cr_2O_3 powders. At a firing process above 1,000°C the mixed powder substitutes to a CTO solid solution. In a high-energy ball mill, the green body gets grind and the particle size controlled to generate the ink. The fine-tuning of the rheological properties of the ink is achieved in a subsequent ball-milling step. In this step, tensides and stabilizers are added to the CTO dispersion [35]. To achieve optimal gas responses, the thick-film layer has to be very porous. With an ink jetting process, this can be realized by different plotting adjustments. One possibility is multiple plotting of single dots on top of each other; alternatively, an array of dots with displaced layers can be plotted. By plotting a threefold CTO structure on

Fig. 3 SEM picture of different plotted CTO dots. *Above*: Ink jetted dots on Pt electrodes with a vertical pitch of 40 μm. This leads to a merging of always two dots in vertical direction. *Below*: Zoom into the ink jetted CTO structure to visualize the particle size [36]

interdigital platinum electrodes on an oxidized silicon substrate, with a pitch of 75 μm, the film thickness reaches 1.5 μm. An SEM picture of a porous annealed CTO thick-film on Pt electrodes can be seen in Fig. 3. Here, the drop spacing in every second vertical direction was set to 40 μm, which leads to a merging of always two adjacent dots. This offers a higher sensitivity due to two connected electrodes. In the horizontal direction, the dot boundaries still have a distance of 6 μm. When decreasing the pitch, the single dots merge and form a closed material layer over the Pt electrodes. The closed layers have straight edges with a reproducible accuracy of 5 μm [36].

The influence of the applied voltage on the piezoelectric printhead is very important; there is a direct link between the applied voltage and drop formation. Lower voltages generate very small and slow droplets; higher voltages induce strong droplet deformations and the creation of small satellites. The droplets can reach velocities up to 15 m/s.

3.3.3 Atmospheric Pressure Chemical Vapor Deposition of CTO

Another method for producing thin-film CTO sensors for ammonia detection is the so-called atmospheric pressure chemical vapor deposition (APCVD) [37]. The APCVD process involves the transportation of volatilized metal precursors over a heated substrate under positive pressure where, following introduction of a gaseous oxygen source, decomposition occurs. Ideally, the desired product is deposited uniformly on the substrate with volatilization of any side products [(2) and (3)]. The first step is the controlled deposition of Cr_2O_3 from chromyl chloride and methanol (2), before introducing the dopant (titanium) using either titanium tetrachloride $TiCl_4$ (3) or titanium tetra-isopropoxide $Ti(OPrI)_4$.

$$4CrO_2Cl_{2(g)} + 4MeOH_{(g)} \rightarrow 2Cr_2O_{3(s)} + 4MeCl_{(g)} + 4HCl_{(g)} + 3O_{2(g)} \quad (2)$$

$$\begin{aligned} 2-xCrO_2Cl_{2(g)} + xTiCl_{4(g)} + 2 + xMeOH_{(g)} \rightarrow Cr_{2-x}Ti_xO_{3(s)} + 2 \\ + xMeCl_{(g)} + 2 + xHCl_{(g)} + (1.5 - 0.5x)O_{2(g)} \end{aligned} \quad (3)$$

The APCVD experiments have to be carried out in a purpose built apparatus comprising a horizontal bed and a cold wall reactor connected by heated stainless steel pipes and valves, through which nitrogen (99.99%; BOC) carrier gas and precursors pass (Fig. 4). All reactions have to be carried out at atmospheric pressure. CrO_2Cl_2, $TiCl_4$/$Ti(OPri)_4$, and ROH (R = Me, Et, H)/Et(O)OEt are sources of chromium, titanium, and oxygen respectively, without further purification. Metal precursors have to be introduced into the gas phase by passing nitrogen through

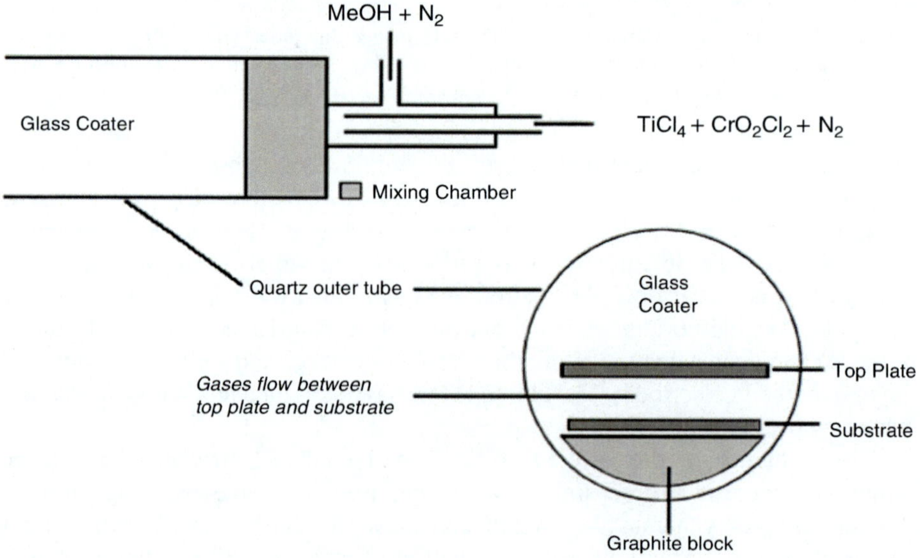

Fig. 4 Design of glass walled reactor (University College London)

bubblers containing the precursors held at constant temperatures. CVD reactions were carried out at 400–550°C for several seconds.

3.3.4 Sol-Gel Process

One main problem for sensor production using screen printing techniques is the synthesis of a printable powder. One possibility of its preparation is the Sol-gel process. This is a wet chemical technique for producing nonmetallic, inorganic particles of high purity out of a chemical solution. Sol-gel is a well-known process and widely used in production and research areas.

The Imperial College London first reported a successful synthesis of $Cr_{1-x}Ti_xO_{3+x/2}$ using a Sol-gel process [28]. With this technique, a nano-crystalline CTO powder with a diameter <1 μm is producible. The basic chromia sol can be prepared by using an aqueous solution of chromium salts by removing the anions. After their removing, the cations are forming Cr–O–Cr linkages. To this solution, the titanium ions can be directly added as a salt. Another opportunity is to integrate the titanium in a separate sol and mix both sols together [30], as performed by the University College London. In the next step, the gel is mixed under the prepared sol. The organic parts of the generated emulsion can be removed by centrifuging and washing with alcohol. The gel powder can be separated by a heating step.

3.3.5 Thin-Film Sensors

To be able to use both the advantage of CTO sensing properties and the advantages of cost-effective microelectronic production, thin-film CTO gas sensors on Si/SiO_2 substrates were developed and their sensing properties examined.

As substrate material 4″ silicon wafers are often used because of their advantages regarding the availability at relatively low costs, the material purity and the excellent surface quality. For the structuring of the gas sensitive metal oxide layer onto the electrodes, a photolithographic mask is used. The polycrystalline p-type CTO is deposited onto the electrodes by reactive sputtering with oxygen pressure as Cr/Ti—sandwich structures given in Fig. 5. The correct Cr/Ti ratio is given by adequate thickness choice. For example, for the deposition of $Cr_{2-x}Ti_xO_{3+z}$ the layers consist of nominal Cr (45 nm)/Ti (15 nm)/Cr (45 nm)/Ti (15 nm)/ Cr (45 nm).

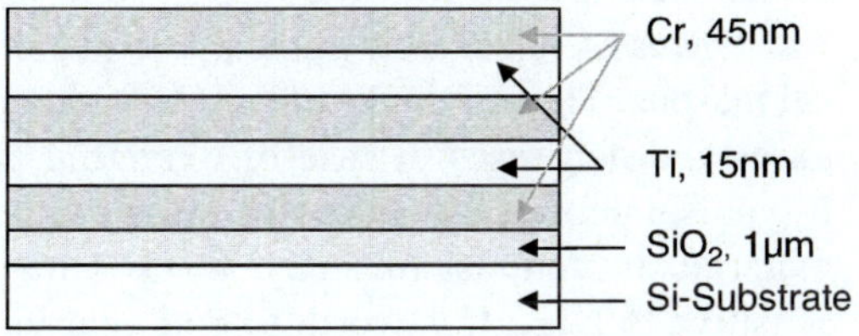

Fig. 5 Schematic view of the sputtered chromium titanium multilayer. The Cr/Ti ratio is given by adequate thickness choice

Fig. 6 SEM picture of a sputtered thin film CTO layer after annealing 1 h at 900°C in synthetic air. The crystals are in the magnitude of 50–200 nm

As deposited, the material is not completely oxidized and still shows a metallic conductivity. Therefore, a reoxidation (2–4 h) at 900°C in synthetic air is necessary. The annealing process causes an alloying of the materials and a formation and growth of crystals. This change in morphology is linked to an increase in the thickness of the layer. For example, annealing at 900°C (4 h) in synthetic air causes a growth of the layer from 160 to 350 nm. The morphological characteristics of the CTO layers can be investigated with an SEM. The polycrystalline film is porous and consists of grains with a size in the range of 50–200 nm (Fig. 6).

3.4 *Gas sensing Behavior*

In order to investigate the humidity and operation temperature dependence of the gas response to different concentrations of NO_2, NH_3, CO, and CH_4, a set of gas measurements were carried out. Figure 7 shows the typical response to trace gases diluted in synthetic air (50% r. H., 25°C) for three CTO thin-film sensors. The basic resistances are all scarcely below 200 kΩ and they differ slightly due to technological reasons. The resistance of the CTO increases when exposed to traces of reducing gases and decreases in case of oxidizing gases. Thus, the material shows typical p-type semiconductor behavior quite close to the characteristics of the thick-film material. In contrast to SnO_2, CTO is less sensitive to CH_4, though it is highly sensitive to NH_3. It is worth to note that in contrast to common SnO_2-sensors, but

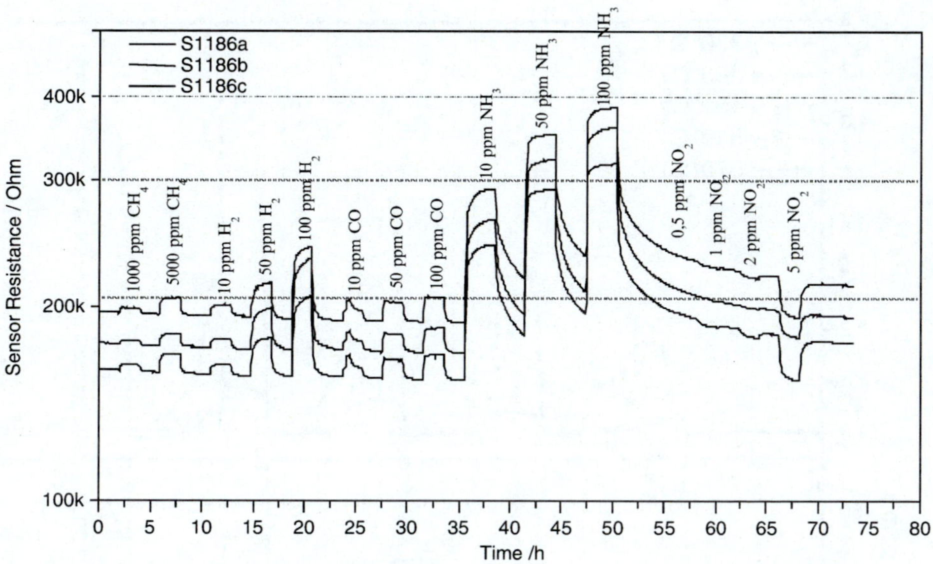

Fig. 7 Response of an evaporated CTO thin-film sensor to different concentrations of NO_2, NH_3, CO and CH_4 in synthetic air at 400°C (50% r.h.)

similar to thick-film CTO, these thin-film CTO-sensors immediately stabilize their baseline resistance.

In the case of NH_3 exposure, the initial response time constants are in range of minutes. In contrast, the recovery time constants are in the range of hours. Switching off the ammonia flow leads to fast resistance decrease followed by a slow recovery process to the equilibrium value. We explain this phenomenon by a superposition of a fast desorption process of the NH_3 molecules away from the hot sensitive layer and a slow desorption of NH_3 from the cold surface of our measurement equipment. The sensor follows this slow process immediately and its response thus shows directly this memory effect.

In order to investigate the temperature dependence of the gas response, a set of gas measurements were carried out. In these experiments, the sensor operation temperature was stepped from 330°C to 570°C. During each 6 h period at constant working temperature, the sensors were exposed 30 min with NH_3 in different concentrations. The sensors were purged with synthetic air 1 h before and 90 min after trace gas exposure. In this case, temperature-dependent sensitivity maxima were found at about 330°C, but with a loss of fast reaction times. In the analyzed temperature range, the NH_3 sensitivity decreases strongly with lower temperatures, but the response times get faster (Fig. 8).

As mentioned above, CTO is particularly of interest due to its reduced cross sensitivity to water vapor. Figure 9 shows the change of the base resistance and NH_3 response of a CTO sensor at 360°C operation temperature as a function of the relative humidity.

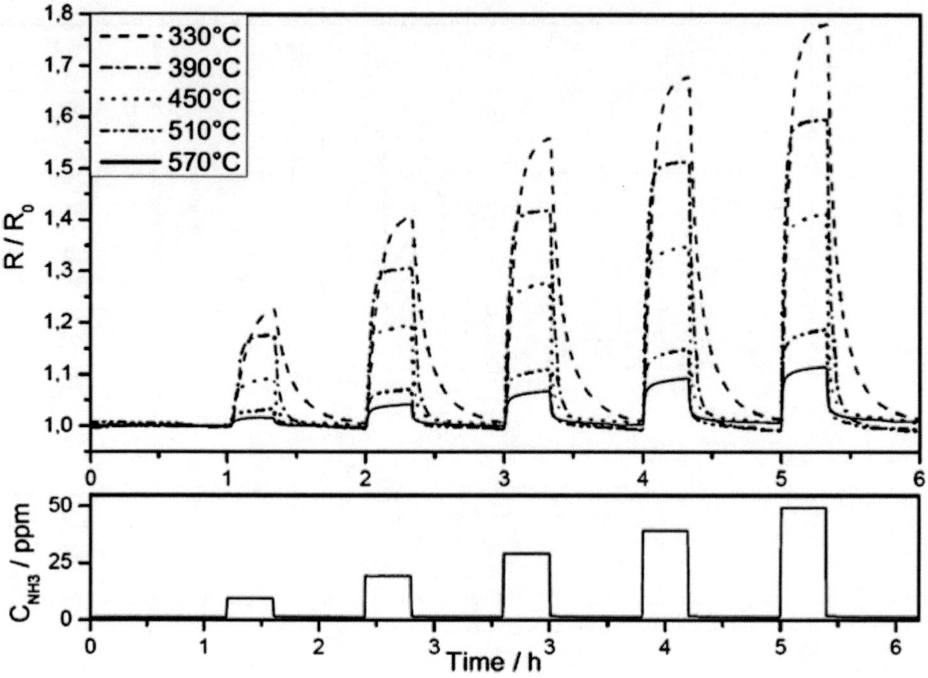

Fig. 8 Gas measurements of an ink jetted CTO sensor to different concentrations of NH_3 with different sensor temperatures. The temperature is varied from 330°C to 570°C. The sensor was exposed to four different NH_3 concentrations between 10 and 50 ppm in 10 ppm steps

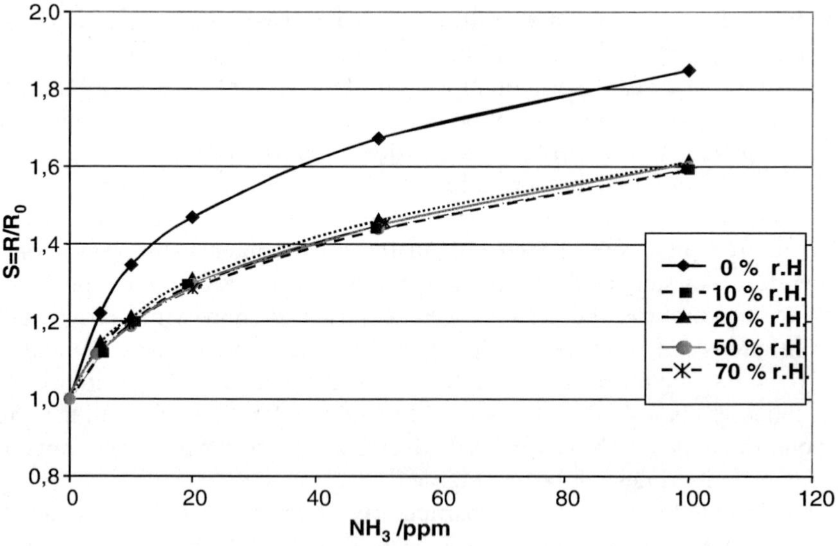

Fig. 9 Normalized response of a CTO thin film sensor versus ammonia concentration at different relative humidity (@ 25°C). The operation temperature of the sensor was 360°C

A remarkable influence of the humidity can be seen between 0% and about 10% humidity. Above 10% the resistance is almost independent of increasing humidity.

4 Chromium Titanium Oxide in Work Function Type Sensors

4.1 Introduction

CTO has already been extensively investigated as conductivity type sensor in the medium temperature range above 300°C. The conductivity response to NO_2 and NH_3 is considerable, whereas the effect of humidity is small compared to SnO_2 films. This has been the motivation to investigate CTO for application as a work function type gas sensor at lower operation temperature compared to resistive sensors. $Cr_{1.8}Ti_{0.2}O_3$ is known as the composition with maximum sensitivity to NH_3 [38]. $Cr_{1.8}Ti_{0.2}O_3$-layers have been have investigated by Kelvin probe and the hybrid suspended gate FET (HSGFET) measurements in order to evaluate the work function shift due to trace gas exposure. The main focus will be on NH_3 as target gas. Both types of measurements revealed comparable and reversible shifts of the work function due to the exposure to ammonia. The work function shift for the reducing NH_3 is—as expected—negative. For a HSGFET, the sensitivity is −30 mV per decade of NH_3 concentration at room temperature. The response time (*t*90) is ~2 min for a concentration of 20 ppm. The sensitivity decreases with increasing temperature. At room temperature, the CTO layers show a low cross sensitivity to humidity on both NH_3 response and sensor baseline signal.

4.2 Experimental

A layer of 300 nm $Cr_{1.8}Ti_{0.2}O_3$ (CTO) was deposited as thin film on 1 μm thick thermal SiO_2 on a p + Si-wafer. At a given O_2 pressure, the film was deposited by alternated reactive sputtering and evaporation of Ti/Cr layers followed by a finishing oxidation and annealing at 900°C in synthetic air. Detailed information about preparation, morphology, and gas-sensing behavior in the medium temperature regime can be found in the previous chapter.

4.2.1 Kelvin Probe

A Kelvin oscillator is a well-established tool for measuring differences in work functions, whose origin is dating back to 1897.

The Kelvin probe consisted of a piezoelectric driven gold mesh as reference electrode (3 mm in diameter) [39]. The probe was installed horizontally in an electrically grounded aluminum chamber (noncorroding alloy of Al) of about

20 cm^3 volume. The sample was mounted on a holder by electrically conductive glue. The holder can be heated by a resistive heater up to 200°C. Temperature was measured by a PT100 and regulated by a temperature controller within a range of 1°C. The distance of the sample to the reference electrode was adjusted in the range of ~0.3 to 1.0 mm, which was limited downward by the amplitude of the vibrating reference electrode. In the vicinity of the Kelvin probe, only electrically conducting materials were used and grounded carefully in order to shield stray capacitances. The gas inlet was near the sample. The gold electrode had been well conditioned during more than 6 months of continuous operation at different temperatures, humidity and with various gases (NO_2, NH_3, H_2, CO) such that reference measurements with LPCVD Si_3N_4 [40] revealed only small contributions of the gold electrode to the contact potential difference signal (CPD) of less than 5 mV.

4.2.2 Suspended Gate Field Effect Transistor Using CTO as Sensitive Layer

The hybrid suspended gate field effect transistor as a gas sensor was explained in detail in [41]. For the sake of simplicity and flexibility in experiments, we dispensed with the fabrication of a micromachined gate as substrate for the gas sensitive film. Instead, we cleaved the CTO covered Si wafer into small samples of about 1 mm × 1 mm. One of these samples is placed on the transducer chip, such that it completely covers only one of the electronically active channels and with the sensitive surface facing this channel. With the sample being the suspended gate, this hybrid structure forms the gas sensitive FET. The sample touches part of the guard ring (100 nm Pt) as depicted in Fig. 10.

The channel of the reference FET remains uncovered. Due to the different thickness of the field oxide and the passivation film on the source and drain lines, a gap of 1.0 mm is formed through which gas species can access the sensitive area. The air gap in the sensitive area itself is 1.2 mm. This ready-to-use gas sensor is glued by a nonconducting epoxide onto a small insulating slab of mica which, in turn, is glued onto a glass substrate with a Pt resistive heater. The slab ensures negligible substrate currents of the device, which is important to avoid a control of the FET by a shifting substrate potential. The heater can be operated to nearly 200°C, which is at the operation limit of silicon-based devices. The sensor is put in a chamber (noncorroding alloy of Al) with gas inlet and outlet on opposite sides to ensure a laminar flow. The volume is about 40 cm^3. Electrical feedthroughs are realized at the bottom of the device. From outside through the top of the chamber, a

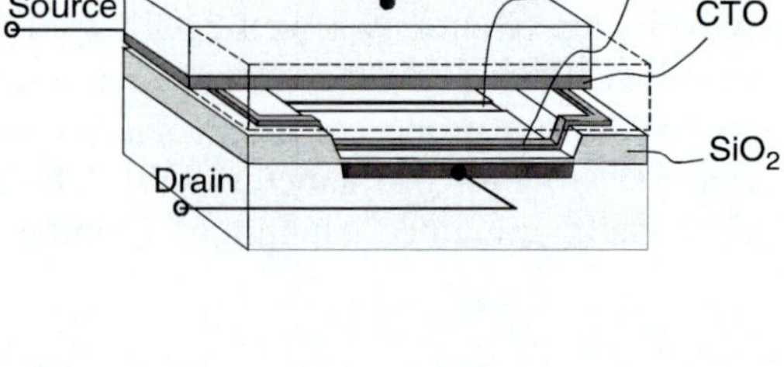

Fig. 10 Schematic view of the HSGFET with part of the gate and CTO shown transparent [41]

guided gib made of Teflon with a flat aluminum tip is placed onto the sample. A weight (100–500 g) is put on the Teflon gib outside the chamber to press the sample as close to the channel as possible to minimize the distance between sample and channel and, hence, to maximize the capacitive coupling of the gate potential to the channel. In addition, a good ohmic contact of the aluminum tip to the sample body is ensured.

4.2.3 Gas Measurements

Gas exposures consisted of three pulses of equal concentration after several hours of stabilization in a constant flow of dry synthetic air. Each pulse lasted 40 min interrupted by 40 and 120 min, respectively. The concentration varied between 13 and 100 ppm for NH_3. Measurements have been carried out at temperatures between RT and 130°C and at relative humidity of 0% and 30%. Changes of work function were measured by both Kelvin probe and HSGFET. Its principle of operation and design is described in [42]. The HSGFET was modified such that CTO rested on parts of the guard ring. The guard ring was set to gate potential. The necessary variation of gate voltage ΔV_G to keep the channel conductivity of the FET constant is taken as sensor signal and interpreted as work function shift $\Delta\Phi/e$.

4.3 *Results and Discussion*

Both methods of measurement revealed similar and comparable results: CTO exhibit negative responses to NH_3 as expected for a reducing gas with values as small as -100 mV. They are fast (e.g., $t90 = 120$ s for 100 ppm/30°C/30% r.h.), reproducible and they relax to baseline after gas exposure (Fig. 11).

In general, Kelvin probe measurements show smaller absolute values and relaxes slower compared to ~12 min for the HSGFET. This can be explained by reactions taking place on the gold reference electrode which compete with the CTO reaction. CTO nearly saturates at concentration levels of ~20 ppm only yielding almost linear increases of absolute signal height in the regime above up to 100 ppm. Figure 12 depicts these signal heights as measured from baseline to the biggest value of response at room temperature. Apparently, the presence of 30% relative humidity is of little effect, especially at higher concentrations of NH_3.

Investigations showed the temperature dependence of the signal heights. It indicates an almost excursive shift of the adsorption curve to positive responses starting at a temperature of about 60°C. The reason for this shift is not clearly understood. Ostrick et al. [43] and Gupta et al. [44] have reported similar temperature behavior on work function GasFETs which they interpreted as a negative temperature coefficient for adsorption of NH_3. But they did not observe a shift to positive signal values. Possibly, similar to the onset of reactions at the reference electrode of a Kelvin probe, temperature-induced quick reactions account for a

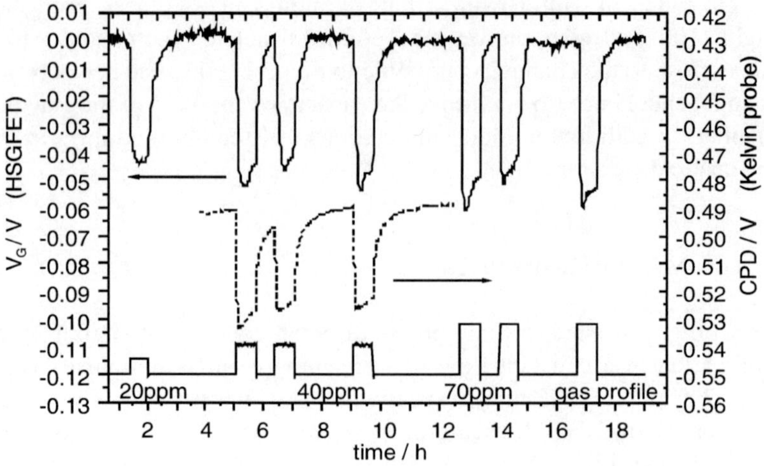

Fig. 11 HSGFET and Kelvin probe signals (*dashed*) of NH_3 exposures at 40°C/30% r.h

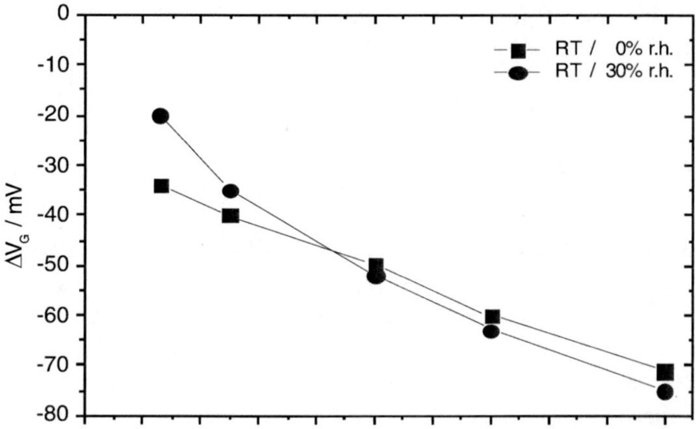

Fig. 12 HSGFET signal heights for various concentrations of NH_3 at room temperature

potential redistribution within the HSGFET. Further on, at 130°C the height of the signal response to NH_3 becomes dependent on the level of background humidity.

Nearly the same value of transition temperature is found when switching from dry synthetic air to 30% relative humidity background level. At $T < 60$°C the response ΔV_G is smaller than 30 mV and relaxes to the previous baseline during the humidity exposure within ~10 min. While at $T > 90$°C the response increases dramatically and a new baseline settles.

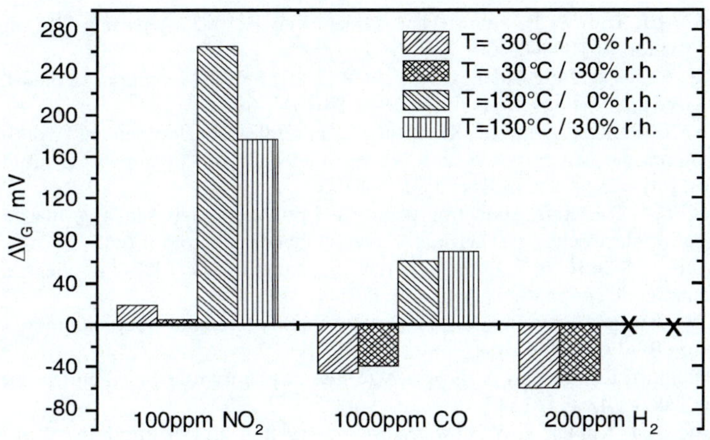

Fig. 13 Cross sensitivity: HSGFET responses to 100 pmm NO_2, 1,000 ppm CO and 200 ppm H_2 at different temperatures (30°C; 130°C) and humidity (0%; 30%)

Figure 13 depicts the sensitivity of CTO to the gases NO_2, CO and H_2 as measured with HSGFET. Most striking is the enormously reduced sensitivity to 100 ppm NO_2 at 30°C.

Acknowledgments This work is partly funded by the European Commission (Glassgas project) by the US National Science Foundation (NSF, DMR-9701699) and by the German BMBF (MISSY-project). The authors wish to thank Dr. M. Burgmair and Prof. Dr. I. Eisle, Universität der Bundeswehr München for the Kelvin probe and HSGFET measurements. We also like to thank very much Dr. Graham A. Shaw and Dr. Peter Smith, University College London for the support of the work.

References

1. Appl M (2006) Ammonia. In: Ullmann's Encyclopedia of industrial chemistry. Wiley-VCH, Weinheim, doi: 10.1002/14356007.a02_143.pub2
2. Kalich J (1980) Harmful gases in the piggery air and its influence on the fattening performance of pigs. J Tierzüchter 32(9):386–388
3. Gieshoff J, Pfeifer M, Schafer-Sindlinger A, Spurk P, Garr G, Leprince T (2001) Advanced urea SCR catalysts for automotive applications. Society of Automotive Engineers, http://www.sae.org/technical/papers/2001-01-0514
4. Mayo N, Harth R, Mor U et al (1995) Electrochemical response to H_2, O_2, CO_2 and NH_3 of a solid-state cell based on a cation- or anion-exchange membrane serving as a polymer electrolyte. Anal Chim Acta 310:139–144
5. Jessel W (2001) Gase – Dämpfe – Gasmesstechnik. Dräger Safety AG, Lübeck
6. Galdikas A, Mironas A, Strazdiene V, Setkus A et al (2000) Room-temperature-functioning ammonia sensor based on solid-state Cu_xS films. Sens Actuators B 67:76–83
7. Sen S, Muthe KP, Joshi N, Gadkari SC et al (2004) Room temperature operating ammonia sensor based on tellurium thin films. Sens Actuators B 98:154–159

8. Connolly EJ, Timmer B, Pham HTM, Groeneweg J (2005) A porous SiC ammonia sensor. Sens Actuators B 109:44–46
9. Moos R, Müller R, Plog C, Knezevic A et al (2002) Selective ammonia exhaust gas sensor for automotive applications. Sens Actuators B 83:181–189
10. Franke ME, Simon U, Moos R, Knezevic A et al (2003) Development and working principle of an ammonia gas sensor based on a refined model for solvate supported proton transport in zeolites. Phys Chem Chem Phys 5:5191–5195
11. Bidan G (1992) Electroconducting conjugated polymers: new sensitive matrices to build up chemical or electrochemical sensors: a review. Sens Actuators B 6:45–56
12. Krutovertsev S, Sorokin S, Zorin A, Letuchy Y, Antonova O (1992) Polymer film based sensor for ammonia detection. Sens Actuators B 7:492–497
13. Cai QY, Jain MK, Grimes CA (2001) A wireless, remote query ammonia sensor. Sens Actuators B 77:614–619
14. Lähdesmäki I, Lewenstam A, Ivaska A (1996) A polypyrrole-based amperometric ammonia sensor. Talanta 43:125–134
15. Lähdesmäki I, Kubiak WW, Lewenstam A, Ivaska A (2000) Interference in a polypyrrole-based aperometric ammonia sensor. Talanta 52:269–275
16. Hirata M, Sun L (1994) Characteristics of an organic semiconductor polyaniline film as a sensor for NH_3 gas. Sens Actuators A 40:159–163
17. Kukla AL, Shirshov YM, Piletsky SA (1996) Ammonia sensors based on polyaniline films. Sens Actuators B 37:135–140
18. Chabuksvar VV, Pethkar S, Athawale AA (2001) Acrylic acid doped polyaniline as an ammonia sensor. Sens Actuators B 77:657–663
19. Osakai T, Kakutani T, Senda M (1987) A novel amperometric ammonia sensor. Anal Sci 3:521–526
20. Trinkel M, Trettnak W, Reininger F, Benes R et al (1996) Study of the performance of an optochemical sensor for ammonia. Anal Chem Acta 320:235–243
21. Klein R, Voges E (1993) Integrated-optic ammonia sensor. Sens Actuators B 11:121–125
22. Yimit A, Itoh K, Murabayashi M (2003) Detection of ammonia in the ppt range based on a composite optical waveguide pH sensor. Sens Actuators B 88:239–245
23. Arnold MA, Ostler TJ (1986) Fiber optic ammonia gas sensing probe. Anal Chem 58:1137–1140
24. Cao W, Duan Y (2005) Optical fiber-based evanescent ammonia sensor. Sens Actuators 110:252–259
25. Capteur Sensors & Analysers Ltd (2000) NH_3 sensor data sheets
26. Moseley PT, Williams DE (1990) A selective ammonia sensor. Sens Actuators B 1:113–115
27. Williams D (1999) Semiconducting oxides as gas-sensitive resistors. Sens Actuators B Chem 57(1–2):1–16
28. Atkinson A, Nartowski AM (2003) Sol–gel synthesis of sub-micron titanium-doped chromia powders for gas sensing. J Sol–Gel Sci Technol 26:793–797
29. Huo L, Zhao H, Gao S, Pokhrel S (2007) Sol–gel derived polycrystalline CTO thick films for alcohols sensing application. Sens Actuators B 120:560–567
30. Parkin I, Williams D, Chabanis G (2001) Microspheres of the gas sensor material $Cr_{2-x}Ti_xO_3$ prepared by the sol emulsion gel route. J Mater Chem 11:1651–1656
31. Gnanasekar KI, Prabhu E, Gnanasekaran T, Periaswami G, Jayaraman V (1999) Preparation and characterisation of Cr2-xTixO3 and its sensor properties. Sens Actuators B 55:175–179
32. Gmelin (1922) Handbuch der Chemie, vol 33, Chrom p 32 published since 1922
33. Brydson R, McBride SP (2004) Analytical transmission electron microscopy and surface spectroscopy of ceramics: the microstructural evolution in titanium-doped chromia polycrystals as a function of sintering conditions. J Mater Sci 39:6723–6734
34. Williams DE, Smith P, Pratt K, Slater B, Catlow CRA, Stoneham AM, Niemeyer D (2002) Experimental and computational study of the gas-sensor behavior an surface chemistry of the solid-solution $Cr_{2-x}Ti_xO_3$ ($x < 0.5$). J Mater Chem 12:667–675

35. Magdassi S (2010) The Chemistry of inkjet inks. World Scientific, Singapore, ISBN-13 978-981-281-821-8
36. Peter C, Kneer J, Wöllenstein J (2011) Inkjet Printing of Titanium Doped Chromium Oxide for Gas Sensing Application. Sensor Letters 9(2):807–811
37. Shaw GA, Parkin IP, Williams DE (2003) Atmospheric pressure chemical vapour deposition of $Cr_{2-x}Ti_xO_3$ (CTO) thin films (< 3 μm) on to gas sensing properties. J Mater Chem 13:2957–2967
38. Wöllenstein J, Plescher G, Kühner G, Böttner H, Niemeyer D, Williams DE (2002) Preparation, morphology, and gas-sensing behavior of $Cr_{2-x}Ti_xO_{3+z}$ thin films on standard silicon wafer. IEEE Sens J 2:403–408
39. Besocke K, Berger S (1976) Piezoelectric driven Kelvin probe for contact potential measurements. Rev Sci Instrum 47(7):840–842
40. Eisele I, Flietner B, Doll T, Lechner J, Leu M (1994) Reliable hybrid GasFETs for work-function measurements with arbitrary materials. Sens Actuators B 22:1994
41. Burgmair M, Eisele I, Doll T (2001) Low power gas detection with FET sensors. Sens Actuators B 78:19–25
42. Burgmair M, Wöllenstein J, Böttner H, Karthigeyan A, Anothainart K, Eisele I (2002) Ti-substituted chromium oxide in work function type sensors: ammonia detection at room temperature with low humidity cross sensitivity. not published in a journal, can be found at http://forschung.unibw.de/berichte/2002/7ut0fvd1htxsoquwk7yl5hkzsvu8aa.pdf
43. Ostrick B, Pohle R, Fleischer M, Meixner H (2000) TiN in work function type sensors: a stable ammonia sensitive material for room temperature operation with low humidity cross sensitivity. Sens Actuators B 68(1–3):234–239
44. Gupta R, Gergintschew Z, Schipanski D, Vyas P (1999) New gas sensing properties of high TC cuprates. Sens Actuators B 56:65–72

Combined Humidity- and Temperature Sensor

T. Bürgler, F. Krogmann, and J. Polak

Abstract Temperature and humidity are very important parameters by measuring gas concentration. Most of the measuring procedures are showing a high temperature dependency; hence, the exact knowledge of the medium temperature is necessary in order to carry out high accurate measurements. While temperature compensation can be found in most gas measuring systems, the humidity of the medium is neglected in many cases. For rough concentration measurement this limitation can be accepted, but for highly accurate measurements humidity acts like an interfering gas.

The measurement of both these parameters can happen via different procedures. For the temperature measurement, an often implemented system is the use of characteristics of the gas measuring procedures (e.g., diode structures at silicon-based semiconductor gas sensors) or present resources in the application (e.g., resistance temperature sensor in processing units). The measurement of the moisture is clearly more difficult to integrate into the systems. Most frequently, type of moisture detection is realized with capacitive humidity sensors. These sensors allow the measurement of the relative humidity by evaluation of a capacitive signal obtained by a humidity-sensitive polymer. The knowledge of the temperature in these sensors is also indispensable for the determination of the absolute humidity respectively of the dew point.

The relative humidity has a strong temperature dependency (especially in the high humidity range). Therefore, it is an advantage to measure the temperature close to the humidity sensor. Thereby measurement mistakes could be inhibited effectively.

In this chapter, the measurement principle of thin film temperature sensors and capacitive humidity sensors, and examples for integration of both these technologies are described.

T. Bürgler, F. Krogmann, and J. Polak (✉)
R & D, Innovative Sensor Technology IST AG, Industriestr. 2, Wattwil CH 9630, Switzerland
e-mail: thomas.buergler@ist-ag.com; florian.krogmann@ist-ag.com; jiri.polak@ist-ag.com

M. Fleischer and M. Lehmann (eds.), *Solid State Gas Sensors – Industrial Application*, Springer Series on Chemical Sensors and Biosensors (2012) 11: 139–150,
DOI 10.1007/5346_2011_7,
Published online: 10 January 2012

Keywords Capacitive polymer humidity sensor, Dew point, Heating humidity sensor, Platinum-temperature sensor

Contents

1 Introduction

"Humidity" is a term that has various meanings and needs to be specified in more detail. So this chapter starts with a discussion of the commonly used terms in this field.

1.1 Relative Humidity

The relative humidity is the ratio of the partial pressure of the water vapor inside the gas to the partial pressure of saturated water vapor [1] [see (1)]:

$$\varphi = \frac{p_d}{p_s} \tag{1}$$

where φ = relative humidity; p_d = partial water vapor pressure; p_s = saturated water vapor pressure.

The relative humidity depends on the temperature, because the saturated water vapor pressure also depends on the temperature. The relative humidity is a dimensionless quantity. However, the relative humidity is often given in percent of the ratio between the partial water vapor pressure and the saturated water pressure.

1.2 Absolute Humidity

The absolute humidity gives the effectively present water quantity in an air volume. It is independent of the temperature of the gas. The absolute humidity can be calculated with (2) [2]:

$$a[g/m^3] = \frac{m_d}{V} \tag{2}$$

where a = absolute humidity; m_d = mass of the water vapor; V = volume.

1.3 Dew Point

If unsaturated air (relative humidity <100%r.H.) is cooled, at first the humidity content and the partial pressure of the water remain constant. However, the relative humidity increases since the saturated water vapor pressure decreases in colder air. Therefore, cold air can absorb less humidity than warm air. The saturated water vapor pressure corresponds exactly to the partial water vapor pressure at the dew point temperature, which means that the saturation limit with 100%r.H. is reached. If the saturated air is cooled under the dew point temperature, the excess humidity condenses out in the form of fog (fine water droplets), and the relative humidity remains at 100%r.H. Water is still not condensed out and the relative humidity is 100%r.H. directly at the dew point temperature (Table 1).

The dew point temperature can be calculated using the Magnus formula [3] (3):

$$tp = \frac{-\ln\frac{p_d}{C_1} \times C_3}{\ln\frac{p_d}{C_1} - C_2} \tag{3}$$

where tp = dew point temperature; p_d = partial water vapor pressure; C_1, C_2, and C_3 = Magnus coefficients.

Table 1 Magnus coefficients

Phase	t (°C)	C1 (mbar)	C2	C3 (°C)
Ice	−50.9 to 0.0	6.10714	22.44294	272.440
Water	−50.9 to 0.0	6.10780	17.84362	245.425
Water	0.0 to 100	6.10780	17.08085	234.175

2 Temperature- and Humidity Sensor

2.1 *Platinum-Temperature Sensor*

Platinum-temperature sensors are established in a wide range of application because of their high accuracy, broad application area, and robustness. Another advantage is that the characteristic of the Platinum-temperature sensor is specified in the norm DIN EN 60751 [4]. For that reason a simple replacement of temperature sensor is possible, as well as the access to widely available standard electronic components for evaluation.

In the past, wire wound platinum resistors were used but in the last decades the thin film temperature sensor succeeded more and more. The reasons for this transition can be in the clearly lower prices, ever smaller designs, and better robustness of the thin film elements.

The measurement of the temperature is realized by a resistance measurement of the platinum. Thereby almost all platinum thin film elements follow the temperature characteristic which is specified in DIN EN 60751.

$$-200 \text{ to } 0^{\circ}\mathrm{C} \quad R(t) = R_0\left(1 + \mathrm{A} \times t + \mathrm{B} \times t^2 + \mathrm{C} \times [\mathrm{t} - 100] \times \mathrm{t}^3\right) \tag{4}$$

$$0 \text{ to } 850^{\circ}\mathrm{C} \quad R(t) = R_0\left(1 + \mathrm{A} \times t + \mathrm{B} \times t^2\right) \tag{5}$$

where R_0 = resistance in Ohm at 0°C; t = temperature according to ITS 90.

The coefficients A, B, and C are specified as

$\mathrm{A} = 3.9083 \times 10^{-3}\ {}^{\circ}\mathrm{C}^{-1}$;
$\mathrm{B} = -5.775 \times 10^{-7}\ {}^{\circ}\mathrm{C}^{-2}$;
$\mathrm{C} = -4.183 \times 10^{-12}\ {}^{\circ}\mathrm{C}^{-4}$.

This characteristic is widespread; however, for automotive a 3,770 ppm/K characteristic and for the Russian market a 3,910 ppm/K characteristic are widely in use, too. Other characteristics (3,750 ppm/K) can nowadays only be found in custom-specific applications.

The platinum-temperature sensors are scaled in Pt100, Pt1000, Pt10,000, or even Pt100k. This description corresponds the electrical resistance of the sensor at 0°C (R_0). It means that a Pt100 sensor has a resistance of 100Ω at 0°C and according to a Pt1000 1,000Ω.

Thin film platinum-temperature sensors consist of a photolithographic structured meander-shaped arranged high-purity platinum layer. The platinum layer is applied with thin film procedure to the ceramic substrate. The structures are patterned with the same procedures as in the semiconductor technology are used.

In order to produce the sensors resistance value very accurately, the requested value is achieved by laser trimming during the production process. Thereby in the design implemented shortcuts are opened as required by a laser trimmer (Fig. 1).

Fig. 1 Platinum-temperature sensor, left and top trim structure, right meander structure

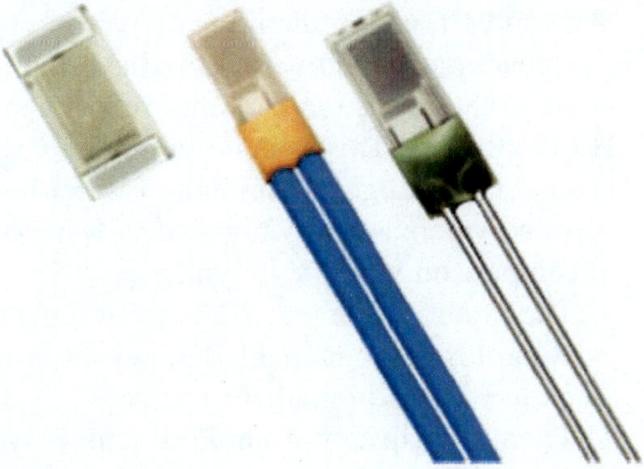

Fig. 2 Platinum-temperature sensors

The meander structures have to be protected after the trim process because the platinum layers are only within a range of 0.1–2 μm. This is done generally via a glass passivation, which is applied with a thick film process.

Sensors can be produced for application temperatures from −200°C to 1,000°C.

The designs of temperature sensors vary very strongly and are mostly coordinated with the application. High temperature sensors are contacted with pure platinum wires but, e.g., SMD sensors for printed boards are usually supplied with solder pads. Also sensor elements with insulating wires are available or also bondable elements (Fig. 2).

2.2 Capacitive Humidity Sensors

As already described in the abstract, most humidity sensors use a moisture-sensitive polymer for determining humidity. The composition of the capacitive humidity sensor is like a plate capacitor [5]. The moisture-sensitive polymer serves as dielectric, which is surrounded by a bottom electrode and a steam-permeable top electrode. The substrate material is ceramic or glass (Fig. 3).

Fig. 3 Composition of a humidity sensor

Another design for capacitive humidity sensors are interdigital structures (IDS), which are covered with a moisture-sensitive polymer. IDS sensors have no top electrode. Therefore, they have disadvantages compared to the sensors with top electrode. Performance of IDS sensors is mostly worse, especially in high humidity application (>90%r.H.) and in chemically aggressive environments. The reason for that can be found in the protective properties of the metal top electrode.

Capacitive humidity sensors have the possibility to detect water in the vapor phase. That means the humidity sensor can be seen as a gas sensor, which reacts to H_2O molecule. This is done via the storage of water molecules into the polymer layer [6]. Although the amount of water inside the polymer is limited, the signal of these polymers is pretty high, since water has a high dielectric constant ($\varepsilon_r = 80.5$) in comparison with the polymer ($\varepsilon_r = 2\ldots4$).

The water vapor exchange between polymer and ambient air occurs via an existing diffusion gradient. The sensor is in the equilibrium with the environment when the diffusion gradient is zero.

The most capacitive humidity sensors, which are available on the market, have a basic capacitance between 30 and 500 pF. The basic capacitance is manufacturer and sensor type specific. The change of the capacitance about the whole humidity range (0–100%r.H.) depends on the basic capacitance respectively on the slope (pF/%r.H.). Typical values for the slope (0–100%r.H.) range between 8 and 100 pF.

The capacitive humidity sensors have no prescribed DIN characteristic in contrast to the above-described temperature sensor. Therefore, the humidity sensor elements have to be calibrated for every different sensor. The complexity of the calibration depends on the desired accuracy and the sensor properties. A one-point calibration is enough for much applications if the humidity sensor has a good linearity (<1.5%r.H.). Minimum a two-point calibration is necessary when the sensor has a bad linearity or a higher accuracy is required (<±2%r.H.). If the system has to achieve a defined accuracy requirement within a certain temperature range, then a calibration at different temperatures is necessary. However, a high accuracy can be achieved with statistic values if the temperature dependency and the behaviour of the sensors respectively the system is repeatable. In this case no calibration is required at different temperatures.

The capacitance of a polymer humidity sensor can be changed via a change of the active sensor area and/or the thickness of the polymer layer. This arises from (6) (calculation of the capacitance of a plate capacitor).

$$C = \varepsilon_0 \times \varepsilon_R \times \frac{A}{d} \tag{6}$$

where

C = capacitance;

ε_0 = dielectric constant of vacuum;

ε_R = dielectric constant of insulating layer;

A = active sensor area;

D = plate distance.

A change of the basic capacitance (capacitance at 23°C and 30%r.H) has also to consequence of a change of the sensitivity respectively the slope in pF/%r.H. It is essential that the higher the basic capacitance of a humidity sensor is, the steeper is the slope.

The polymer layer expands (heating) and contracts (cooling) at a change of the temperature. Therefore, the distance d (6) also changes as well as the capacitance of the sensor. That means a capacitive humidity sensor with a polymer layer as dielectric is temperature dependent. This temperature dependency has to be compensated.

3 Humidity Sensors in Applications

Using capacitive humidity elements, there are some constrains that decide whether an accurate measurement can be achieved. The most relevant points from the application point of view are described in the following.

The most critical point for all application is the knowledge of the exact temperature, which has a direct influence of the measurement accuracy. The first reason is that the capacitive signal of the polymer has a temperature dependence—means that at the same level of humidity but different temperatures there is a different capacity read-out. Thus, a temperature compensation is necessary in order to provide an accurate humidity value.

The second reason is given by the calculation of the dew point. The measured humidity is often used to calculate the dew point, so that the exact temperature is essential. Looking at the Magnus formula (3), the temperature accuracy directly influences the dew point accuracy, especially for high humidity levels. In spacious systems, a close distance between temperature-sensing position and humidity-sensing is therefore advantageous, since high temperature gradients can often be observed.

Many inaccuracies are often not influenced by the sensing element but due to the assembly and electronics. The read-out capacities are mainly in the range of a few tens to hundreds of picofarad. In that range, the smallest leakage currents and capacity variations in the electronic circuits can influence the measurement dramatically. Also the used components (e.g., microcontrollers, capacitors, or resistors) show temperature drift effects which makes the design of a good electronics very challenging.

Moreover, the most electronics heat them self up. This can generated locally hotter places, which can effect the humidity sensor, too. Example: Due to self heating of the electronics the humidity sensor has a increased temperature of 0.5°C with respect to the ambient temperature. This yields in an measurement error of 2.8%r.H. at 20°C and 90%r.H.

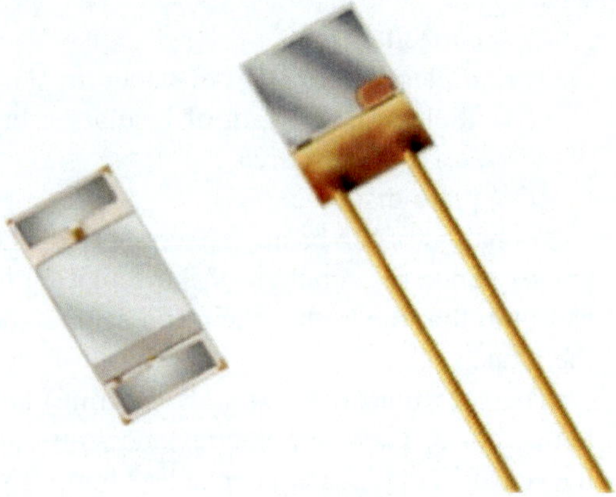

Fig. 4 Humidity sensors

A different problem can be found in the used materials for the circuit board. Most materials absorb humidity at high humidity levels, which they desorb at lower humidity levels or at higher temperatures. This can generate a microclimate over the circuit board, leading to a measurement error or an increased response time of the whole system. These problems can be dramatically reduced by not using SMD elements but sensor elements with wires. The short distance between the circuit board and the element helps to stabilize the climatic environment at the sensor element. The other main advantage using wired elements can be achieved by having the possibility to hermetically seal and separate the electronics from the sensor. On one hand, this brings a better measurement accuracy in application of high humidity and condensation, and on the other hand, in aggressive environments it is the only possibility to achieve long-term stable electronics (Fig. 4).

A different issue is degradation effects of the polymer or drift behaviors due to cross-sensitivity to different gasses. Ammonia from pig farms, chlorine from swimming pools, or methane out of biogas are examples for possible contaminations. The resistance against these influences are one key parameter which chooses the correct element for the application. Also, many of these effects can be annealed by heating the elements to higher temperatures. The drift is mostly due to absorption of the contamination gas inside the polymer. By heating up to higher temperatures (e.g., 130°C), the weak van-der-Waal Forces between gas and polymer break and release the gas to the ambient.

4 Combined Sensor: Humidity + Temperature Sensor/Heater

One system, which is getting more and more popular, is the combined temperature and humidity sensor. It combines the possibility of a temperature measurement directly at the place of the humidity measurement, as well as the possibility to heat the element from time to time.

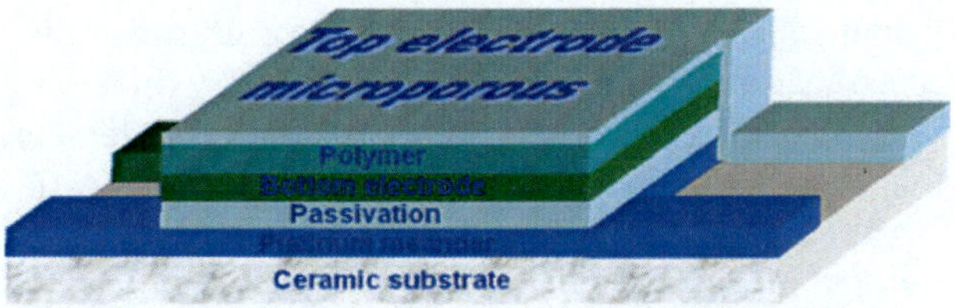

Fig. 5 Composition of a combined sensor

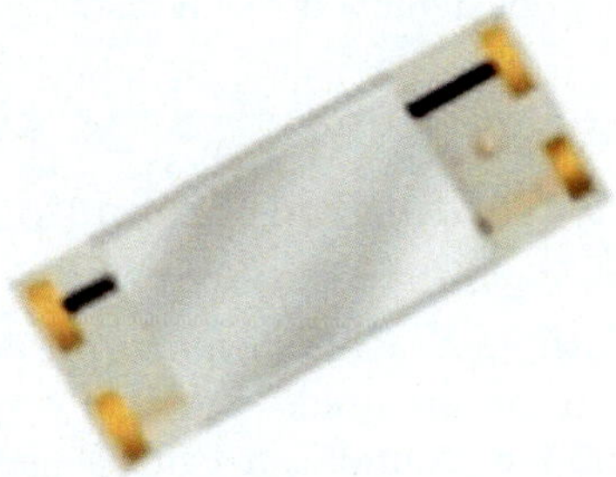

Fig. 6 Combined sensor—P14 SMD thermo

The sensor consists of the two sensor principle described previously one over the other. On a ceramic substrate a standard platinum sensor is constructed, afterwards the humidity sensing element is fabricated on top of the glass passivation layer of the temperature sensor. Such a system is shown in Fig. 5.

Using such a technology, both the elements are in a distance of only some tens of microns, but have the same advantages like the individual sensors: accurate, robust, and standardized temperature sensing as well as a robust humidity sensor with a water permeable top cover. One advantage is for sure the good integration of the systems and therefore the low area need in space-critical assemblies (Fig. 6).

Due to the close connection between the humidity-sensing polymer and the temperature-sensing platinum, the thermal coupling is nearly ideal. The distance between the both is in the range of 10–50 microns, only separated by the glass passivation layer of the temperature sensor and the bottom electrode of the humidity sensor.

The platinum meander structure can also be used as a heater. By applying higher current, self-heating can be used to heat the polymer. Since the characteristics of the platinum meander can simultaneously be measured, the exact knowledge of the temperature at every time is ensured, allowing a precise control of the heating temperature and suppressing overheating of the polymer.

As described previously, the sensor can be exposed to contaminating gasses which can be driven out of the polymer by heating. This normally shows a drift of the sensor signal due to diffusion or absorption of molecules. By periodically heating the elements to temperatures over 100°C, the molecules are driven out and the long-time stability and long-time accuracy can be increased strongly.

Since the temperature is very well known while heating, it can also be used for self-calibration or sensor verification. By heating the element to 150°C from room temperature, the humidity at 1 bar atmospheric pressure at even 100%r.H. is less than 1%r.H. In case of aggressive or abrasive environment, this enables a sensor verification. Using an element with a very linear or good defined characteristic, this technique can even be used to calibrate the element while being in the application.

Another main advantage can be achieved in case of condensation on the element. After exposed to water droplets, the polymer needs some time to desorb the water stored inside the polymer. This leads to higher response times after condensation on the element. For recovering the elements faster, the heating can be used to cast out the water out of the polymer, so that an equilibrium between ambient and polymer can be achieved much faster.

5 Optimized Technology on Customer Request

Different requirements can be defined depending on which application the humidity sensors are used.

For example, it can be necessary that a certain basic capacitance has to be provided. This is usually given by the used electronics. Because of the evaluation, electronic has to be defined exactly in a picofarad range. The adaptation of the basic capacitance can occur via a change in the thickness of the polymer layer or the dimension of the electrodes.

Another user-specific adaptation can be realized in the field of the response time of the sensor. The most capacitive humidity sensors have a response time in a range of 5–10 s. The response time can be reduced clearly (<1.5 s) via adapted choice of the top electrode. An example for such application is in the field of weather balloons.

The choice of the connection is another key parameter. SMD elements are very easy to integrate in circuit board designs and low cost in handling, but often show inaccuracies and drift effects to the used materials. By choosing sensor element with wires, this effect can be reduced. Depending on the application, the wire material can be chosen from a wide range. Starting from pure nickel wires, many other materials such as silver, silver-, gold-, or platinum-cladded nickel wires are possibly used. In some applications, very long wires up to several meters are necessary to protect the electronics from the influence of the measuring ambient. Using such elements, hermetically sealed electronics can be realized, which are placed in temperature-controlled conditions.

Last, it should be stated that before starting the development for a specific application, the temperature and humidity range and accuracy should be defined accurately. Especially in dew point application, this is most critical. Due to the strong dependence of the dew point, a sensor can show very accurate measurement results at moderate temperatures, and at high temperatures it can lead to very high deviation. Using an element with ±2%r.H. accuracy and a temperature accuracy of 0.5°C, the calculation of the dew point can be achieved with an accuracy of ±1.5°C at 30%r.H. When the temperature increases to 90°C, the accuracy with the same element can only be in the range ±5°C. Therefore, a close collaboration between sensor and system manufacture is necessary to finish those projects successfully.

References

1. Babin SM (2000) Water vapor myths: a brief tutorial. http://fermi.jhuapl.edu/people/babin/vapor/index.html
2. Schaumburg H (1992) Sensoren. Teubner, Stuttgart
3. Sonntag D (1990) Important new values of the physical constant of 1986, vapour pressure formulations based on the IST-90 psychrometer formulae. Z Meteorol 70(5):340–344
4. IEC 60751 (2008) Industrial platinum resistance thermometer and platinum temperature sensors (norm)
5. Hauptmann P (1990) Sensoren – Prinzipien und Anwendungen. Carl Hanser, München
6. Osada Y, De Rossi DE (2000) Polymer sensors and actuators. Springer, Berlin

Gas Sensor Investigations in Characterizing Textile Fibres

N. Felde and D. Kohl

Abstract The response of tin oxide and tungsten oxide sensor elements to the volatiles of heated polymer fibres of textile fabrics was investigated. As an example, polyethylene terephthalate (PET) fibres covered with finish were chosen. Both the contributions to the sensor signal from the fibre and the finish are discussed. All compounds originating from the fibres contain at least one aromatic ring, the dominating components are benzoic acid, phenol, and 1-phenyl-1,2-propandion. The finish contributes with several aliphatic aldehydes and ketones of chain lengths from 5 to 19 carbon atoms. Rings with five carbon atoms are also due to the finish. The investigated tin oxide sensor and one of the tungsten oxide sensors respond only to one decomposition compound of the fibres, namely to 1-phenyl-1,2-propandion. The second investigated tungsten oxide sensor does not respond to any PET decomposition product. However, all sensors react with high signals to decomposition compounds of the finish.

Keywords Gas chromatography, Phenyl-propandion, Polymer pyrolysis, Tin oxide sensor, Tungsten oxide sensor

Contents

N. Felde and D. Kohl (✉)
Institute of Applied Physics, University of Giessen, Heinrich-Buff-Ring 16, Giessen 35392, Germany
e-mail: Kohl@physik.uni-giessen.de

M. Fleischer and M. Lehmann (eds.), *Solid State Gas Sensors – Industrial Application*, Springer Series on Chemical Sensors and Biosensors (2012) 11: 151–172,
DOI 10.1007/5346_2011_4, © Springer-Verlag Berlin Heidelberg 2012,
Published online: 10 January 2012

1 Introduction

The detection of complex gas mixtures by gas sensors gains growing interest. Fields of application are food flavour analysis, air quality systems, solvent mixtures, and fire detection. For sensor evaluation, the high resolution gas chromatography/selective odorant measurement by multisensor array (HRGC/SOMMSA) approach has been established [1]. An essential part of this method is the use of a gas-chromatographic column ending in a split with the sensors working in parallel to a reference detector. In an earlier study, we presented results of sensor evaluation for the detection of specific organic smoke compounds by using the HRGC/SOMMSA approach [2]. Smouldering fires of wood were investigated in that paper. Smoke was sampled by different kinds of enrichment on charcoal and SPME fibres. It was shown that the reaction of an SnO_2 sensor to beechwood smoke is primarily due to 2-methoxyphenol and 2,6-dimethoxyphenol derivates with para-substituted alkyl and alkenyl groups, which originate in hardwood lignin. Both types of compounds occur in relatively high concentrations during smouldering and both are sensitively detected by SnO_2 sensors.

A similar experimental arrangement was used here to investigate a frequently used polyester (PES) fibre, polyethylene terephthalate (PET, some brand names are Diolen, Trevira, Terylene for fibres, and Hostaphan or Mylar for films, e.g. for food packaging). The robust fibres, often simply called polyester, are very abrasive resistant and have only a low humidity take up.

2 Experimental Conditions and Results

2.1 Samples and Apparatus

The following four Dupont PET fibres (M. Witschas, 2003, company DuPont, private communication) were investigated:

1. 1.6-38-158-NSD (monomers ethylenglycol, terephtalic acid, 0.08% weight finish)
2. 1.7-12-157-NSD (monomers ethylenglycol, terephtalic acid, 0.20% weight finish)
3. 1.6-38-702-WSD (monomers ethylenglycol, terephtalic acid, 2-sodium-sulfoisophthalic acid, 0.08% weight finish)
4. 3.3-60-483-WSD (monomers ethylenglycol, terephtalic acid, unknown phosphor organic component 1–5%, 0.08% weight finish)

A phosphor organic component replaces 1–5% of the terephthalic acid for flame retardation.

To get volatiles from the PET samples, four temperatures, 200°C, 280°C, 385°C, and 485°C, were adjusted. Volatiles with masses above 50 were evaluated by a Saturn GC-MS. Details of the method are given in reference [3]. Results at 200°C were taken with the static head space system of the GC-MS. At higher temperatures, a separate furnace was used, and the volatiles were enriched by solid-phase microextraction (SPME). Especially SPME, enrichment with 100 μm polydimethyl siloxane (PDMS) non-bonded delivered well-resolved chromatograms at higher molecular masses. Analytes with retention times below 27 min are suppressed (active charcoal can be used for light masses with retention times below 20 min).

At 285°C, 15 peaks (count rate around 20 kc or higher) appear and are identifiable by the NIST database [4]. In addition, nonanal with a low count rate of 10 kc was characteristic for PET and excited sensor response. The masses of identified molecules are between 142 (nonanal) and 226 (pentadecanal). All identified compounds belong to the aliphatic aldehydes (up to C15, partly with C=C bonds in the chain) or to the aliphatic ketones (C11–C13). Volatiles at 385°C and 485°C appear within a similar retention time range (20–41 min) and within the same mass range. However, completely different classes of compounds are observed. At 385°C 22 peaks appear, 17 of them could be identified, compare Fig. 1. At 485°C, 21 of 30 peaks (>20 kc) could be identified. The lowest mass observed is 94 (phenol); the heaviest is 206 (terephthalic acid, methyl vinyl ester). All compounds contain at least one aromatic ring (exceptions are undecanal and tetradecanal at 385°C). In nearly all cases, oxygen atoms appear within the side chains of the ring-forming ester-, aldehyde-, keton-, or alcohol groups.

The retention time range for PET 1.7-12-157-NSD volatiles at 385°C can be extended by the use of a 75 μm carboxen/PDMS cross linked SPME fibre to 3–43 min (lower enrichment factor below 20 min), compare Fig. 1. This figure shows also that the 100 μm PDMS fibre enriches aliphatic aldehydes relatively weakly. Therefore, the results from 75 μm carboxen/PDMS were taken for further evaluation.

The chromatograms of the four different PET samples showed the same peaks with only minor variance in height. Therefore, in the following, only results of fibre type 1.7-12-157-NSD are discussed.

Figure 2 shows results for all investigated temperatures. Only major peaks, typical for compound classes, are tagged by numbers. As mentioned above, qualitative differences occur only between 200°C/280°C and the higher temperatures.

The PET fibres are coated either with 0.08% weight or with 0.20% weight of finish, and the peak heights at 200°C of the aliphatic compounds vary correspondingly. Especially by washing with soap, these peaks shrink. At 485°C, the peaks of aliphatic compounds are reduced by a higher ratio, too.

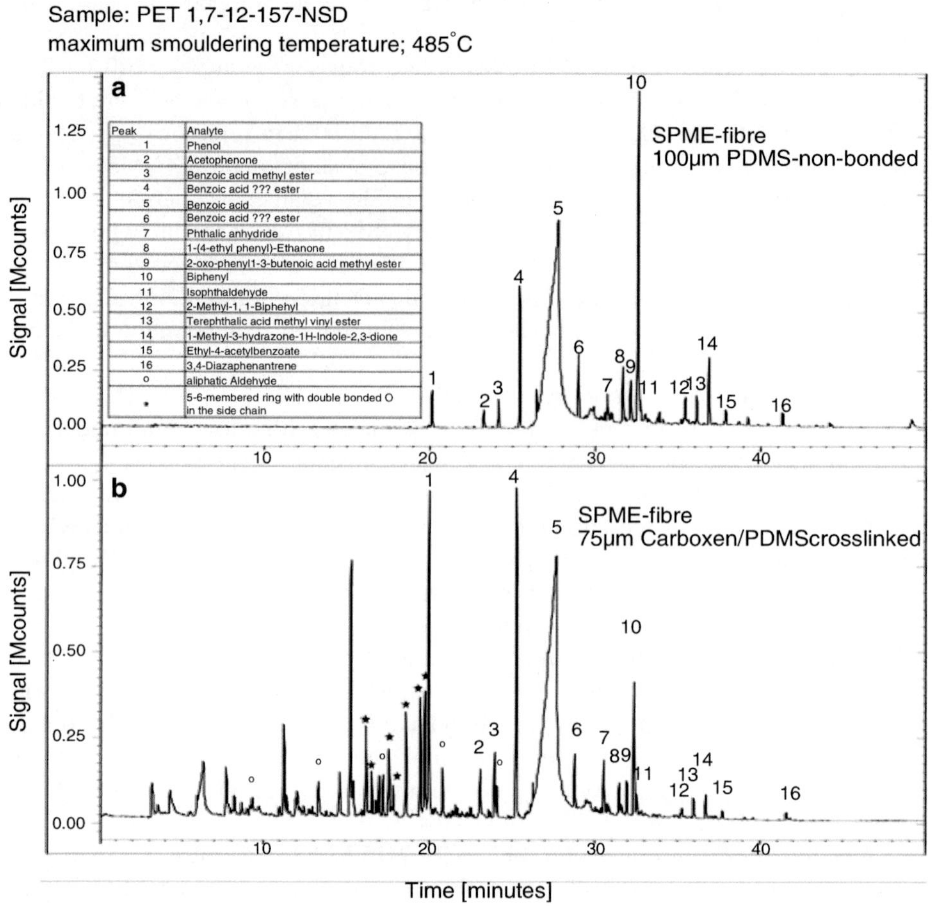

Fig. 1 Comparison of chromatograms obtained from different enrichment techniques. ???, The peak is small and is presumably due to benzoic acid. However, it cannot be excluded that it is caused by an ester

2.2 Sensor Elements

Commercial tungsten oxide (W330 and GGS 5330) and tin oxide (GGS 1330) thick-film sensor elements made by UST (Geschwenda, Germany) are applied for volatile detection. A further investigated tin oxide sensor (GGS 7330) is not reported, because it responded irreproducible. All sensors were operated at 350°C.

Since the absolute conductance change of oxidic sensor elements reacts too slowly to be compared with mass spectrometer peaks, the first derivative is usually sketched [1]. Thereby, residual conductance changes from the peaks preceding in time have lesser influence on an actual peak. But it has to be kept in mind that products of a preceding absorption may still be present on the sensor surface when the next compound is arriving at the column outlet. Maxima of the first derivative of the

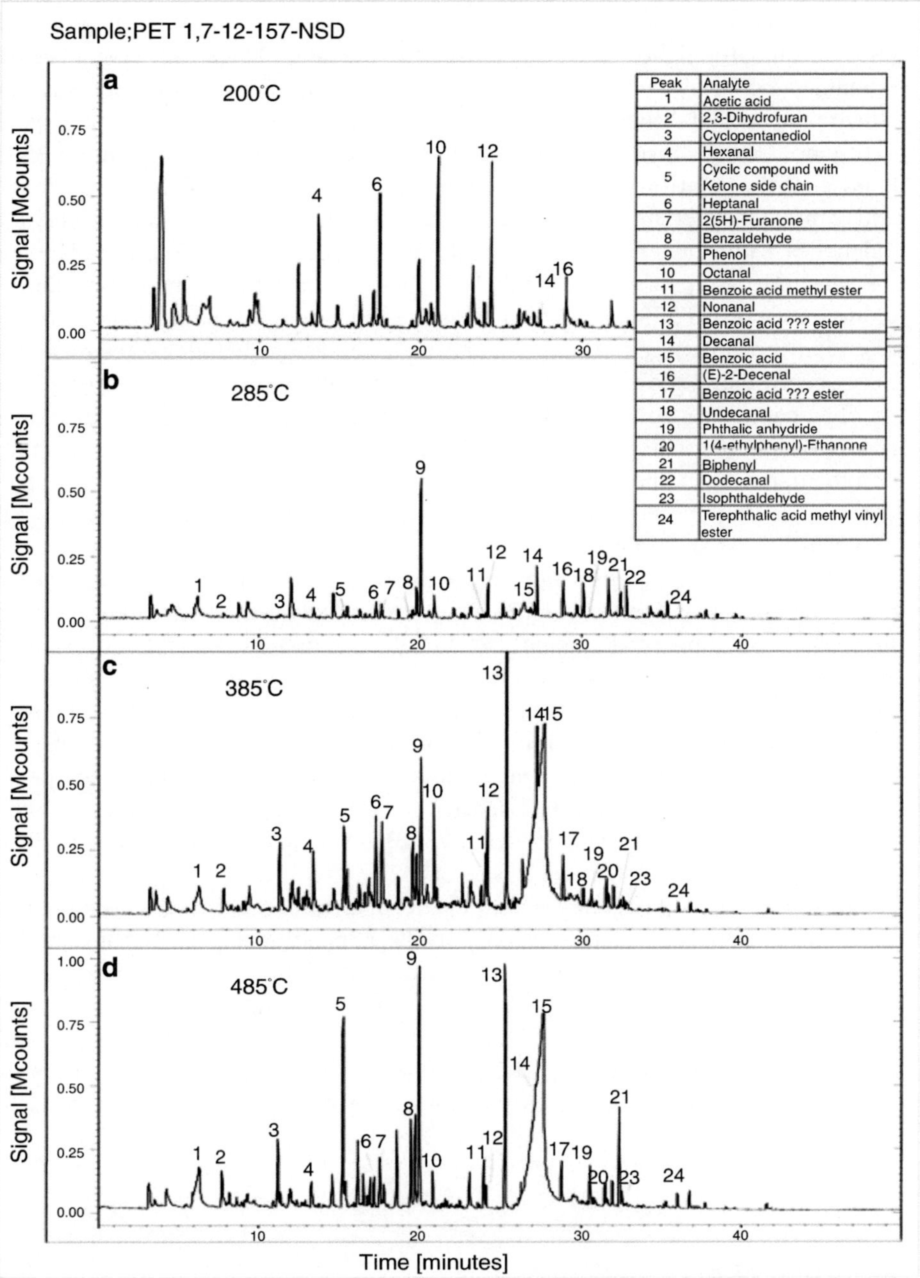

Fig. 2 PET decomposition compounds found with MS at various temperatures. ???, The peak is small and is presumably due to benzoic acid. However, it cannot be excluded that it is caused by an ester

sensor signal follow the mass spectrometer maxima with a delay of 1 or 2 s up to about 25 amu, for higher masses typically delays of about 15 s are observed. For some aliphatic alcoholes (e.g. heptanol, octanol) and phenol, the sensors react even more slowly; the special difficulties with phenol are discussed explicitly in reference [3].

Figure 3 shows results of a tungsten oxide sensor (W330, temperature 350°C) exposed to the volatiles of PET. The upper curve was measured first, the two others 1 and 2 months later in order to get some information concerning stability. The aldehydes with peaks 8, 12, 18, 20, and 22 are detected in a stable way. However, the sensitivity to light masses (peak 3–5) changes drastically. The second tungsten oxide sensor, GGS 5330, behaves similar for the aldehydes, some other peaks with retention times between 10 and 25 min are decreasing. Parallel screening by the mass spectrometer showed that the compound distribution remained nearly constant.

The tin oxide sensor GGS 1330 varies in sensitivity, compare peaks 10 and 12 in Fig. 3. Between retention times of 15 and 20 min, some peaks do not show up in all chromatograms. So, the sensor signal is reproducible only for a part of the detected compounds. In the following, only peaks which appear stable in all chromatograms of a sensor type are regarded.

Both tungsten oxide sensors are very sensitive with a high signal to noise ratio and have rather similar properties (Fig. 3). They react mainly to medium and heavy masses in the chromatogram, predominantly to aliphatic compounds. As a reference, the mass spectrometer peaks tagged separately for aliphatic and cyclic compounds are shown in the upper part of Fig. 4. The traces for 385°C and 485°C with the W330 sensor as an example show that in spite of a multitude of cyclic compounds in the mass spectrometer chromatogram the sensor reacts almost exclusively to aliphatic compounds. The GGS 5330 has an additional sensitivity to 1-phenyl-1,2-propandion. Both tin oxide sensors (initial sensitivity for GGS7330) are about one order of magnitude less sensitive than the tungsten oxide sensors. The GGS 1330 reacts mainly to low and medium masses. The signal is less dependent on molecule structure as for tungsten oxide. The GGS 1330 reacts not only to the aliphatic compounds but also rather strongly to some of the cyclic compounds. Table 1 shows the observed tungsten and tin oxide response to various classes of identified compounds.

Finally, it should be mentioned that the investigations of smouldering polyamide fibres have shown that heterocyclic compounds occurring in high concentrations, e.g. nitriles, amides, and amines, do not evoke relevant reactions on the sensor elements used above.

3 Discussion

3.1 Thermal Decomposition of PET

The majority of identifiable decomposition compounds are aromatic rings of five or six elements with ketones, esters, carboxyl, or aldehyde side groups.

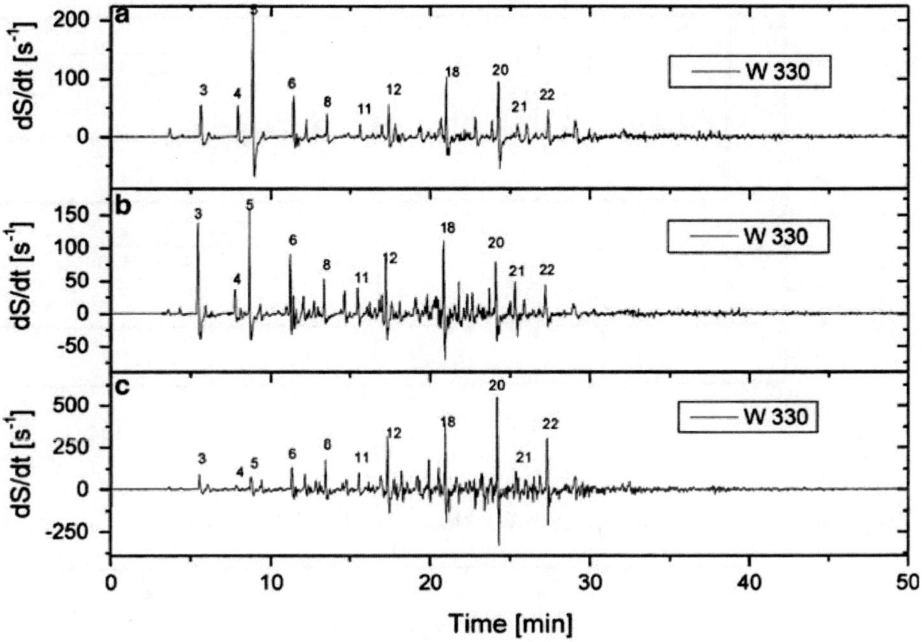

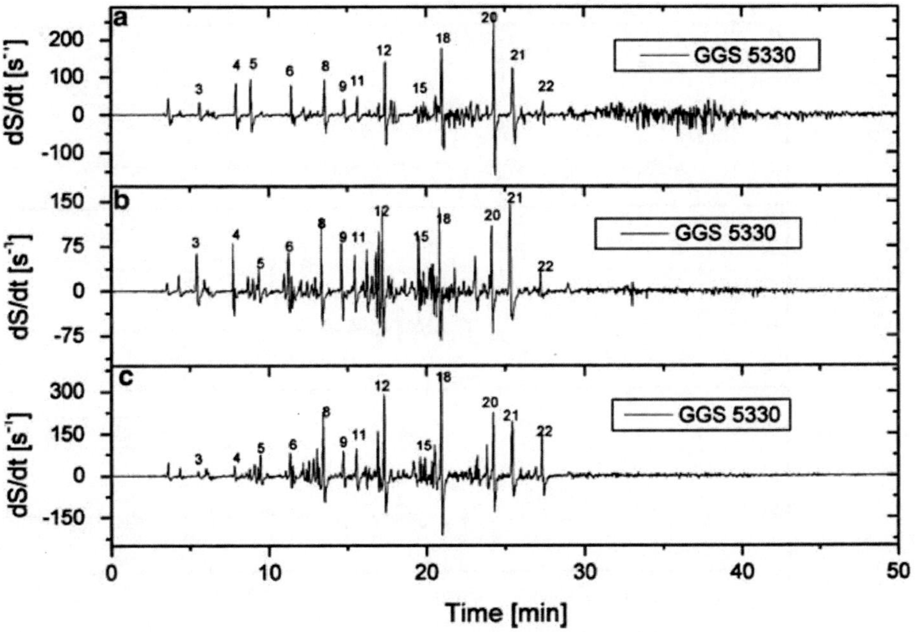

Fig. 3 (continued)

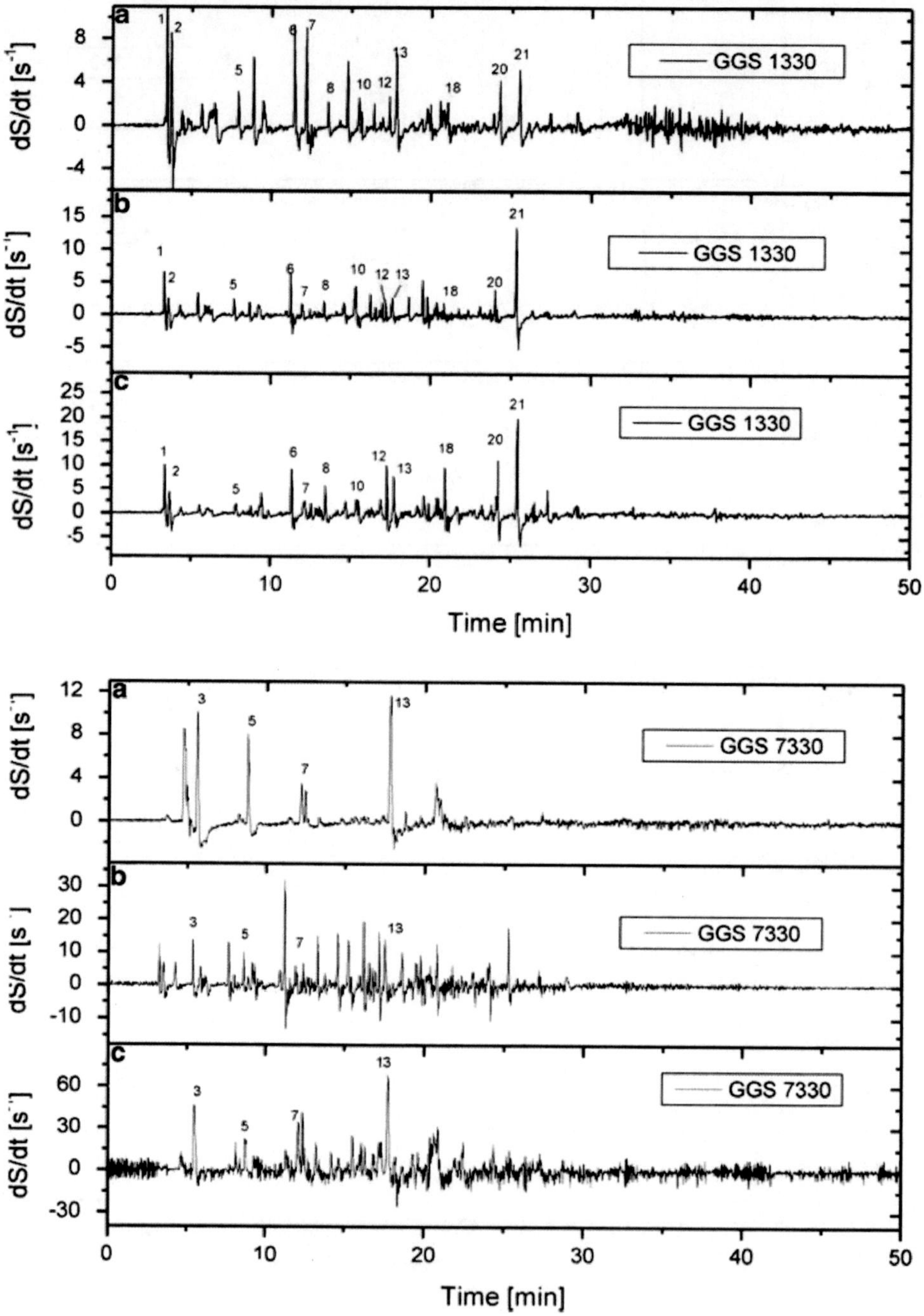

Fig. 3 Chromatograms of tungsten oxide (W 330, GGS 5330) and tin oxide (GGS 1330; GGS 7330) sensors exposed to the volatiles of PET smouldering at about 400° C. *Graph A*: first chromatogram taken, *Graphs B and C*: chromatograms taken 1 and 2 months later. *Identified peaks*: 3, 2-Butenal; 4, 2,3-Dihydrofuran; 6, 1,2-Cyclopentanediol, Hexanal; 9, 2-Cyclopentene-1-one; 10, cyclic compound with ketone side chain; 12, Heptanal; 13, 2(5H) Furanone; 14, 3-Methyl-2,5-furandione; 15, Benzaldehyde; 16, 2H-Pyran-2-one; 17, Phenol; 18, Octanal; 19, Benzoic acid methylester; 20, Nonanal; 21, 1-Phenyl-1,2-propanedione; 22, Decanal

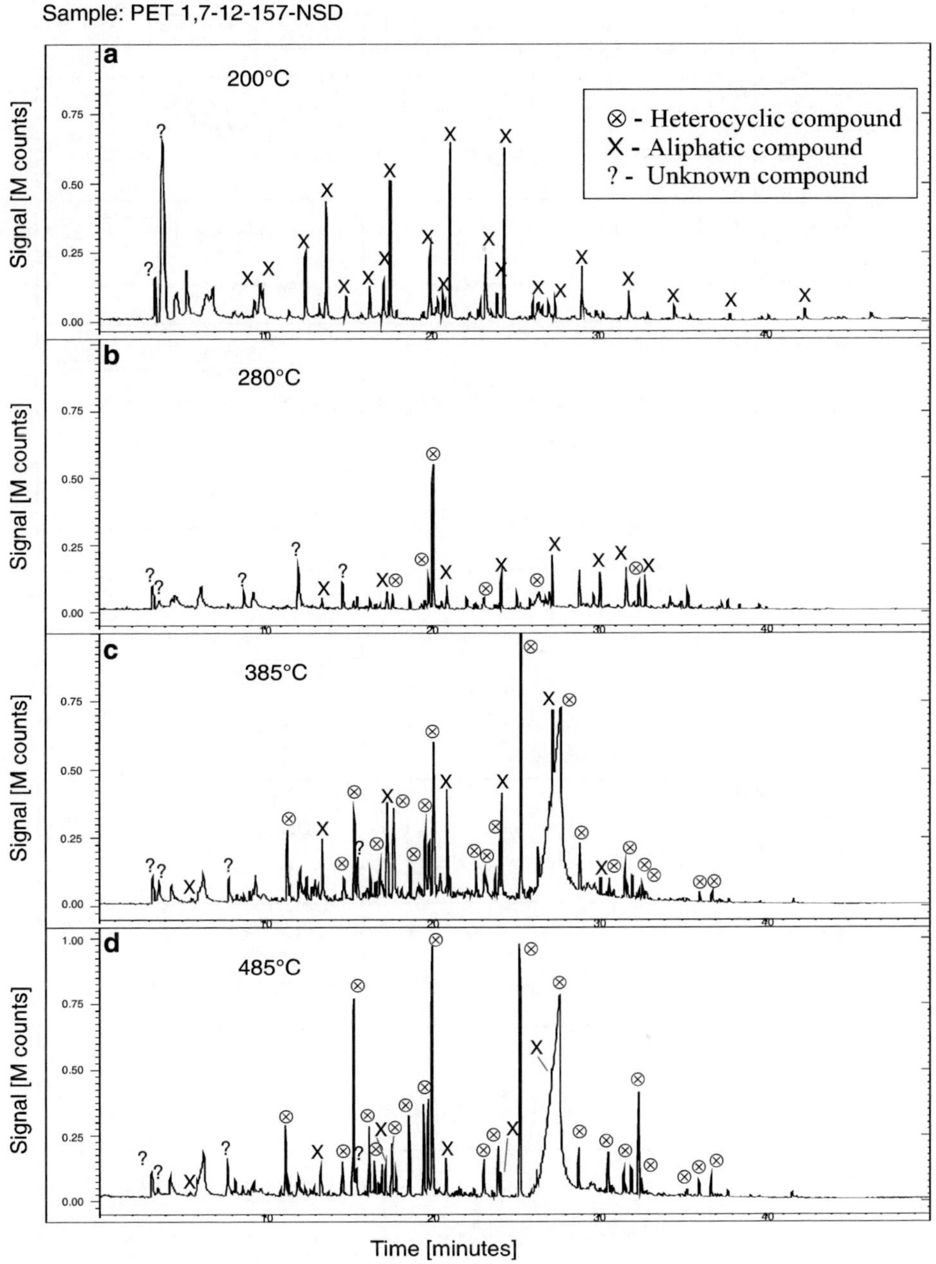

Fig. 4 (continued)

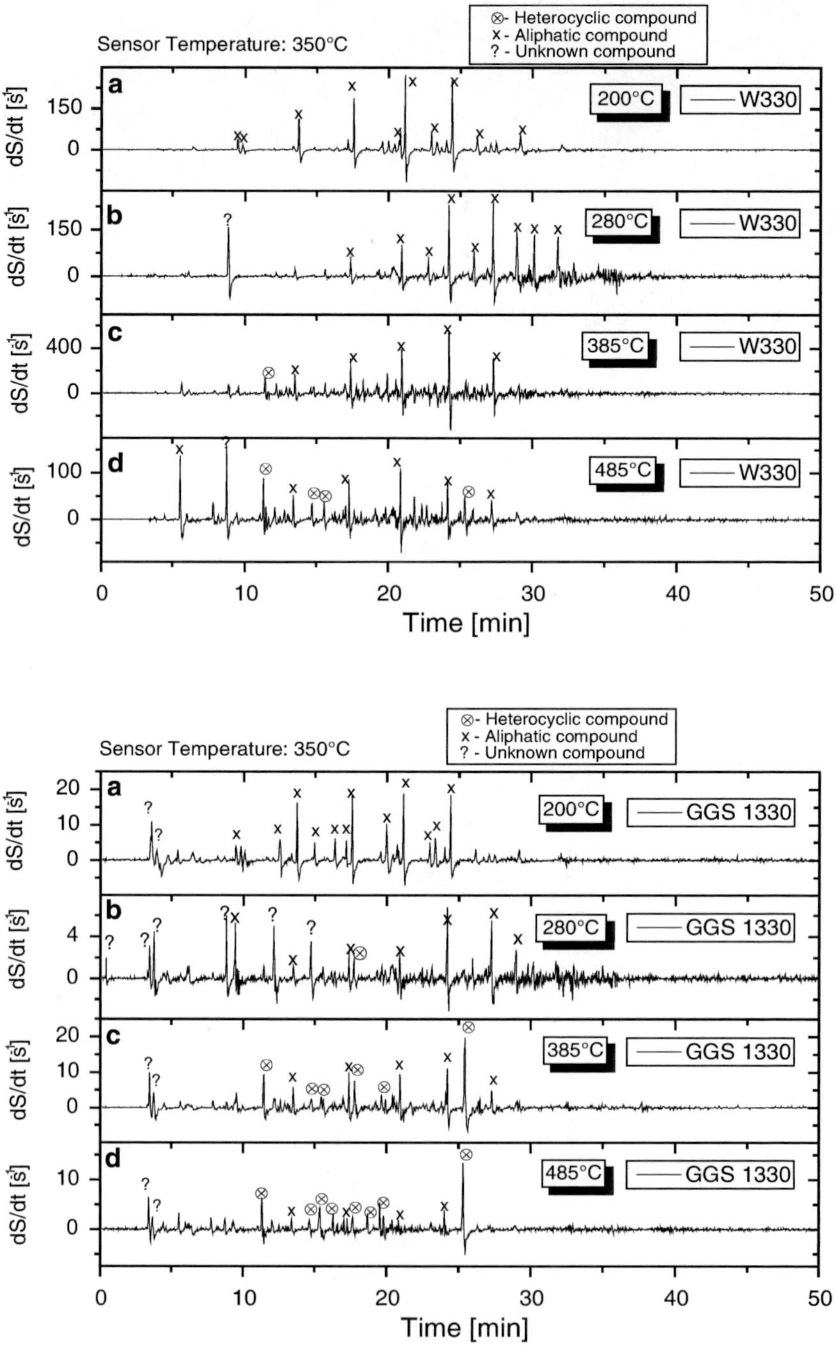

Fig. 4 Chromatograms of the mass spectrometer (to be taken as reference), a tungsten oxide (W 330) and a tin oxide (GGS 1330) sensor: aliphatic and cyclic decomposition compounds of a PET sample. Enrichment by 75 μm Car/PDMS

Table 1 Sensitivity of tungsten oxide (W330 and GGS 5330) and tin oxide (GGS 1330) thick-film sensor elements operated at 350°C to compounds formed during thermal decomposition of PET fibres

Retention time (min)	Decomposition compound	W330 tungsten oxide	GGS5330 tungsten oxide	GGS1330 tin oxide
Aliphatic aldehydes				
13.4	Hexanal	∅	++	+
17.3	Heptanal	+	++	+
20.9	Octanal	++	++	∅
24.1	Nonanal	+	++	+
27.4	Decanal	?	?	–
Aromatic carbonic acids/-ester				
24	Benzoic acid, methyl ester	––	––	––
25.3	1-Phenyl-1,2propanedione	––	∅	∅
27.7	Benzoic acid	––	––	––
28.8	Benzoic acid, ??? ester	––	––	––
32	2-Oxo-4-phenyl-3-butenoic acid, (methyl ester)	––	––	––
36	Terephthalic acid, methyl vinyl ester	––	––	––
37.8	Ethyl-4-acetylbenzoate	––	––	––
Aromatic compounds with keton-, aldehyde-, or hydroxy side groups				
19.5	Benzaldehyde	––	––	
20	Phenol	?	?	?
23.1	2-Methyl-phenol	––	––	––
30.7	Phthalic anhydride	––	––	––
31.5	1-(4-Ethylphenyl)-ethanone	––	––	––
32.6	Isophthalaldehyde	––	––	––
Aromatic hydrocarbons				
36.7	1-Methyl-3-hydrazone-1H-indole-2,3-dione	––	––	––
41.2	3,4-Diazaphenantrene	––	––	––
Aromatic hydrocarbons				
8.2	1,5-Hexadiyne/benzene	––	––	––
17	Styrene/1,3,5,7-cyclooctatetraene	––	––	––
32.4	Biphenyl	––	––	––
Rings with 5 or 6 atoms including oxygen in the ring or in the side chains				
7.7	2,3-Dihydrofuran	––	––	
11.2	*trans*-1,2-Cyclopentanediol	+	+	+
14.6	2-Methyl-2-cyclopentene-1-one/ spirohexan-5-one	––	∅	––
15.3	Cyclic decomposition product with a ketone group	––	––	–
16.2	2-Cyclopentene-1,4-dione	∅	––	––
17.6	2(5H)-Furanone	––	––	+
17.8	*p*-Benzoquinone	––	––	––
18.6	3-Methyl-2,5-furandione	––	––	––
19.8	2H-Pyran-2-one	––	––	––

Five categories: ++, very strong signal; +, strong; ∅, medium; –, weak; ––, very weak or no signal; ?, Peak not stable during ageing; ???, The peak is small and is presumably due to benzoic acid. However, it cannot be excluded that it is caused by an ester

At low temperatures, near 280°C, phenol is most intense; at higher temperatures, intensities of benzoic acid and 1-phenyl-1,2-propandion increase (Fig. 2).

Literature offers only sparsely information with respect to decomposition products and mechanisms. Aseeva [5], Ortner [6] and Witschas (M. Witschas, 2003, company DuPont, private communication) mention benzoic acid, terephthalic acid, and benzene as gaseous products of heated PET. Ortner lists also alcohols (e.g. glycol), aldehydes (especially acetaldehyde) and ketones in his publication. In a high vacuum pyrolysis of PET at 350°C, Lüderwald et al. [7] identify a variety of ion masses in the range from 42 to 920 amu. The authors propose several decomposition reactions. The most frequently occurring mechanism, shown in Fig. 5, is a splitting of the ester group (*cis*-elimination). Thereby a carboxyl end group is formed which splits off carbon dioxide.

Aseeva [5] describes the PET decomposition by a complex mechanism with molecular and radical constituents. The main reaction path is again based on a *cis*-elimination with a lot of consecutive decomposition and transformation steps leading to a great number of decomposition products. Especially benzoic and terephthalic acid, and their anhydrides are mentioned there.

Lüderwald et al. [7] explain the formation of terephthalic acid by splitting of two carbon–oxygen bonds (compare Fig 5). None of the mentioned papers proposes an idea for the formation of benzoic acid. Because during the PET decomposition in vacuo Lüderwald et al. [7] do not detect the ion of the benzoic acid molecule at $M/Z = 122$, we assume that benzoic acid is not formed by a thermal decomposition reaction, but rather by a subsequent oxidation of a benzene ring having lost his side groups. Such a process can explain also the occurrence of phenol and benzaldehyde.

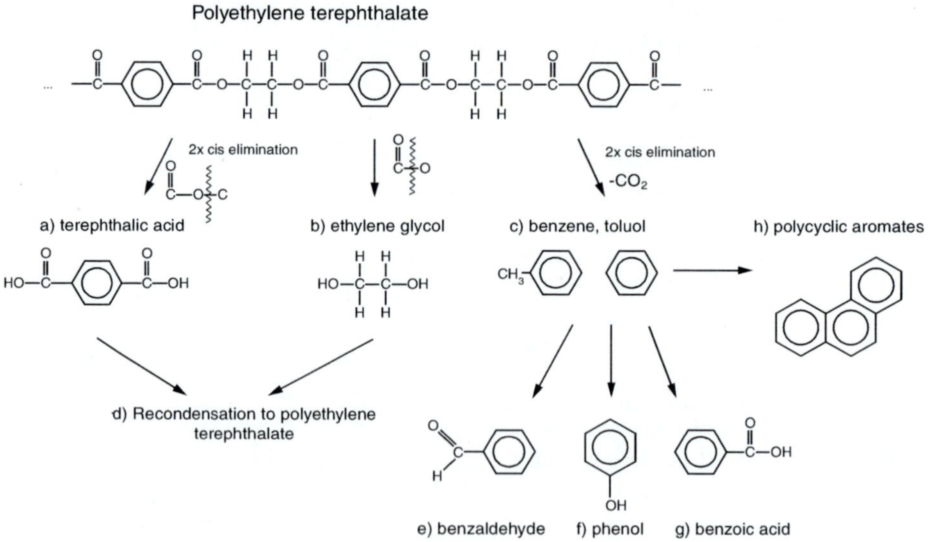

Fig. 5 Reaction paths during thermal decomposition of PET. Formation of carboxyl and vinyl groups by *cis*-elimination followed by split off of carbon dioxide from the carboxyl group. A lot of recondensate accumulates on the walls and is well visible

Besides benzoic acid, several esters of benzoic acid are identified, too. These compounds may be due to a reaction of benzoic acid and possibly its anhydride with alcohol components arising from the decomposition mechanisms shown in Fig 5. Such a consumption of alcoholic compounds to esters would explain the very weak peaks of alcohols, glycols, and benzoic acid anhydride in the chromatogram. A high reaction probability between decomposition components is visible by the intense formation of repolymerized species on the walls of the oven and in the enrichment tubes, Fig. 5. Correspondingly, terephthalic acid and glycol do not occur in the mass spectra.

After 25.3 min, a sensor relevant component is observed which belongs to the five strongest peaks. The mass spectrum contains M/Z = 105 (basis fragment), 77 and 51 as main fragments.

Lafferty [8] designates M/Z = 77 and 51 as main fragments of methyl benzoyl groups or of a phenyl group. In any case, an aromatic ring is present. The sequence M/Z = 105, 77 and 51 is also contained in two species with only slightly different elution times, benzoic acid methyl ester and benzoic acid ethyl ester. Therefore, it can be guessed that the unknown species is a benzene ester, too. From the relative position to the mentioned neighbouring esters, it can be derived that the ester group features a short chain length. Al-Flahi [9] identified such a compound as one of the main constituents (4.2%) in SPME GC/MS of the drug khat. His mass spectrum of 1-phenyl-1,2-propandion resembles strongly the spectrum found here. The molecular ion with M/Z = 148 is extremely weak in the spectrum, compare Fig. 6.

Summing up the model proposals all identified aromatic compounds can be explained by decomposition reactions followed by a reaction of these products with each other and with oxygen. The stable aromatic ring [10] does not break at the temperatures used.

Grause [11] investigated PET recycling by burning PET material at 730°C and analysed decomposition products with a GC-MS apparatus. The products are very similar to the products found in our experiments, compare Tables 2 and 3. The high values for aromatic compounds, benzene and toluol, should be due to the high combustion temperatures. Under his experimental conditions benzoic acid and phenol will be oxidized to carbon dioxide and benzene.

3.2 *Thermal Decomposition of Finish*

Aliphatic chains with 4–15 carbon atoms are identified in our spectra. Aldehyde, ketones, and alcohols are mentioned in the literature as decomposition products of PET, [6, 11]. However, there are some arguments that in the low-temperature spectra (200°C and 285°C), these compounds are not formed during the thermal decomposition of PET. We observed that the aliphates appeared already at 200°C. However, the thermal decomposition of PET needs about 300°C [6]. Further, the peaks of the aliphates are reduced relatively to other peaks after cleansing. They are also reduced for the fibres with lower coverage of finish. If the volatiles are

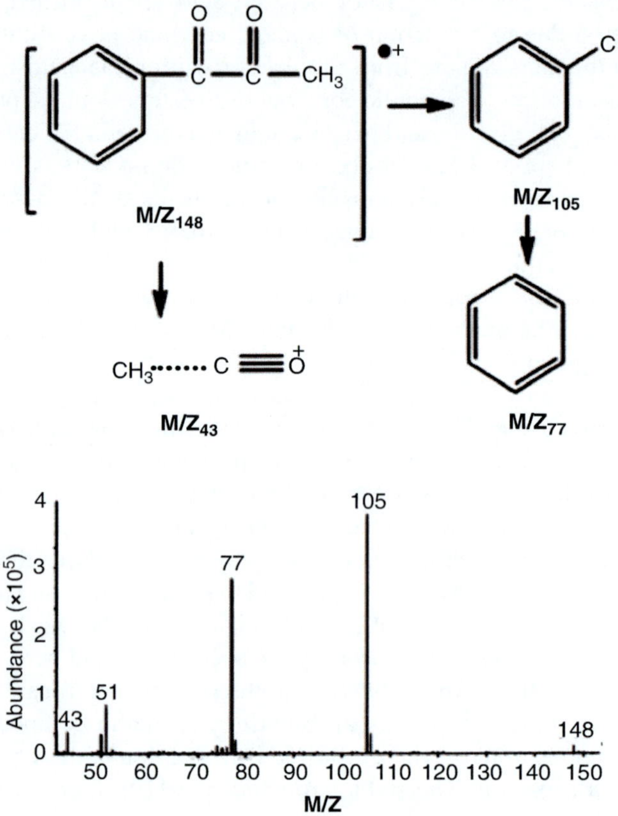

Fig. 6 Structural formula (Al-Flahi [10]) of 1-Phenyl-1,2-propandion and its mass spectrometer fragmentation products. Below the mass spectrum of 1-Phenyl-1,2-propandion taken by Al-Flahi [10] is reproduced

collected only at temperatures above 465°C, then the aliphates are not found. These findings are consistent with the experience that long-chain aliphates are not derivable from thermal decomposition of polymers. The formation of longer chains is energetically not favourable.

The PET samples of DuPont, we used, are coated with finish. These fats and oils (M. Grünwald, 2006, HTL Dornbirn (Innsbruck), private communication) were found to decompose to about 38% methylketones (C3–C17), to about 36% aldehydes (C3–C17) and to about 3% alcohols (C4–C14) [12].

Another group of compounds, not mentioned in the literature on PET decomposition, are cyclic compounds with five ring atoms, e.g. furanes or cyclopentanes. They cannot be formed from phenylic rings with six atoms, because rings with more carbon atoms are energetically favoured [13]. A break of the bond in the main chain of PET can only deliver aliphatic compounds with five or six atoms. One or two oxygen atoms (dioxane) are included in any case. However, these oxygen-containing compounds were not observed. The identified rings with five atoms

Table 2 Compounds found during the combustion of PET at 730°C (M. Grünwald, 2006, HTL Dornbirn (Innsbruck), private communication)

Combustion of PET	Run 1 w[%]	Run 2 w[%]
Aliphatic compounds	0.02	0.90
Aromatic compounds	72	45
Benzene	45	5.1
Toluol	3.5	2.0
Ethylbenzene	0.81	0.71
m-Xylol	0.15	0.13
Styrol	6.2	7.3
Phenylethene	0.15	0.16
C3-benzene	0.71	2.0
C-benzene	0.49	0.96
Inden	1.3	1.7
Naphthalene	0.38	0.65
Methylnaphthalene	0.31	0.51
Biphenyl	11	20
Methylbihenyl	0.72	1.3
Diphenylmethane	0.46	0.65
Fluoren	0.27	0.43
Other aromatic compounds	0.46	1.5
Not identified	0.95	0.09
Oxygen-containing compounds	25	47
Acetone	0.35	2.3
Cyclopentanon	0.16	0.11
Acetophenone	15	27
4-Methylacetophenone	2.7	4.6
Benzaldehyde	0.66	0.81
Other aldehydes and ketones	0.23	2.4
Ether	0.04	–
Aliphatic alcohols	0.02	–
Phenol	1.1	1.4
Other phenols	0.04	0.07
Benzoic acid methylester	0.32	0.46
Other ester	0.01	0.55
Benzoic acid	4.2	7.6
Other organic acids	–	0.05
Oxygen-containing heterocycles	–	0.13
Other compounds	2.1	0.11
Thiophen	0.75	–
Benzonitrile	1.3	–
Nitrogen heterocycles	–	0.04
Halogenated HC	–	0.03
Water	0.45	6.3

can only be formed from two short aliphatic compounds. Since these compounds are not observed by other authors during PET decomposition, it is likely that they are formed during the thermal decomposition of the finish.

Table 3 Identified cyclic compounds during heating of PET in air

Retention time (min)	Decomposition compounds	M/Z (amu/e)
Aromatic carbonic acids and ester		
24	Benzoic acid, methyl ester	136
25.3	1-Phenyl-1,2-propanedione	148
27.7	Benzoic acid	122
28.8	Benzoic acid or ??? ester	
32	2-Oxo-4-phenyl-3-butenoic acid, (methyl ester)	176 (190)
36	Terephtalic acid, methyl vinyl ester	206
37.8	Ethyl-4-acetylbenzoate	192
39.3	Benzoic acid phenyl ester	182
Aromatic compounds with keton-, aldehyde-, or hydroxy side groups		
19.5	Benzaldehyde	106
20	Phenol	94
23.1	2-Methyl-phenol	108
30.7	Phthalic anhydride	148
31.5	1-(4-Ethylphenyl)-Ethanone	148
32.6	Isophthalaldehyde	134
38.7	Benzophenone	182
Aromatic compounds with nitrogen atoms		
36.7	5-Ethyl-1H-Indole-2,3-dione	175
41.2	3,4-Diazaphenantrene	180
44.2	4-Phenylbenzylhydrazide	212
Aromatic hydrocarbons		
8.2	1,5-Hexadiyne/Benzene	78
17	Styrene/1,3,5,7-Cyclooctatetraene	104
32.4	Biphenyl	154
42.4	Phenantrene	178
44.3	*o*-Terphenyl	230
49.2	*m*-Terphenyl	230
Rings with five or six atoms including oxygen in the ring or in the side chains		
7,7	2,3-Dihydrofuran	70
11,2	*trans*-1,2-Cyclopentanediol	102
14,6	2-Methyl-2-cyclopentene-1-one/spirohexan-5-one	96
16,2	2-Cyclopentene-1,4-dione	96
17,6	2(5H)-Furanone	84
17,8	*p*-Benzoquinone	108
18,6	3-Methyl-2,5-furandione	112
19,8	2H-Pyran-2-one	96

???, The peak is small and is presumably due to benzoic acid. However, it cannot be excluded that it is caused by an ester

3.3 Sensor Response

Any changes of conductance of a semiconducting oxide surface are due to a redox process of the gas molecule with surface oxygen or are due to a charge transfer of an ionosorbed molecule [14]. The amount of charge transfer is given by the Mulliken charge [15]. The type of oxidation, partial with more or less sensor selectivity or total with lower selectivity, depends on the type of oxygen species involved, the acidity of the oxide [16, 17], and eventually on added catalytically active noble metal clusters [18].

The partial oxidation of hydrocarbons is frequently described as Mars-van-Krevelen mechanism [16–18]. Carbon of an adsorbed molecule reacts with nucleophilic surface lattice oxygen (O^{2-}), thereby the sensor conductance increases. Later, the reduced surface site is reoxidized by oxygen from the ambient air, accompanied by a return of the conductance to its initial value.

The acidity of an oxide surface depends on several parameters. First, it depends on the ionic/covalent character or difference of electronegativity in the metal/oxygen bond (oxidation number of the metal cation); second, it depends on the radius of the metal cation; and third, it depends on the coordination number of the metal cation in the lattice structure.

Tungsten oxide has a more covalent metal–oxygen bond because of the higher electronegativity of tungsten (Pauling electronegativity 2.36). It has medium-to-strong acid Lewis and Brönsted sites [16]. There are no nucleophilic or basic properties. Tungsten trioxide as an acid metal oxide reacts with water or adsorbed hydroxyl groups (in humid air) to tungsten acid or more precisely to a surface tungsten acid which is sketched in Fig. 7 [19]. The high electronegativity of the tungsten surface atom attracts the electrons of the oxygen atom in the hydroxyl group. So the basic (nucleophilic) property of oxygen nearly disappears.

Tin oxide with a Pauling electronegativity of 1.96 has an ionic bonding with a medium-to-weak Lewis acidity. Correspondingly, the basic character is medium to strong. In the presence of water (humidity of ambient air), hydroxide is formed on the surface, Fig. 7. Hydrogen atoms can be more strongly adsorbed on the nucleophilic oxygen atoms.

For selective oxidation, covalent oxides are preferred; ionic oxides tend to total oxidation [16].

3.3.1 Strong Sensor Response to Decomposition Products of Finish

As described, already at 200°C, finish decomposes to aliphatic aldehydes and ketones. All investigated sensors respond with a medium-to-strong conductance increase to these compounds. Aliphatic aldehydes are easily partially oxidized to carbonic acids. Especially strong and selective react acidic surfaces, e.g. WO_3 or CrO_3 [16]. A detailed description is given by Lercher [20]. In this acid-catalysed reaction, in a first step, the aldehyde is protonized, as shown in Fig. 8. In a second step, a nucleophilic attack of the deprotonized "surface tungsten acid" to the electrophilic carbon atom forms a "surface tungsten acid ester". Such a surface group can finally desorb as a carbonic acid, compare Fig. 8.

For more basic oxides, e.g. SnO_2, the electrophilic carbon can be directly attacked by the surface oxygen atom, as shown in Fig. 8. A dissociative adsorption follows, and finally carbonic acid desorbs again [21]:

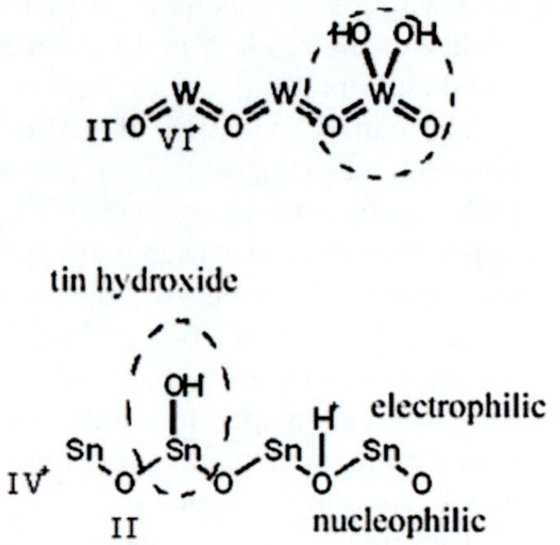

Fig. 7 Reaction of a tungsten- and tin oxide surfaces with water (humid air) to surface tungsten acid and to tin hydroxide, respectively

$$CH_3CHO_{ads} + O_{exposed} \rightarrow CH_3COO_{ads} + V_O.$$

The reaction rate is limited in both cases by the decomposition of the surface intermediate to carbonic acid [22]. This reaction step is faster in the acid-catalysed oxidation. Therefore, the WO_3 is more sensitive than SnO_2 to aldehydes. Generally spoken, a less acidic oxide (higher electronegativity of metal atom) is less sensitive to aldehydes.

The more acid tungsten oxide is in comparison with tin oxide also more sensitive to ketones. But the reasoning is different from that for aldehydes. Because of the position of the carbonyl group in ketone, an oxidation by one of the above reactions is not possible [3]. However, if some humidity is present, then a reaction including an adsorbed hydroxyl group is possible. In reference [21], the adsorption of acetone reacting by its carbonyl group with the adsorbed hydroxyl group to desorbing formic acid is described. That is in accordance with Lercher [20] who used the weakly basic acetone to evaluate the strength of acidity of mixed oxides. Brönsted acidity of a hydroxyl group increases with Lewis acidity on a metal oxide [16]. So the higher response of WO_3 on the more acid tungsten oxide is explained.

3.3.2 Weak Sensor Response to Decomposition Products of PET Fibres

Only one of the many compounds of PET decomposition identified by the mass spectrometer detector evokes a significant reproducible conductance change. Namely 1-phenyl-1,2-propandion affects one of the tungsten oxide sensors, GGS 5330, and the tin oxide sensor GGS 1330 (Fig. 3). Other significant conductance changes found there are not caused by decomposition products of PET. Since the side chain of 1-phenyl-1,2-propandion is similar in structure to acetone (Fig. 9), detection mechanisms can involve reactions for acetone described in the preceding section.

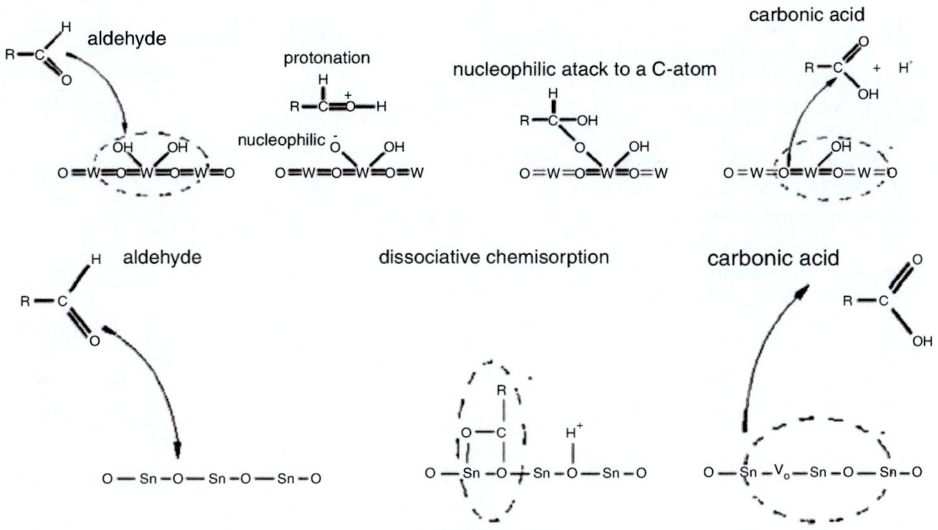

Fig. 8 Nucleophilic attack (acid-catalysed oxidation) of an aliphatic aldehyde on tungsten oxide. Electrophilic attack of an aliphatic aldehyde on tin oxide

So we will proceed to a discussion of the remaining PET decomposition products. None of the sensors responds to the highest mass spectrometer peak, benzoic acid. The reaction to another prominent peak, phenol, is only weak and so slow that it is difficult to measure at the outlet of the column. Such result is surprising on a first glance, but it is understandable. The majority of the PET decomposition products are formed at 300°C–480°C in air. That means that these products are stable in air against further oxidation at these temperatures. Since the sensor elements were operated at 350°C, only compounds formed already at lower temperatures and such compounds for which the sensor surface has a stronger catalytical activity can show up.

Benzoic acid is emitted in high concentrations from PET but does not cause a conductance change for any of the sensors used. Busca has shown by infrared spectroscopy [17] that benzoic acid on V–Ti metal oxide-based catalysts is oxidized to CO_2 and benzene. A minimum temperature of 400°C is necessary for this oxidation. If the reaction path on the investigated sensor surfaces would be similar, then the sensor operation temperature would have been too low for detection.

Toluol and benzene on metal oxides are typically oxidized to benzaldehyde and phenol [14, 16, 17, 23]. Both, benzaldehyde and phenol, can be further oxidized on metal oxides, to benzoic acid and chinones [17, 24, 25]. Activation temperatures for the partial oxidation of phenol to chinon are not reported in literature. However, Eberheim [25] found tin oxide sensors, operated at 270°C, selective to phenolic compounds. By the points given above, benzaldehyde and phenol should be more reactive as benzoic acid. The spectra showed small responses of the investigated sensors, but the reproducibility was lower in comparison to the aliphatic aldehydes and ketones [3].

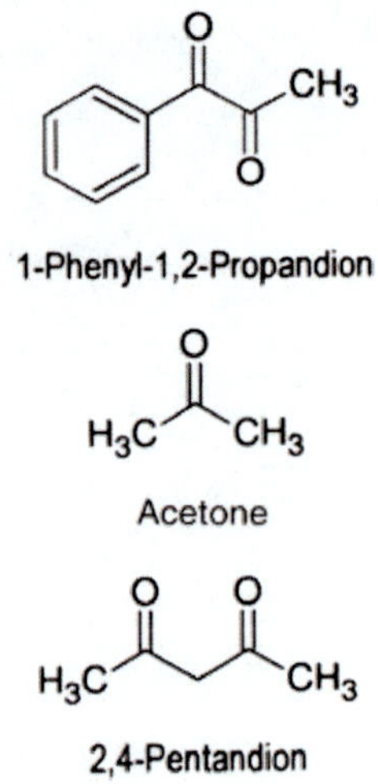

Fig. 9 1-Phenyl-1,2-propandion, acetone and 2,4-pentandion. The first compound is detected by the tungsten oxide sensor (GGS 5330) and the tinoxide sensor (GGS 1330). Acetone reacts by its carbonyl group with hydroxyle adsorbed on the oxide surface causing a sensor conductance increase [21] [17]

It is intelligible why a cyclic aldehyde (benzaldehyde) is less sensitively detected than an aliphatic aldehyde regarding the nucleophilic attack to the carbon atom sketched in Fig. 8. In case of an aliphatic aldehyde, the formal positively charged carbon atom of the carbonyl group is attacked. In case of the cyclic aldehyde, the positive charge is delocalized within the ring structure and thereby less accessible for a nucleophilic attack.

4 Conclusion

Heated PET fibres produce a spectrum of volatiles which is dominated by compounds of the oily and fatty coatings (finish). The formation of these compounds, aliphatic aldehydes and ketones, starts at 200°C and is completed at 450°C. The three investigated sensor types are sensitive to these compounds, and reaction mechanisms are discussed.

The cyclic decomposition products of the PET fibre are easily detectable by the mass spectrometer detector. However, the sensor reaction mechanisms of oxidic semiconducting sensors prefer aliphatic compounds relative to cyclic compounds as shown by experimental results and by considering reaction mechanisms. The cyclic PET decomposition components lose the competition with the aliphatic avivage components on the sensor surface.

There is one possibility to improve the situation for oxidic sensors. The fibres can be heated up in the absence of air (pyrolysis) to get higher amounts of oxidizable products for detection by the sensors. Alternatively, it would be interesting to use a metal oxide sensor, operated at higher temperature, e.g. gallium oxide at 800°C.

References

1. Hofmann T, Schieberle P, Krummel C, Freiling A, Bock J, Heinert L, Kohl D (1997) Sensors and Actuators B Chemical, http://www.journals.elsevier.com/sensors-and-actuators-b-chemical/ 41:81–87
2. Kohl D, Eberheim A, Schieberle P (2005) Thin Solid Films 490:1–6
3. Felde N (2005) Gassensor-Detektionsmechanismen für Schwelgase von PA-Granulat und PET-Fasern. Doctoral thesis, Giessen, Germany
4. National Institute of Standards and Technology, NIST 98 MS Library Database, http://www.nist.gov/srd/upload/Ver20Man.pdf
5. Aseeva RM, Zaikov GE (1981) Combustion of polymer materials. Hanser Publishers, München
6. Ortner J, Hensler G (1995) Beurteilung von Kunststoffbränden. Bayrisches Amt für Umweltschutz, http://www.lfu.bayern.de/luft/doc/kunststoffbraende.pdf
7. Lüderwald I, Urrutia H (1976) Makromol Chem 177:2079–2091
8. Mc Lafferty FW, Turecek F (1993) Interpretation von Massenspektren. Spektrum Verlag, Heidelberg
9. Al-Flahi A, Zou J, Yin X (2004) J Zhejiang univ sci 5(4):428–431
10. Mortimer CE (1986) Chemie. Georg Thieme Verlag, Stuttgart
11. Grause G (2003) Chemisches Recycling von Polyethylenterephthalat und Polymethylmethacrylat. Doctoral thesis, Hamburg
12. Belitz H-D, Grosch W, Schieberle P (2001) Lehrbuch der Lebensmittelchemie. Springer, Heidelberg
13. Vieweg R, Müller A (1966) Kunststoff Handbuch Vol 6 (Polyamide). Carl Hanser Verlag, München
14. Paul RK, Badhulika S, Mulchandani A (2011) Appl Phys Lett 99(3). doi: 10.1063/1.3614544 (Article number 033103)
15. Kohl D (2001) J Phys D Appl Phys 34:R125–R149
16. Busca G, Finocchio E, Ramis G, Ricchiardi G (1996) Catal Today 32:133–143
17. Busca G (1996) Catal Today 27:457–496
18. Centi G, Perathoner S (1998) Catal Today 41:457–469
19. Mortimer CE (1986) Chemie – das Basiswissen der Chemie. Georg Thieme Verlag, Stuggart
20. Lercher JA, Gründling C, Eder-Mirth G (1996) Catal Today 27:353–376
21. Kohl D (1996) Semiconductor and calorimetric sensor devices and arrays. In: Kress-Rogers E (ed) Handbook of biosensors and electronic nose: medicine, food & the environment. CRC Press, Boca Raton
22. Carey FA, Sundberg RJ (1995) Organic chemistry. Wiley-VCH, Weinheim
23. Yoo JS (1998) Catal Today 41:409–432
24. Vollhardt KPC, Schore NE (1988) Organische Chemie. Wiley-VCH, Weinheim
25. Eberheim A (2003) Qualifizierung von Halbleiter-Gassensoren für die Detektion spezifischer organischer Rauchgaskomponenten. Doctoral thesis, Giessen, Germany

New Approaches for Exhaust Gas Sensing

R. Moos

Abstract Steadily increasing emission standards for passenger cars and heavy duty vehicles combined with the need for fuel efficiency lead to novel powertrain concepts, for example to leanly operated gasoline direct injection engines, or to novel exhaust gas aftertreatment concepts such as Lean NO_x Traps (LNT), ammonia selective catalytic reduction catalysts for NO_x reduction (SCR), or even to a combination of both. Also, diesel particulate filters (DPF) are in series production.

To control these novel exhaust gas aftertreatment systems and to monitor on-board the proper operation of these systems (on-board diagnosis, OBD), novel exhaust gas sensors are required or are at least be very helpful. Since the development of exhaust gas sensors has always to be seen in interaction with the corresponding exhaust gas aftertreatment systems, novel types of exhaust gas sensors have gained in importance just recently, when the time was ripe for novel exhaust gas aftertreatment concepts. This article reports on several types of NO_x sensors and ammonia sensors.

Additionally, a very recent novel concept is presented. Here, the catalyst itself works as a sensing device that gives directly information on its own status. The readout can be wirebound (demonstrated for LNT and SCR) or even be wireless by applying radio frequency techniques. It will be shown that this allows to detect the oxygen loading degree of three-way catalysts very precisely. It can be also applied to determine the ammonia loading of SCR catalysts and the soot loading of DPF.

As a conclusion, these novel methods may provide a future alternative for low emission-aiming engine control as well as for OBD of low emission vehicles with novel exhaust gas aftertreatment systems. However, it is clear that all novel sensors or systems do not only have to meet the technical requirements but also have to be very inexpensive, reliable, and cost effective.

R. Moos (✉)
Bayreuth Engine Research Center, University of Bayreuth, 95440 Bayreuth, Germany
e-mail: Functional.Materials@Uni-Bayreuth.de

M. Fleischer and M. Lehmann (eds.), *Solid State Gas Sensors – Industrial Application*, Springer Series on Chemical Sensors and Biosensors (2012) 11: 173–188,
DOI 10.1007/5346_2011_6, © Springer-Verlag Berlin Heidelberg 2012,
Published online: 10 January 2012

Keywords Ammonia sensor, DPF, Lambda-probe, LNT, NO_x sensors, On-board diagnosis, SCR, Soot sensor

Contents

1 Introduction

Steadily increasing fuel costs in conjunction with the pressure on the automotive industry to reduce CO_2 emissions lead to booming market shares for Diesel propelled passenger cars. Since they are operated leanly, nitrogen oxides (NO_x) removal by the use of conventional three-way catalysts (TWC) in an engine that is controlled by lambda sensors, for example [1] is not possible. At the same time, for both passenger cars and for commercial trucks—the latter are almost completely Diesel-operated—NO_x emission limits have been strongly reduced.

Well-known for NO_x removal of coal power plants is the ammonia SCR process. It has been adapted for automotive applications. For heavy duty vehicles, such systems are already in serial application, and recently, they have been serialized for passenger cars as well.

Leanly operated direct injection gasoline engines with NO_x storage catalysts (often denoted as lean NO_x traps, LNT) were developed also and have been introduced recently. Here, NO_x is adsorbed and stored in the form of nitrates during a lean phase. Just right before the LNT storage capacity is exhausted and the LNT begins to let pass NO_x, nitrate reduction occurs in a short rich phase. In the meantime, the LNT concept can also be found in the aftertreatment of Diesel engine exhausts, as well as the SCR system is discussed for leanly operated gasoline engines. Even the combination of both NO_x removal technologies, where ammonia is formed during the short rich phase, is in serial application [2].

Both novel concepts, ammonia SCR catalyst and LNT, require novel exhaust gas sensors that supplement the almost 40 years old lambda-probe.

Ammonia SCR technology allows reducing NO_x selectively in lean atmospheres. A urea–water solution (AdBlue™) is injected into the exhaust. There, it decomposes to ammonia and CO_2. In the SCR catalyst, NO_x reacts with the formed ammonia to nitrogen and water. NO_x conversion depends on many parameters in the SCR catalyst such as temperature, oxygen content, NO and NO_2 concentration, space velocity, and on the amount of previously stored NH_3 [3]. Since in contrast to power plant applications transient operations with steadily varying exhaust conditions and with nonconstant NO_x concentrations prevail in automobiles, a closed-loop controlled system would be beneficial. By utilizing ammonia sensors, the NO_x conversion could be increased to 93% in the European transient test [4]. Despite ammonia breakthroughs downstream of the SCR catalyst should be avoided, about 10 ppm NH_3 can be tolerated. An ammonia exhaust gas sensor could detect ammonia breakthroughs that stem from overinjection of AdBlue™ or from a catalyst failure. In short: a sensor-based closed-loop control system offers the possibility to effectively monitor the catalyst state (OBD) and to achieve a high conversion rate at the same time [4, 5].

To control these exhaust gas aftertreatment systems as well as for a reliable OBD, suitable exhaust gas sensors become increasingly important. Since both exhaust gas aftertreatment processes store gases, it might also be helpful for an optimized control of these systems to obtain direct knowledge of the degree of catalyst loading.

The *serial development* focuses in two directions. Selective exhaust gas sensors such as NO_x sensor or an ammonia sensor have already been serialized or are announced to enter the market soon. These sensors base upon the platform of planar lambda sensors (cf. [6] for planar lambda sensor technology). They take advantage the high degree of maturity of planar zirconia sensors and the long experience in manufacturing and operating these devices.

In an early stadium of *research* are novel and promising concepts to measure directly the degree of catalyst loading without the indirect determination of an exhaust component. For that reason, electrical properties of catalyst coatings are measured during operation and the catalyst state is derived in-situ from these data.

Since this book is concerned with *gas* sensors, another novel trend, soot sensors, is not considered here. The reader is referred to [7] for an overview on exhaust-suitable methods for diesel particulate detection. Some interesting results on conductometric soot sensors are given in [8] and [9].

2 Sensors for Lean NO_x Traps

During a lean phase, nitrogen oxides are stored as nitrates. Before the storage capacity of the LNT is exhausted, a short rich phase is applied and the nitrates get reduced. Usually, it is not possible to operate such systems open-loop controlled. Therefore, a NO_x sensor downstream of the LNT is helpful to avoid overloading the LNT by detecting NO_x breakthroughs in an early state. An additional lambda-probe is required to ensure that the LNT is regenerated only as long as necessary, since otherwise CO or hydrocarbon breakthroughs may occur.

Besides the direct measurement of the exhaust gas component NO_x or of the normalized air-to-fuel ratio λ, it would be beneficial for an optimized control to know the degree of catalyst loading. In the last years, initial work is conducted in research labs to measure directly—without the detour to the exhaust gas concentration—the state of an exhaust gas catalyst.

2.1 NO_x Sensors

In the middle of the 1990s, a threefold exhaust sensor device was proposed [10]. It bases upon the planar zirconia technology. With this sensor not only NO_x but also the oxygen concentration in the lean exhaust can be detected. An additional binary lambda signal is also provided. The sensor is well documented in the literature [11]. It shows only very small cross-sensitivities to components that are usually present in the exhaust (CO_2, O_2, CO, HC, H_2, SO_2) [10]. The remaining cross-sensitivity for ammonia can be used for SCR purposes (see below) [12]. Due to its two pumping chambers and the electronic circuitry to deal with very small currents, the sensor is relatively complex and costly. At the moment, it is in discussion whether the sensor accuracy suffices to meet the requirement for future OBD [13]. Very recently, a similar NO_x was announced from a second supplier [14].

Two separate review papers on exhaust gas NO_x sensors were published recently [15, 16]. Both give an overview on further opportunities to detect NO_x directly in the exhaust. If one reads these papers critically, one gets the impression that all the novel attempts require much more research work and it is not clear whether they may ever reach the technical ripeness for serial application.

2.2 Sensors for Directly Determining the Loading Degree of an LNT

This complete novel approach is described in [17, 18]. By in-situ measuring the electrical impedance of the coating material itself, the state of an LNT (i.e., NO_x loading, state of regeneration, state of sulfurization, and thermal aging) can be determined.

On an electrically insulating ceramic plate, interdigital electrodes are applied by thick- or thin-film technology. A passivated heater on the backside can also be used as a resistance thermometer. A coating, which is identical with the ceramic LNT monolith coating, is applied on top of the interdigital electrodes.

A sketch of a typical test in synthetic exhaust is shown in Fig. 1. The sensor is heated to the same temperature at which an LNT would operate, for example 350°C, and the sensor impedance is determined at some kHz. Three different gas mixtures ("A," "B," "C") are applied consecutively. It starts with the rich gas composition "C," for example $\lambda \approx 0.8$, followed by a lean gas composition without NO, $\lambda \approx 2.1$; "A." Then the lean gas composition "B," which is composed like "A" but with NO, is applied.

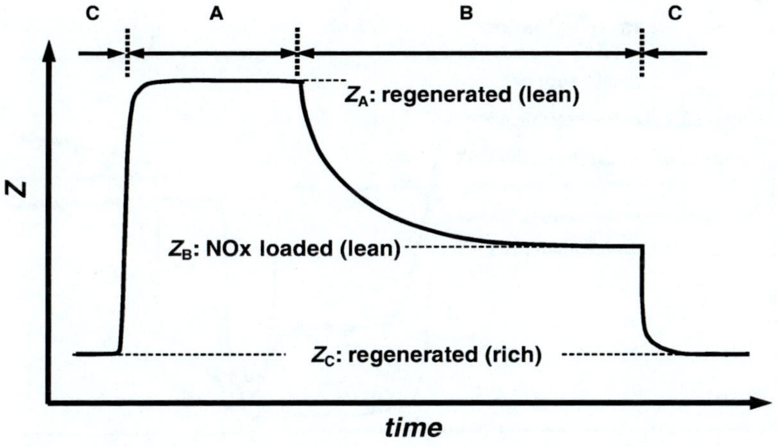

Fig. 1 The electrical impedance of LNT materials at 350°C. Simplified after Ref. [17]

As can be seen, the electrical impedance Z takes three different values, denoting three different states of the LNT: Z_C (regenerated, rich), Z_A (regenerated, lean), and Z_B (NO_x loaded, lean). Let us start our considerations at the lean regenerated point, Z_A. As soon as NO is added to the exhaust, the catalyst coating stores NO_x and the impedance decreases as the NO_x loading increases. The impedance reaches Z_B, the value for "NO_x loaded, lean." Since sensor material and LNT material are identical, it is clear that the catalyst loading degree can be determined directly with the help of the electrical measurement. Of course, the temperature dependence of the electrical impedance has to be corrected [17].

The precise end of the regeneration phase can also be determined. Figure 2 shows the course of a regeneration phase conducted in a synthetic gas test bench. Two status sensors were employed, one upstream and the other downstream of an LNT. The LNT was NO_x loaded in lean exhaust (for details, see [19]). At $t' = 0$, the regeneration started ($\lambda \approx 0.75$). The signal of a wide band lambda-probe (also called UEGO-sensor or linear lambda-probe) downstream of the LNT mirrors the behavior. During regeneration, the UEGO-sensor remains at the stoichiometric value of $\lambda \approx 1$. The sensor upstream of the LNT represents the first part of the LNT. As its impedance shows, it is almost instantaneously regenerated. After about 6 s, the regeneration is completed and lambda downstream of the LNT reaches again the feed gas value ($\lambda \approx 0.75$). Just at this point, the status sensor downstream of the LNT switches to the value of the "regenerated" state, indicating that also the last part of the LNT is regenerated. The status sensor signal is much more pronounced than the signal of the UEGO-sensor. At $t' = 8$ s, the regeneration process is finished. The feed gas is switched back to lean (with NO). The first part of the catalyst, indicated by the first sensor, switches almost immediately to the lean value. It lasts some seconds until the last part of the catalyst is also oxygen loaded and the sensor switches to the value "regenerated in lean exhaust." This occurs

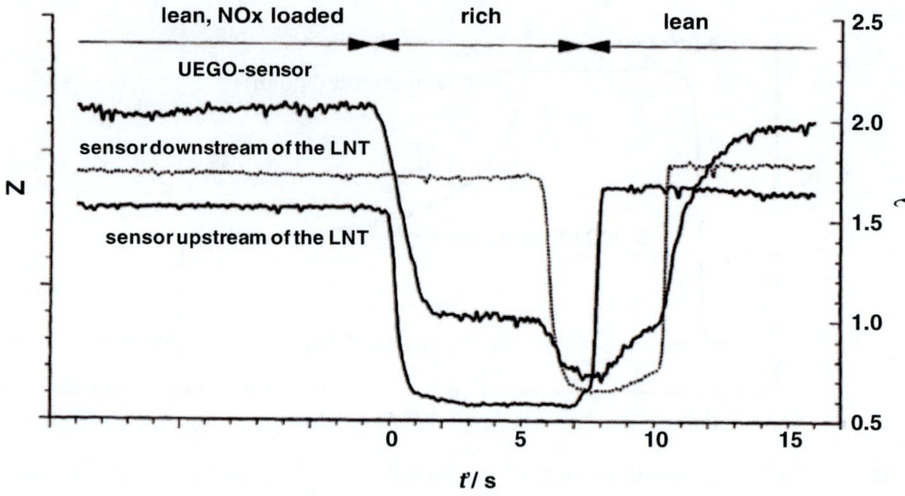

Fig. 2 Regeneration of an LNT. Comparison of a sensor upstream of an LNT (representing the first few millimeters of the coating), a sensor downstream of an LNT (representing the last millimeters of the catalyst coating), and a wide band lambda-probe installed downstream of an LNT. For more details see text and Ref. [19] (from [19])

simultaneously with the UEGO-sensor signal downstream of the LNT, which also reaches the lean value of the feed gas. Again, the signal is much more pronounced.

It has been additionally shown that both sulfurization and thermal aging can be distinguished [17]. Dynamometer results are very promising [17].

3 Sensors for NH_3–SCR–deNOx

As written above, suitable exhaust gas sensors become increasingly important for the control ammonia of SCR systems. An ammonia exhaust gas sensor can detect an ammonia breakthrough as the result of an overinjection of urea water solution or of a catalyst defect.

3.1 NH_3 Sensors

Up to now, no ammonia exhaust gas sensor has been serialized. In the following, sensors in the R&D state are reviewed. Only sensor principles are considered that have—at least from the author's point of view—the potential to meet the technical requirements for long-term stability in the exhaust. Additional requirements are high sensitivities, low cross sensitivities, reproducible sensor signals, and a low-cost sensor manufacturing technology. Several principles are proposed in the literature, comprising

- Conductometric sensors based on semiconducting metal oxides
- Zeolite-based sensors
- Mixed potential sensors
- NO_x sensors with a pronounced ammonia cross sensitivity

Conductometric sensors based on semiconducting metal oxides are applied for the detection of many different gases. $Cr_2Ti_{2-x}O_{7-2x}$-based ammonia sensors are in use for nonautomotive applications since they can be produced reproducibly in low-cost film technologies [20]. Like most conductometric semiconducting gas sensors that base on surface-effects, they do not withstand very well the high exhaust temperatures. A more promising candidate for exhaust applications is a recently presented conductometric ammonia exhaust gas sensor based on tungstated zirconia, that is on an acidic mixture of ZrO_2, Y_2O_3 and WO_3 [21]. The sensor film is applied on interdigital electrodes and covered with a porous protection layer. The long-term stability depends on the yttria content of the functional film [22]. Optimized sensors may be applied in the exhaust, since they retain their good ammonia sensitivity even after high temperature aging at 600°C for 100 h in air.

Zeolite-based sensors were originally suggested by Daimler and later developed by Delphi [23, 24]. Measurand is the ammonia-dependent electrical impedance of a zeolite functional film, which is determined in a sensor setup very similar to the above-mentioned conductometric ammonia sensor. At the sensor operation temperature of 420°C, exhaust components such as CO, CO_2, HC, O_2, and NO do almost not affect the sensor impedance. The remaining small water cross sensitivity can be compensated. Despite the facts that engine tests were very promising and the sensors are stable up to 600°C, it seems that they will not be serialized since the long-term stability in real exhaust is not fully given and since baseline drifts might occur [25].

Mixed potential sensors are very promising for the detection of ammonia in the exhaust. Here, a potential difference forms between two different electrodes that are separated by solid electrolyte yttria stabilized zirconia, YSZ. General details on the theory of mixed potential sensors can be found in [26, 27]. At the three phase boundary of each electrode (gas, electrode, solid electrolyte), a potential that depends on the surrounding gas atmosphere composition forms. Hence, the catalytic properties of the electrode materials determine the sensitivity of the sensor and its cross sensitivity to interfering gases.

In the *serial development* stage is a sensor that has been originally presented by Wang et al. in Ref. [28]. The sensing electrode contains SCR-active oxides of molybdenum, vanadium, and/or tungsten. It is doped with materials like bismuth and with other stabilizers to improve long-term stability [28]. An oxidation process and a reduction process compete with each other at the electrodes:

$$\frac{1}{2}\,O_2 + 2e^- + V_O^{\bullet\bullet} \rightarrow O_O \tag{1}$$

and

$$\frac{2}{3}\,NH_3 + O_O \rightarrow H_2O + \frac{1}{3}\,N_2 + 2e^- + V_O^{\bullet\bullet} \tag{2}$$

wherein $V_O^{\bullet\bullet}$ denotes doubly positively charged oxygen vacancies and O_O stands for lattice oxygen.

As a result, the formed potential difference between both electrodes, U, depends on the ammonia partial pressure:

$$U \propto \frac{RT}{2F}\left(\frac{2}{3}\ln p_{NH_3} - \frac{1}{2}\ln p_{O_2} - \ln p_{H_2O}\right). \tag{3}$$

The sensor shows only a very slight cross sensitivity toward other exhaust components [29], mainly toward NO_2. At a first glance, one obtains from (3) a cross-effect of oxygen and water. Because oxygen and water in the exhaust are inversely correlated by the air-to-fuel ratio λ, water and oxygen interferences cancel out each other in parts. Dynamometer tests are very promising [30]. The sensitivity remains constant even after aging at 800°C for 570 h [31]. Compared to conductometric and impedimetric sensors, it is a major advantage of the mixed potential principle that the baseline $U(0\ \text{ppm}\ NH_3) = 0$ V is very stable.

Another mixed potential ammonia sensor *research* approach further improves the above-described setup. The sensor contains a thick-film heater on an alumina substrate and a YSZ solid electrolyte. Porous gold electrodes are in contact with the exhaust. In contrast to the above-described setup, one of the electrodes is covered by a porous catalytically active layer. Therefore, the functions of the classical active electrode for mixed potential sensors (electrical conductivity, high selectivity, and

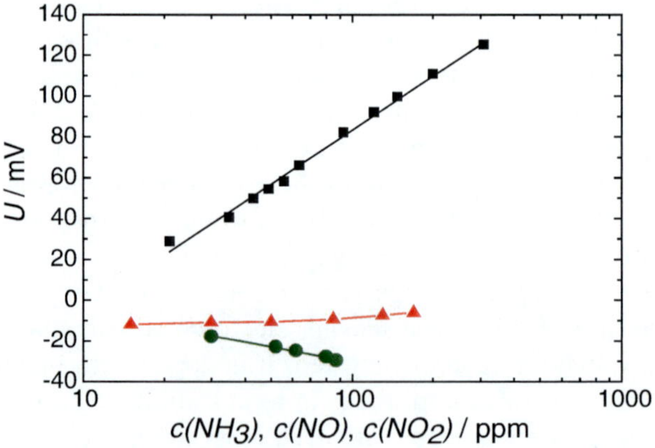

Fig. 3 Semi-logarithmic sensor characteristics towards NH_3, NO, and NO_x of an mixed potential ammonia sensor as described here. Sensor temperature is 550°C (from Ref. [34] with permission from Elsevier)

long-term stability) are separated. Commercially available catalyst materials can be used. In addition, a smaller design is possible.

Results with respect to sensitivity and cross-sensitivity toward NO and NO_2 are shown in Fig. 3. As porous catalyst layer, a commercial and exhaust gas-proven SCR formulation was selected (V_2O_5–WO_3–TiO_2) [32]. The sensor was measured at 565°C. The semilogarithmic plot reveals a behavior in accordance to (3) with a slope of 88 mV per decade NH_3. While the NO influence is negligible, NO_2 has a significant effect, however with a slope of only −20 mV per decade NO_2. Since usually either NH_3 or NO_2 are present in the exhaust, the cross-effect of NO_2 might not be detrimental. The lambda cross-sensitivity is low and can be neglected for $\lambda > 1.3$ (more than ca. 4.5 vol.% exhaust oxygen). Dynamometer tests show promising results. Details of this sensor can be found in [33–35].

To conclude the mixed potential section, a recent approach shall be discussed [36]. NiO/Au sensing electrodes in combination with platinum reference electrodes are reported to be sensitive and selective. The sensor is operated at 800°C and behaves linear with a slope of approx. 0.4 mV per ppm NH_3. Cross-effects toward H_2O, NO, or CO are negligible. A small effect of hydrocarbons remains. NO_2 has an opposite effect of approximately −0.05 mV per ppm NO_2. The sensor performs best under constant oxygen concentration (7%) [36].

NO_x sensors with a pronounced ammonia cross sensitivity can also be applied for control and monitoring of the SCR system [10], for example if one implements a cross sensitivity factor [37]. This behavior can easily be understood if one considers that at the high sensor operation temperatures a part of the ammonia gets oxidized according to

$$2\ NH_3 + \frac{5}{2}\ O_2 \rightarrow 2\ NO + 3\ H_2O \tag{4}$$

The formed NO leads to an additional sensor signal. It is a major advantage of this sensor that it is already commercially available for NO_x sensing purposes in the exhaust (see Sect. 2.1).

3.2 Sensors for Directly Determining the Ammonia Storage Degree of an SCR Catalyst

The maximum NO_x conversion of an ammonia SCR catalyst does depend not only on temperature but also on the amount of previously stored ammonia [38]. Especially at low temperatures, high conversion requires a completely ammonia loaded SCR catalyst. Under these conditions, ammonia slip may occur. One can either determine the actual degree of ammonia loading by a model that is based on the ammonia storage capacity of the SCR coating in conjunction with engine operation data and a NO_x balance. Probably more accurate would be a direct measurement with a loading sensor. It is the basic idea to use a part of a catalyst material itself as a sensor [39, 40]. Similar to the LNT loading sensor, the electrical impedance of a catalyst material film is dependent on the amount of stored ammonia. The basic setup of such a sensor reminds to the impedimetric or conductometric ammonia sensors. A film of catalyst material is applied on an alumina substrate with an integrated heater and platinum interdigital electrodes. Basic tests were conducted in synthetic exhaust with 5% O_2, 1% H_2O and N_2 using a zeolite SCR material. Whereas at the beginning the catalyst coating is free of stored ammonia, during the loading mode at constant temperature, ammonia is added to the feed gas and is stored in the film. At the subsequent measuring mode, the temperature is increased and stored ammonia desorbs and/or gets oxidized. The changing conductivity of the sensitive film is a measure for the stored amount of ammonia [39, 41].

4 Radio Frequency-Based Catalyst Gauging

It has been shown in the previous sections that by measuring the electrical properties of the catalyst coating itself, one can observe directly and in situ the state of LNT and SCR catalysts. A comprehensive overview on this topic is given in [42]. However, these methods require that the coating is applied to planar electrodes which are contacted by leads. A wireless method would not only be more elegant but would also avoid problems of mounting the planar substrates into the catalyst. Therefore, several attempts were conducted to gauge directly the state of catalysts directly by observing the penetration of the material by radio frequency waves. This is possible in principle for all catalysts that store gases but will be demonstrated here for TWC only.

Today, gasoline operated vehicles are equipped with TWC. Since the TWC converts reducing components such as carbon monoxide or hydrocarbons, and oxidizing components such as nitrogen oxides best at $\lambda \approx 1$, λ is measured by lambda-probes upstream of the TWC and is readjusted if required. Besides catalytically active noble metals, the TWC contains ceria–zirconia mixed oxides, which are capable to store oxygen. These ceria-based materials store oxygen under lean atmospheres ($\lambda > 1$) and release oxygen under rich conditions ($\lambda < 1$). Details of the catalyst function can be found, for example in [43]. As a result of the ability of the TWC to store oxygen, an oscillating air-to-fuel ratio λ in the feed gas is damped and the lambda-value downstream of the TWC remains close to 1. With a second lambda-probe downstream of the TWC, the oxygen storage capacity can be measured, and indirect information can be obtained whether the TWC still converts as required (OBD).

Under oxygen excess, the trivalent cerium in Ce_2O_3 gets oxidized to Ce^{4+} (CeO_2) and vice versa in case of rich gas. Since the conductivity of the TWC-washcoat component ceria–zirconia changes with the surrounding oxygen partial pressure [44], a conductivity measurement of the TWC might provide information on the oxygen loading of the TWC, very similar to the behavior of the LNT as shown above. In simple terms, the higher the oxygen partial pressure, the higher the amount of oxidized ceria and the lower the conductivity of the washcoat.

This is the starting point of a novel approach to determine the oxygen loading of a TWC in-situ by a contactless measurement with the help of radio frequency (RF) but without lambda-probes. The interactions of the TWC-material with RF-waves reflect changes in the electrical TWC-material properties, and therefore the oxygen loading of the catalyst can be determined [45]. Due to the metallic catalyst housing, the system acts as a cylindrical cavity resonator. Resonance frequencies and losses at the resonance frequencies of the electromagnetic waves depend on the electrical material parameters of the inserted TWC. A change in the conductivity of the TWC-material changes the magnitude of the electromagnetic waves [46].

In an initial study, the conductivity of the TWC coating was investigated. For that purpose, interdigital electrodes (IDE) were coated with an almost standard washcoat and were analyzed in synthetic exhaust with changing λ in a test rig [47].

TWC-monoliths with the same coating in a metallic housing were measured with the RF-system at low space velocities. The schematic assembly is illustrated in Fig. 4. Both the gas mixture and the catalyst were heated to about 410°C. An automatic vector network analyzer (VNA) was used to detect resonance frequency shifts or an attenuation of the electromagnetic wave. Electromagnetic waves with the amplitudes a_1 and a_2, respectively, are applied to the two RF antennas (marked by RF in Fig. 4). The back-scattered waves were gauged at these two ports and the parameters b_1 and b_2 determined. According to (5), S parameters can be defined as reflection coefficients (S_{11}, S_{22}) or transfer functions (S_{12}, S_{21}).

$$\begin{pmatrix} b_1 \\ b_2 \end{pmatrix} = \begin{pmatrix} S_{11} S_{12} \\ S_{21} S_{22} \end{pmatrix} \begin{pmatrix} a_1 \\ a_2 \end{pmatrix}. \tag{5}$$

The transfer function S_{21} describes the transmission of energy from one port to the other through the interior of the cavity resonator. Details of the measuring technique can be found in [48].

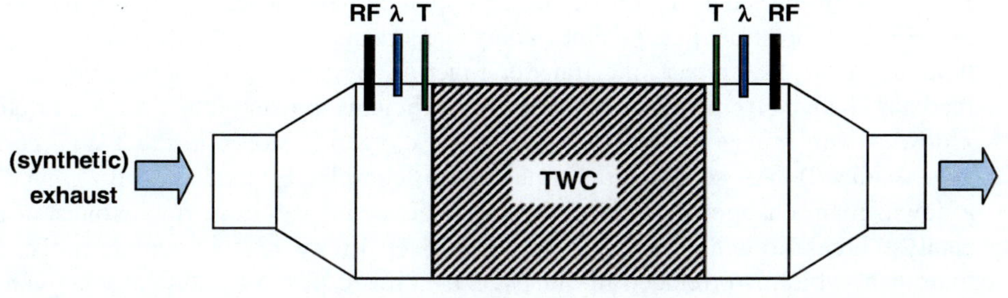

Fig. 4 Schematic setup for tests of the radio frequency-based catalyst gauging. See Fig. 5 for results

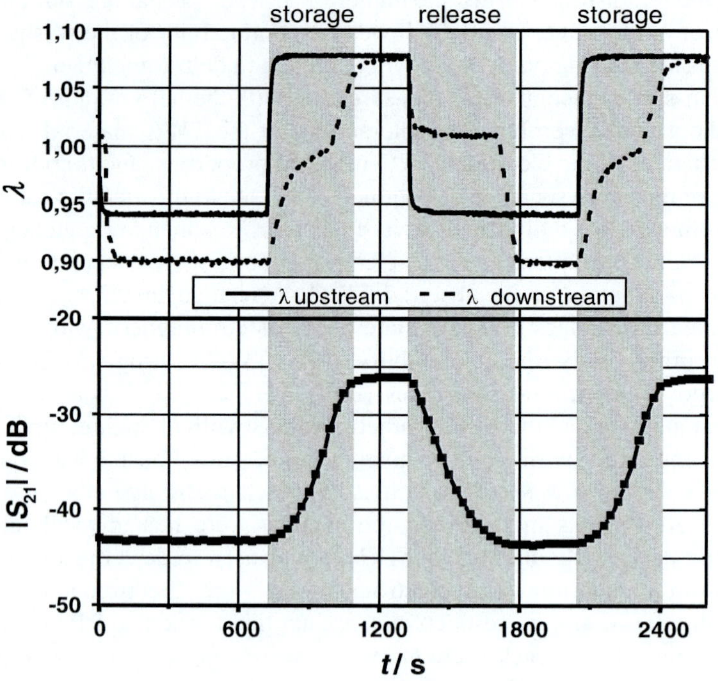

Fig. 5 Air-to-fuel ratio, λ, and transmission factor $|S_{21}|$ at a fixed frequency during lean-rich witches (modified from Ref. [50])

In general, the cavity resonator is perturbed when the TWC changes its conductivity as a function of oxygen loading and the characteristics of the resonator S-parameters mirror the oxidation/reduction state of the catalyst [46].

In an initial measurement, several spectra of $|S_{21}|$ in the range of 1–2 GHz are taken. Starting at reducing conditions, oxygen is added till ceria is completely oxidized [49]. A significant change at the resonance peaks in the amplitude and in the resonance frequency position occurs. Some pronounced frequencies at which $|S_{21}|$ increases strongly with the oxygen stored in the TWC coating were found.

Figure 5 shows the transmission amplitude $|S_{21}|$ at the fixed frequency of 1.27 GHz (modified from [50]). In the upper part of the diagram, the lambda-probe signals upstream (solid line) and downstream (dashed line) of a TWC are plotted. In the lower part, the transfer function $|S_{21}|$ in dB is shown. When the feed gas changes from rich ($\lambda < 1$, oxygen deficient) to lean ($\lambda > 1$, oxygen rich), λ downstream moves toward to $\lambda \approx 1$, since oxygen is stored in the TWC (ceria gets oxidized). As soon as the catalyst is "completely loaded" with oxygen, λ downstream changes to lean (differences between λ-values in rich exhaust after catalyst breakthrough are due to the cross-sensitivity of the lambda-probe to hydrogen, which is produced in the TWC as a result of the catalytically activated water–gas shift reaction). When changing the feed gas to rich, λ downstream decreases again, since the previously stored oxygen is released.

The RF-Signal shows a gradient as long as the catalyst stores or releases oxygen (gray shaded areas in Fig. 5) and a constant value occurs as long as the catalyst is in a constant state. These results suggest that it is possible to infer from an RF-signal to the instantaneous oxidation state of the catalyst. The analysis provides an integral measurement of the device, so that a control of the entire catalyst is possible. Compared to λ-probe-based air-to-fuel control, the RF approach could be advantageous with regard to future OBD for better monitoring of exhaust aftertreatment components. The effect of changing transmission coefficients could be reproduced and is a reliable indication of the oxygen storage in the TWC [51]. Furthermore, it is shown that cross-effects of water, carbon monoxide, carbon dioxide, and gas flow rate can be neglected [52]. Measurements were also conducted in engine test benches and had proven the applicability of the RF approach under real exhaust conditions [53].

The stability of the resonance frequency positions are presently under study as well as the temperature-dependency and cross-sensitivities toward several components that occur in the exhaust. Besides TWC characterization, the transfer to other catalyst types such as lean NO_x traps opens up a wide spectrum of possible OBD applications. Initial attempts for measuring the NO_x loading of LNT were successful [56] as well as ammonia storage determination in SCR catalysts [54] or soot load detection in DPF [55].

5 Conclusion

The advent of novel combustion concepts like lean burn operating gasoline direct injection required novel exhaust gas aftertreatment concepts. It pushed the development of NO_x sensors and NH_3 sensors. It is shown how the development of exhaust gas sensors interacts always with the progress in exhaust gas aftertreatment systems. Therefore, novel types of sensors will become increasingly important, the more the emission regulations get tightened.

As a perspective on the farer future, it might be that catalyst state sensors or the more sophisticated RF state gauging may supplement or replace conventional exhaust gas sensors.

References

1. Moos R (2006) Automotive exhaust gas sensors. In: Grimes CA, Dickey EC, Pishko MV (eds) Encyclopedia of sensors, vol 1. American Scientific, Stevenson Ranch, CA, USA, pp 295–312
2. Weibel M, Waldbüßer N, Wunsch R, Chatterjee D, Bandl-Konrad, B, Krutzsch, B (2009) A Novel Approach to Catalysis for NO_x Reduction in Diesel Exhaust Gas. Top Catal 52:1702–1708. doi: 10.1007/s11244-009-9329-7

3. Kröcher O, Devadas M, Elsener M, Wokaun A, Söger N, Pfeifer M, Demel Y, Mussmann L (2006) Investigation of the selective catalytic reduction of NO by NH_3 on Fe-ZSM5 monolith catalysts. Appl Catal B Environ 66:208–216. doi:10.1016/j.apcatb.2006.03.012
4. Aneja R, Flathmann K, Savonen C, Tindall T (2004) SCR potential and issues for heavy duty applications in the USA. In: 10th Diesel Engine Emission Reduction (DEER) Workshop, Coronado, CA, USA
5. Hammerle R (2003) Urea SCR and DPF system for diesel sport utility vehicle meeting Tier II Bin 5. In: 9th Diesel Engine Emissions Reduction (DEER) Workshop, Newport, RI, USA
6. Baunach T, Schänzlin K, Diehl L (2006) Sauberes Abgas durch Keramiksensoren. Physik J 5:33–38
7. Riegel J, Klett S (2008) Sensors for modern exhaust gas aftertreatment systems. In: Proceedings of 5th international exhaust gas and particulate emissions forum, Ludwigsburg, Germany, pp 85–96
8. Weigel M, Roduner C, Lauer T (2010) Particle-filter onboard-diagnosis by means of a soot-sensor downstream of the particle-filter. In: 6th international exhaust gas and particulate emissions forum, Ludwigsburg, Germany, pp 62–69
9. Hagen G, Feistkorn C, Wiegärtner S, Heinrich A, Brüggemann D, Moos R (2010) Conductometric soot sensor for automotive exhausts: initial studies. Sensors 10:1589–1598. doi:10.3390/s100301589
10. Kato N, Nakagaki K, Ina N (1996) Thick film ZrO_2 NO_x sensor. SAE Technical Paper 960334. doi: 10.4271/960334
11. Zhang H, Pfleger C, Lemire B (2000) Integration Eines smart NOx-Sensors im Abgasstrang für die Benzindirekteinspritzung. In: 21. Internationales Wiener Motorensymposium (4.-5. Mai 2000). Fortschrittberichte VDI, vol 12, nr. 420, pp 288–311
12. Hofmann L, Rusch K, Fischer S, Lemire B (2004) Onboard emissions monitoring on a HD Truck with an SCR system using NOx sensors. SAE Technical Paper 2004-01-1290. doi: 10.4271/2004-01-1290
13. Kim YW, Van Nieuwstadt M (2006) Threshold monitoring of urea SCR systems. SAE Technical Paper 2006-01-3548. doi: 10.4271/2006-01-3548
14. Sasaki H, Scholl D, Parsons M, Inagaki H, Shiotani K, Visser J, Zawacki G, Kawai T, Teramoto S, Kubinski D (2010) Development of an Al_2O_3/ZrO_2-composite high-accuracy NO_x sensor. SAE Technical Paper 2010-01-0041. doi: 10.4271/2010-01-0041
15. Zhuiykov S, Miura N (2007) Development of zirconia-based potentiometric NOx sensors for automotive and energy industries in the early 21st century: what are the prospects for sensors? Sens Actuators B Chem 121:639–651. doi:10.1016/j.snb.2006.03.044
16. Fergus JW (2007) Materials for high temperature electrochemical NOx gas sensors. Sens Actuators B Chem 121:652–663. doi:10.1016/j.snb.2006.04.077
17. Zimmermann C (2007) Neuartiger Sensor zur Bestimmung des Zustandes eines NOx-Speicherkatalysators. In: Moos R, Fischerauer G (eds) Bayreuther Beiträge zur Sensorik und Messtechnik, vol 2. Shaker, Aachen
18. Moos R, Zimmermann C, Birkhofer T, Knezevic A, Plog C, Busch MR, Ried T (2008) Sensor for directly determining the state of a NOx storage catalyst. SAE Technical Paper 2008-01-0447. doi: 10.4271/2008-01-0447
19. Birkhofer T, Leye H, Knezevic A, Moos R, Plog C, Ried T, Voigtländer D (2003) Method for identifying the state of an NOx storage catalyst. US Patent specification, US 6,619,108
20. Moseley PT, Williams DE (1989) Gas sensors based on oxides of early transition metals. Polyhedron 8:1615–1618
21. Satsuma A, Shimizu K, Hattori T, Nishiyama H, Kakimoto S, Sugaya S, Yokoi H (2007) Polytungstate clusters on zirconia as a sensing material for a selective ammonia gas sensor. Sens Actuators B Chem 123:757–762. doi:10.1016/j.snb.2006.10.011
22. Nishiyama H, Kakimoto S, Inoue R, Yokoi H, Ishida N, Oshima T, Sugaya S, Imaeda K, Hattori T, Satsuma A (2008) Ammonia sensor. US Patent specification, US 7,341,694

23. Chen DK, Bloink RL, Valdes CA, Ker EL, Zhang J (2006) Ammonia sensor element, heater, and method for making the same. US Patent specification, US 7,069,770
24. Moos R, Müller R, Plog C, Knezevic A, Leye H, Irion E, Braun T, Marquardt K-J, Binder K (2002) Selective ammonia exhaust gas sensor for automotive applications. Sens Actuators B Chem 83:181–189. doi:10.1016/S0925-4005(01)01038-3
25. Birkhofer T, Knezevic A, Müller R, Plog C (2005) Method for producing and managing a sensor. US Patent application US 2005/0130338
26. Guth U (2008) Gas sensors. In: Bard AJ, Inzelt G, Scholz F (eds) Electrochemical dictionary. Springer, Berlin, pp 294–299
27. Zosel J, Westphal D, Jakobs S, Müller R, Guth U (2002) Au–oxide composites as HC-sensitive electrode material for mixed potential gas sensors. Solid State Ionics 152–153:525–529. doi:10.1016/S0167-2738(02)00355-7
28. Wang DY, Symons WT, Farhat RJ, Valdes CA, Briggs EM, Polikarpus KK, Kupe J (2006) Ammonia gas sensor. US Patent specification, US 7,074,319
29. Wang DY, Yao S, Cabush D, Racine D (2007) Ammonia sensor for SCR NOX Reduction. In: Proceedings of 13th diesel engine-efficiency and emissions research (DEER) conference, Detroit, MI
30. Wang DY, Yao S, Shost M, Yoo J, Cabush D, Racine D, Cloudt R, Willems F (2008) Ammonia sensor for closed-loop SCR control. SAE Technical Paper 2008-01-0919. doi: 10.4271/2008-01-0919
31. Weisgerber V, Wang DY, Symons W, Cabush D (2006) Ammoniaksensoren für die Regelung von SCR-Systemen mit Harnstoffeinspritzung. In: Tille, T. (ed.) Sensoren im Automobil, p. 49–58. Expert, Renningen
32. Schedel H, Fischer S, Ballmert B (2004) Durability of extruded homogeneous SCR catalyst. SAE Technical Paper 2004-01-0075. doi: 10.4271/2004-01-0075
33. Schönauer D, Moos R, Wiesner K, Fleischer M (2007) Selektiver Ammoniakabgassensor auf Mischpotentialbasis. In: Gerlach G, Hauptmann P (eds) 8. Dresdner Sensor-Symposium, 10.-12. Dezember 2007. Dresden, pp 11–14
34. Schönauer D, Moos R, Wiesner K, Fleischer M (2009) Selective mixed potential ammonia exhaust gas sensor. Sens Actuators B Chem 140:585–590. doi:10.1016/j.snb.2009.04.064
35. Schönauer D, Nieder T, Wiesner K, Fleischer M, Moos R (2011) Investigation of the electrode effects in mixed potential type ammonia exhaust gas sensors. Solid State Ionics 192:38–41. doi:10.1016/j.ssi.2010.03.028
36. Elumalai P, Plashnitsa VV, Fujio Y, Miura N (2008) Stabilized zirconia-based sensor attached with NiO/Au sensing electrode aiming for highly selective detection of ammonia in automobile exhausts. Electrochem Solid State Lett 11:J79–J81. doi:10.1149/1.2971171
37. Schär CM, Onder CH, Geering HP, Elsener M (2003) Control of a urea SCR catalytic converter system for a mobile heavy duty diesel engine. SAE Technical paper 2003-01-0776. doi: 10.4271/2003-01-0776
38. Hammerle R (2002) Urea SCR and DPF system for diesel sport utility vehicle meeting Tier II Bin 5. In: Proceedings of 8th diesel engine emissions reduction (DEER) workshop, San Diego, CA, USA
39. Kubinski DJ, Visser JH (2008) Sensor and method for determining the ammonia loading of a zeolite SCR catalyst. Sens Actuators B Chem 130:425–429. doi:10.1016/j.snb.2007.09.007
40. Binder K, Marquardt K-J, Braun T, Knezevic A, Busch MR, Moos R, Plog C (2004) Method and apparatus for determining the storage state of an ammonia-storing SCR catalyst. US Patent specification, US 6,833,272
41. Moos R, Schönauer D (2008) Recent developments in the field of automotive exhaust gas ammonia sensing. Sens Lett 6:821–825. doi:10.1166/sl.2008.509
42. Moos R (2010) Catalysts as sensors—a promising novel approach in automotive exhaust gas aftertreatment. Sensors 10:6773–6787. doi:10.3390/s100706773
43. Twigg MV (2007) Progress and future challenges in controlling automotive exhaust gas emissions. Appl Catal B Environ 70:2–15. doi:10.1016/j.apcatb.2006.02.029

44. Boaro M, Trovarelli A, Hwang J-H, Mason TO (2002) Electrical and oxygen storage/release properties of nanocrystalline ceria–zirconia solid solutions. Solid State Ionics 147:85–95. doi:10.1016/S0167-2738(02)00004-8
45. Moos R, Spörl M, Hagen G, Gollwitzer A, Wedemann M, Fischerauer G (2008) TWC: lambda control and OBD without lambda probe—an initial approach. SAE Technical paper 2008-01-0916. doi: 10.4271/2008-01-0916
46. Fischerauer G, Spörl M, Gollwitzer A, Wedemann M, Moos R (2008) Catalyst state observation via the perturbation of a microwave cavity resonator. Frequenz 62:180–184. doi:10.1515/FREQ.2008.62.7-8.180
47. Reiß S, Wedemann M, Moos R, Rösch M (2009) Electrical in-situ characterization of three-way catalyst coatings. Top Catal 52:1898–1902. doi:10.1007/s11244-009-9366-2
48. Fischerauer G, Spörl M, Reiß S, Moos R (2010) Mikrowellengestützte Aufklärung elektrochemischer Vorgänge in Katalysatoren und verwandten Systemen. Technisches Messen 77:419–427. doi:10.1524/teme.2010.0066
49. Reiß S, Moos R, Wedemann M, Spörl M, Nerowski A, Fischerauer G (2009) RF-probing of automotive catalysts. In: Sensor 2009, 14th international conference on sensors, technologies, electronics and applications, 26–28 May 2009, Nürnberg, Germany. doi: 10.5162/sensor09/v2/b7.1
50. Moos R, Wedemann M, Spörl M, Reiß S, Fischerauer G (2009) Direct catalyst monitoring by electrical means: an overview on promising novel principles. Top Catal 52:2035–2040. doi:10.1007/s11244-009-9399-6
51. Reiß S, Spörl M, Hagen G, Fischerauer G, Moos R (2011) Combination of wirebound and microwave measurements for in situ characterization of automotive three-way catalysts. IEEE Sens J 11:434–438. doi:10.1109/JSEN.2010.2058798
52. Reiß S, Wedemann M, Spörl M, Fischerauer G, Moos R (2011) Effects of H_2O, CO_2, CO, and flow rates on the RF-based monitoring of three-way catalysts. Sens Lett 9:316–320. doi:10.1166/sl.2011.1472
53. Reiß S, Spörl M, Fischerauer G, Moos R (2009) Realabgastauglichkeit einer HF-gestützten Automobilabgasdiagnose. In: Gerlach G, Hauptmann P (eds) 9. Dresdner Sensor-Symposium, Dresden, 7–9 Dec 2009, pp 263–266
54. Reiß S, Schönauer D, Hagen G, Fischerauer G, Moos R (2011) Monitoring the ammonia loading of zeolite-based ammonia SCR catalysts by a microwave method. Chem Eng Technol 34:791–796. doi:10.1002/ceat.201000546
55. Fischerauer G, Förster M, Moos R (2010) Sensing the soot load in automotive diesel particulate filters by microwave methods. Meas Sci Technol 21:035108. doi:10.1088/0957-0233/21/3/035108
56. Fremerey P, Reiß S, Geupel A, Fischerauer G, Moos R (2011) Determination of the NO_x loading of an automotive lean NO_x trap by directly monitoring the electrical properties of the catalyst material itself. Sensors 11:8261–8280. doi: 10.3390/s110908261

Technology and Application Opportunities for SiC-FET Gas Sensors

A. Lloyd Spetz and M. Andersson

Abstract The development of SiC-FET gas sensors has proceeded for about fifteen years. The maturity of the SiC material and a deeper understanding of the transduction mechanisms and sensor surface processes behind the sensitivity to a number of target substances have recently allowed the development of market-ready sensors for certain applications. Some examples presented below are a sensor system for domestic boiler control, an ammonia sensor for control of the SCR (selective catalytic reduction) and SNCR (Selective Non-Catalytic Reduction) NO_x abatement processes as well as other more or less market-ready applications. In parallel, the basic research continues in order to reach more demanding markets/new applications and also to possibly lower the production costs of the sensors. Therefore, current research and future challenges are also treated, such as the development of new types of conducting ceramics for ohmic contacts to SiC in order to increase the operation temperature beyond the present state of the art.

Keywords Catalytic metals, Combustion control, Exhaust gases, FET sensors, Flue gases

Contents

A. Lloyd Spetz (✉)
Department of Physics, Chemistry and Biology, Linköping University, 581 83 Linköping, Sweden
e-mail: spetz@ifm.liu.se

M. Andersson
Department of Physics, Chemistry and Biology, Linköping University, 581 83 Linköping, Sweden

SenSiC AB, Isafjordsgatan 39B, 164 40 Kista, Sweden
e-mail: mikan@ifm.liu.se; mike.andersson@sensic.se

M. Fleischer and M. Lehmann (eds.), *Solid State Gas Sensors – Industrial Application*, Springer Series on Chemical Sensors and Biosensors (2012) 11: 189–214,
DOI 10.1007/5346_2011_5, © Springer-Verlag Berlin Heidelberg 2012,
Published online: 10 January 2012

1 Introduction

The first gas sensor utilizing a field effect device was invented by I. Lundström, M. S. Shivaraman, and C. M. Svensson in the mid-1970s [1]. This device (a Metal Oxide Semiconductor Field Effect Transistor, MOSFET) was based on the semiconductor silicon (Si), for a recent review, see [2]. By applying a gate contact material (commonly a catalytic metal) capable of interacting with one or more gaseous substances, such that the electric field over the insulator changes upon the interaction, devices with gas sensing properties and a simple readout can be constructed. The basic design of a gas-sensitive transistor device (based on SiC) is given in Fig. 1. A voltage applied between the source and drain contacts causes a current to flow through the channel region. The gas-interacting gate material is placed on top of the insulating layer over the channel region, whereby gas-influenced changes in the electric field from the gate contact to the semiconductor will modulate the channel current. Normally, however, the externally applied gate bias in order to keep a constant current through the device constitutes the sensor output.

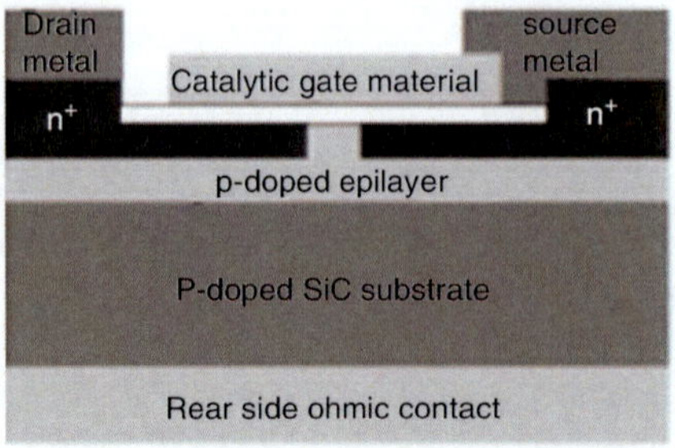

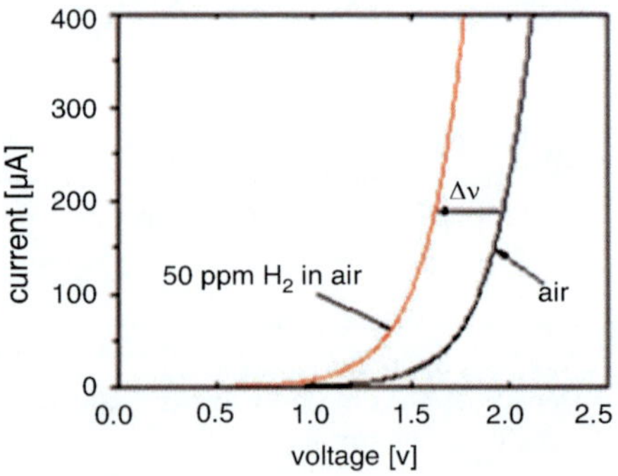

Fig. 1 *To the left*: Schematic drawing (cross section) of a Silicon Carbide Field Effect Transistor (SiC-FET) device with the gate connected to source. The bias is applied between the drain contact and the gate-source contact. *To the right*: The IV characteristics of the SiC-FET device. The sensor signal is the voltage at a constant current, see Fig. 6

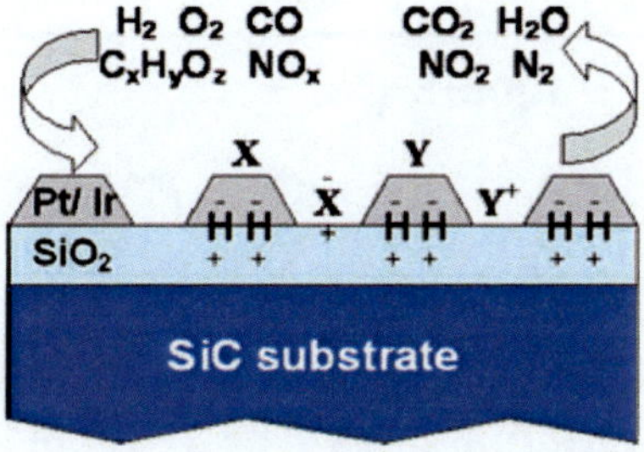

Fig. 2 Schematic representation of the suggested gas sensing principles of SiC-FE gas sensors. Hydrogen atoms adsorbing at the metal/insulator interface (e.g., creating polarized –OH groups on the oxide surface) and possibly other charged or polarizable compounds (or reaction products) adsorbing at the three phase boundaries between metal, insulator, and the gas phase or directly on bare patches of oxide, influences the number of mobile charge carriers in the semiconductor material so changing the I/V-characteristics of the transistor device

The research and development of Si-based FET sensors has over the years generated a lot of knowledge as well as a number of important principles governing their sensing characteristics, results which have also been possible to transfer to later generations of FET sensor devices. An example of such a result is the use of porous catalytic films as the gate contact in FET devices [3–5], see Fig. 2, which was investigated and found required for the sensitivity to gases such as ammonia for Pt- and Ir-gate Si-based sensors, but in fact has become the main structure of the sensing layers in many gas-sensitive FET devices.

The access to SiC material of improved quality from about 1995 started a new era for the FET devices since operation temperatures above 200°C was possible to reach. The wide band gap and chemical inertness of SiC and related materials permits their use in field effect devices intended for applications involving elevated temperatures (up to 800°C) and/or aggressive/corrosive environments, for example car exhaust or flue gases. In 1997, the operation of a gas sensor at 1,000°C for 20 min was published [6], see Fig. 3. The failure was likely not due to poor semiconductor properties at this temperature, but rather contact problems, which is dealt with in Sect. 6. Two recent reviews about the SiC-FET development are found in [2, 7]. Since differences in selectivity and sensitivity in the sensors response to different gaseous substances can be gained by the choice of gate material, its structure and the operating temperature, the SiC-based devices provide more degrees of freedom regarding choice of materials and surface reactions owing to their wider range of operating temperatures. The gate material can be a catalytic metal, for example platinum, rhodium, and iridium or combinations of these, but a number of metal oxides can also be used as catalytic layers in SiC transistor devices, where the gate material does not need to conduct large currents.

In this chapter, a review of important findings regarding the detection mechanisms is given and different operation principles of the FE devices based on MOS capacitors, Schottky diodes and transistors are discussed. In Sect. 5, a SiC-based sensor system for combustion control in domestic boilers, currently

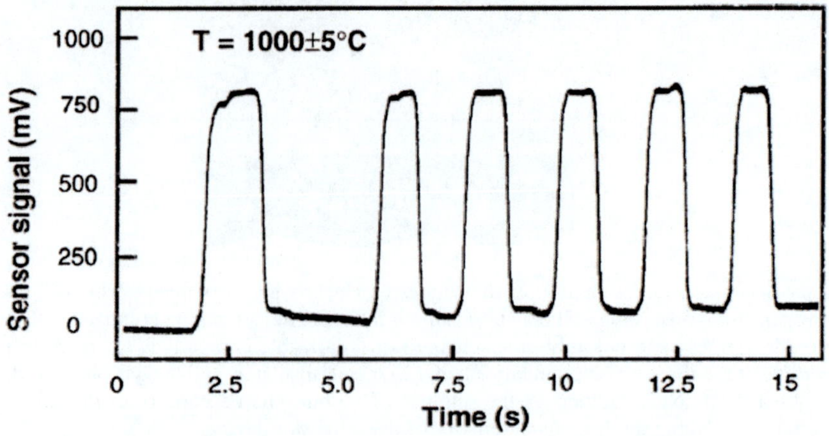

Fig. 3 Pt (100 nm)/TaSi$_x$ (10 nm)/SiO_2 (100 nm)/4H–SiC capacitor device. Response to 0.5% propane pulses in 0.7% oxygen/argon at 100°C. The sensor signal is the voltage at a constant capacitance close to the flat band voltage. The sensor survived for about 20 min [6]. Copyright Wiley-VCH verlag GmbH & Co. KGaA. Reproduced with permission

commercialized by a spin-off company [8], is also presented. This kind of company structure is probably beneficial for the early stage of commercialization of research ideas, since the important close collaboration between researchers, engineers, and end users of the technology, which is the leading theme of this book, is established more naturally. Some interesting results from the ongoing research within the VINNOVA VINN Excellence Center, FunMat (Functional nanoscale Materials), regarding advanced contact materials to SiC for improved high temperature stability are treated in Sect. 6. The VINN Excellence center involves active collaboration between researchers and different industrial partners. This structure of the collaboration between the university and industry has proven to be very efficient providing a possibility for researchers, engineers, and end users to work together.

2 Detection Mechanism of FET Gas Sensors

In order to proceed and tailor devices for specific applications, a more detailed understanding of the detection mechanism of FET gas sensors is needed. One important step has been the development of a model where the active sites for the detected species are located on the insulator side of the metal–insulator interface [9]. Thus, in respect of a thick, homogeneous catalytic gate contact, hydrogen dissociatively adsorbs on the metal surface, rapidly diffuses through the metal, and adsorbs at the metal/insulator interface in the form of polarized hydroxyl (−OH) groups on the oxide surface. The strong dipole moment of the OH-groups at the interface changes the electric field in the insulator and thus the current through the channel of the transistor device.

In the case of a porous gate contact, the importance of the three phase boundaries, where the gas, metal, and insulator simultaneously are in contact, for the sensitivity to certain substances has been established, see Fig. 2 [3, 10]. It is believed that, in addition to hydrogen adsorbing at the oxide side of the gate contact/insulator interface, molecular adsorbents of nonhydrogen containing species, such as CO and NO as well as for example ammonia (NH_3) and various hydrocarbons, contribute to the sensor signal by undergoing redox reactions removing oxygen anions from the metal surface and exposed areas of the oxide [11, 12]. Polarized or charged molecules and/or reaction species might also spill over onto bare patches of oxide and add to the hydrogen induced polarized layer in affecting the current–voltage- or capacitance–voltage-characteristics of the field effect devices, see also Figs. 1 and 2.

Extensive studies in ultrahigh vacuum were performed on Pd gate Si FET sensors [2, 13] and later expanded to Pt gate Si FET devices [14]. From these studies, some interesting differences between Pd and Pt gate FET devices were found. It seemed like the interface between the Pt metal and the Pd metal, respectively, and the SiO_2 insulator had a quite different amount of sites for the hydrogen detection. Further research showed that the structure of the metal in contact with the SiO_2 may be very different according to deposition method and temperature and following different annealing procedures [15, 16]. The number of active sites on the insulator very likely has an inverse dependency on the number of contact points between the metal and the insulator. This becomes even more relevant in view of experimental research, which suggested that the hydrogen atoms adsorbed on the metal surface in fact spill both out onto the exposed insulator surface but also under the metal at the metal insulator border in a porous metal film [2].

Ultrahigh vacuum studies of the hydrogen oxygen interaction in Pt gate SiC-FET devices were also recently performed by Kahng et al. [17], which confirmed the oxygen hydrogen interplay on the catalytic metal surface also at 527°C. They also investigated the effect of sulfur contamination of these devices at 527°C [18]. Sulfur reduced the hydrogen response by about 70% but the sulfur was effectively removed by oxygen and had no detrimental effect on the device gas sensing behavior.

In this respect, it is also interesting to compare the FET and the HSGFET devices, that is FET devices with a suspended gate, regarding the active sites for the gas detection [19]. Here, hydrogen spillover to the oxide does not occur due to the air gap of about 1 μm between the gate metal and the insulator. In this case, the influence on the semiconductor underneath the insulator definitely originates from gas-induced changes of the work function, due to charge transfer to or from the suspended gate material itself during adsorption/reaction processes. For the normal MOSFET devices (also the SiC-based FETs), the gate metal is applied directly on top of the gate insulator, which gives rise to a relatively high voltage change due to gas molecules interacting with the gate area. In comparison, the response of the suspended gate FET gas sensor is somewhat lower. However, the separate processing of the gate and subsequent merging with the rest of the device has other clear advantages, such as large freedom in the choice of gate material besides the processing advantages [19, 20]. Furthermore, a new type of device has been presented where the charging of the suspended gate is transmitted by a floating gate to a small FET-device with a short channel. This minimizes the loss of the sensing signal due to weak coupling over the air gap as described in a recent status report of solid state gas sensor research in Germany [21].

The gas molecules that adsorb on the sensor surface might decompose and/or react with other adsorbed molecules on the catalytic metal if the activation energy is overcome. The temperature dependence is therefore very strong, see Fig. 4 [22]. Dissociation of ammonia and chemical reactions on a Ru surface has been modeled

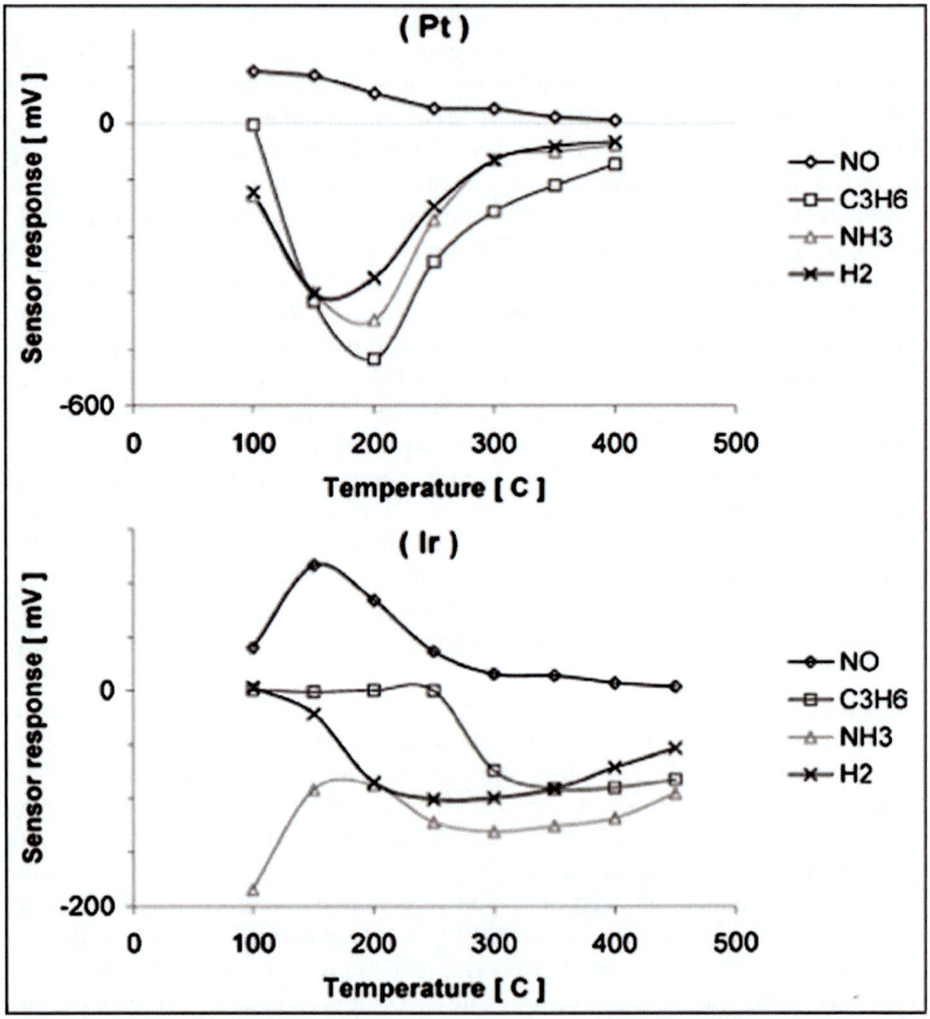

Fig. 4 The response versus temperature of Pt gate (*above*) and an Ir gate (*below*) SiC-FET sensor to NO, CO, C_3H_6, NH_3, and H_2 [22].Copyright Plenum Publishing Corporation. Reproduced with permission

[23, 24]. This work is important and will continue in order to get information about reaction paths on the sensor surface, which is different for different catalytic metals or metal oxides, related to the structure of the material and varies with temperature. A very informative spectroscopic method for investigation of the detection mechanism is represented by the DRIFT (Diffuse Reflectance Infra-red Fourier Transform) spectroscopy. This method can be used with powder samples and we have investigated Pt and Ir impregnated SiO_2 as model surfaces for the SiC-FET devices [10, 25]. From such studies, the creation of OH groups on the SiO_2 surface, originating from decomposition of hydrogen or ammonia [10, 23], has been observed during gas exposure. It was also shown that the OH groups in some way seem to have a connection to the metal border, since increasing the concentration of impregnated metal up to a certain value increased the number of OH groups on the SiO_2 surface [26]. Water is continuously produced from two adsorbed OH groups, OH_s, again forming a free oxygen surface site, O_s:

$$2OH_s \rightarrow H_2O_g + O_s \tag{1}$$

This of course lowers the sensor response but at the same time provides a reaction path for the back reaction, which renders the sensor reversible.

The chemical reactions on the catalytic metal surface also involve oxygen, for example in the water production but also in the detection of hydrocarbons, which are oxidized to CO_2 and H_2O in the overall reaction:

$$C_xH_y + (x+y/2)O_2 \rightarrow xCO_2 + y/2H_2O \tag{2}$$

Similarly, NH_3 consumes oxygen through oxidation and disappears as NO_2 and H_2O:

$$4NH_3 + 7O_2 \rightarrow 4NO_2 + 6H_2O \tag{3}$$

Therefore, oxygen plays an important role in the detection mechanism as also investigated by several other researchers [12, 17, 27]. Schalwig et al. [12] suggested that oxygen in fact also spills over to the oxide surface from the catalytic metal and forms negatively charged anions with a strong influence on the number of mobile carriers in the semiconductor. When these anions are consumed in oxidation reactions, this will influence the sensor signal.

While the response to molecules like H_2, NH_3 and HC is fairly well understood, work is in progress to reveal the details of the mechanism for detection of nonhydrogen containing molecules such as CO and NO/NO_2 [28]. Regarding the CO detection of porous Pt gate SiC field effect capacitors, several mechanisms seem to be involved, but one likely contribution to the response is an increased sensitivity to background hydrogen, a substance which is always present at sub-ppm levels in all gas mixtures/atmospheres. The response of Pt gate FET sensors toward CO has been shown to correlate well with the CO coverage on (or the exclusion of oxygen from) the Pt metal contact. The binary switch in sensor signal, resulting from a decrease in the CO/O_2 ratio or an increase in temperature, corresponds to the transition between a CO covered and an oxygen dominated Pt surface, as also known from the field of CO oxidation over Pt catalysts [29–31]. This transition is accompanied by a switch from a large to a small sensor response and occurs both in the case of a porous and a thick, dense Pt gate contact.

Theoretical prediction indicates that the Pt/MgO material system when applied to FE sensor devices should infer a very low sensitivity to hydrogen [32]. The insensitivity of dense Pt gate sensors toward CO at both higher (>300°C) and lower (<100°C) temperatures or at low CO/O_2 ratios and the overall insensitivity of dense Pt sensors with a gate oxide of magnesium oxide (MgO) toward CO give indications on the response, at least partly, being mediated through an increased sensitivity to background hydrogen. From the results on the study of the CO response characteristics of porous Pt/SiO_2 and porous Pt/MgO sensors, it seems likely that one or more processes other than the increased sensitivity to background hydrogen contribute to the CO response of devices with porous metal gate contacts, one candidate being the removal of oxygen anions from the surface of exposed oxide areas.

3 Processing and Electrical Operation of FET Devices

Standard protocols are now available for processing of SiC-based chemical gas sensors including a gate oxide of reasonable quality, which is important for capacitive and transistor devices. An increased oxide thickness in 4H–SiC capacitor devices tended to reduce the interface state density in work by Nakagomi et al. [33]. For gas sensors based on SiC Schottky diodes, it was shown that an interfacial layer of oxide is necessary in order to get a gas response, for example, to hydrogen [34, 35]. This is again a proof that the response to H_2 emanates from active sites on the oxide surface. The SiC is today commercially available in up to 4″ wafers. We have recently processed SiC field effect transistor devices on a 4″ wafer, which holds about 2,000 sensor devices, see Fig. 5. Since the layout is not yet optimized and we expect the available wafer size to grow in the future, this will reduce the cost of the sensors considerably. Thus, like for most chemical sensors, it is the mounting and packaging of the sensor that constitute the main cost per device.

Schottky diode FE devices have been processed by Neudeck et al. [36] on atomically flat 4H SiC grown as webbed cantilevers with platinum as a combined gas sensing contact and heater. The Pt sensing layer and heater was 2,700 Å thick and stabilized sequentially through annealing in air up to 400°C. The Schottky diode was operated in forward bias as a hydrogen sensor and showed stable results and a high current gain for cycled operation in air/0.5% H_2 in air at 150°C for more than 250 h.

Schottky diodes based on Metal–Reactive insulator–Silicon carbide, MRISiC, devices have also been investigated [37–40], and sometimes show a shift to hydrogen of several volts. This is attributed to resistivity changes of the metal oxide in series with the Schottky diode field effect.

As already mentioned, the detection of gas molecules in FE devices is due to the formation of electric dipoles or charging of the gate area giving rise to a change in

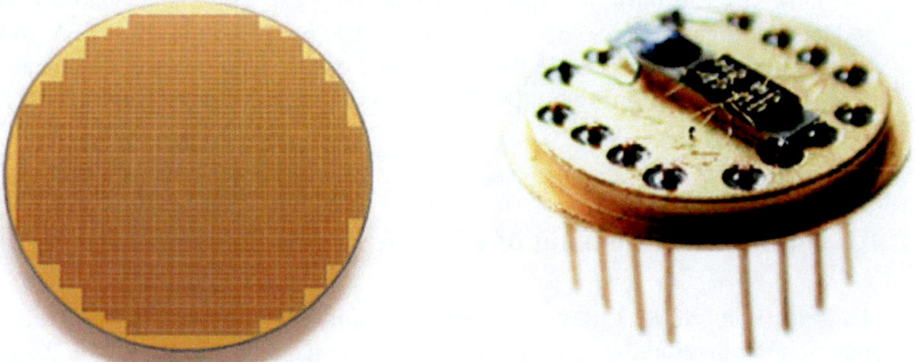

Fig. 5 *To the left*: In total about 1,000 SiC-FET sensors processed at MC2, Chalmers, Sweden, on a three-inch SiC wafer. *To the right*: To facilitate electrical characterization and gas measurements, the MISiCFET sensor devices are glued to a heater substrate mounted on a 16-pin header with gold-plated contacts. The transistors' ohmic contacts are connected to the contacts on the header through gold wires and a Pt100 temperature sensor for temperature control is also mounted on the heater substrate

the electric field in the gate insulator, which in turn influences the number of mobile carriers in the semiconductor. Therefore, the gas response of an FE device results in a change in the current through a transistor, a change of capacitance for a capacitive device and a current change in a Schottky diode. In the SiC-FET device, source and gate (or drain and gate) are connected to form a two-terminal device with electrical characteristics similar to that for a diode. The gas response is preferably measured as the voltage change at a constant current, see Fig. 1, which is the most relevant measurement since the interacting gases charge the gate area, that is changing the built in potential at the gate/insulator interface, described in (4) [41]

$$V_{\mathrm{GD}} = (I_{\mathrm{DS}}2\mathrm{L}/\mu_{\mathrm{e}}C_{\mathrm{ox}}W)^{1/2} + V_{\mathrm{T}} \quad (4)$$

where V_{GD} is the voltage drop between the gate-source and drain contacts, I_{DS} is provided by a constant current generator, μ_{e} is the electron mobility in the MOSFET channel, C_{ox} is the capacitance of the metal-oxide-semiconductor structure, W the width and L the length of the channel, respectively, and V_{T} is the threshold voltage. Changes in V_{T} due to gas interaction with the gate area are monitored as the sensor signal. This has also proven to be a much more long-term stable and reproducible measurement method as compared to the use of a constant bias voltage where the current gain is measured. For the same reason, it is recommended to measure bias at a constant current also in Schottky diode chemical gas sensors. Similarly, it is more reliable to measure voltage at a constant capacitance in the case of capacitive devices. Ghosh et al. showed that the long-term stability increases when a capacitance level close to midgap is chosen, which means that a fairly low capacitance level on the CV-curve is recommended [42]. The same device based on Pd

(150 nm)/AlN (50 nm)/SiC was operated in both capacitive and Schottky diode mode as hydrogen sensor with a response for 100 ppm H_2 in N_2 of 0.35 V in both cases [43]. However, we argue that the transistor device is the best choice among these three devices offering both a very stable and reliable device operated by DC current and simple electronics.

4 Tailor-Made Sensing of FET Devices

The selectivity of FET devices depends mainly on the gate material and the operation temperature, see Fig. 4. The knowledge gained so far about the detection mechanism has supervised new innovative gate material combinations for tailor-made sensors. Platinum and iridium as the gate materials in FET devices were characterized for the response to typical combustion gases, C_3H_6, CO, NO and to NH_3 and H_2, in the temperature range 100–450°C. The comparison showed differences especially in respect to the temperature dependence of the response to these gases, see Fig. 4 [22]. For example, Ir operated at 300°C responds mainly to NH_3 and H_2, and was suggested as gate material in sensors intended for the SCR application, see Sect. 5, since hydrogen is not present in diesel exhaust [44, 45].

The surface of magnesium oxide, MgO, has been theoretically predicted as a surface where hydrogen atoms decomposed from hydrogen containing gases do not form OH groups, and experimental work with Pt/MgO model sensors support this view. This knowledge has been used to design two stacks of sensing layers for the gate of SiC-FETs. Preliminary tests performed on capacitor devices indicate fairly good selectivity to oxygen in the range 1–20% at 300°C in the presence of CO, hydrogen and propene, a model hydrocarbon. The oxygen sensitivity of SiC/SiO_2/MgO/LaF_3 devices showed very small cross-sensitivity to CO and propene, which was decreased even further for the SiC/SiO_2/MgO/IrO_2 sensor structure [32].

Palladium and platinum have also been compared as gate materials in SiC Schottky diodes for detection of hydrogen and methane in air at 400–600°C [46]. The Pd–SiC sensor showed both higher sensitivity and faster initial adsorption as compared to the Pt–SiC sensor. Depletion mode SiC-FETs with the gate consisting of oxide with a top layer of $In_xV_yO_z$ showed sensitivity to NOx at 275–325°C with no interference of O_2 or D_2, while an operation temperature of 25–100°C was beneficial for selective D_2 detection [47]. For the low temperature operated suspended gate FET sensors IrO_2 was shown as a NO_2 sensitive material and TiN as NH_3 sensing material with low cross sensitivity to humidity [19, 48, 49].

5 Results

Switching the semiconducting material from silicon to silicon carbide for the FET gas sensors opened up a large number of new applications at temperatures where silicon FETs are not reliable any longer. Later FET gas sensors in other wide band gap materials were also demonstrated, see Sect. 6, but silicon carbide is by far the most mature in terms of material quality and established processing technology.

5.1 Applications at an Ambient Temperature Above 500°C

The high end record for silicon carbide was demonstrated already in 1997 with a MOS capacitor device sensing hydrocarbons at 1,000°C, see Fig. 3 [6]. The device was operated for 20 min before failure, most likely due to restructuring of the electrical contacts, both ohmic contacts and the gate metal.

5.1.1 Syngas Control in a Power Plant

Loloee et al. have used very reliable MOS capacitors based on Pt–SiO_2–6H SiC structures (n-type) operated at 1 MHz at a sensor operating temperature of 620°C [50]. The sensor is operated using constant capacitance measurements and has been demonstrated, for example, for syngas control in a power plant. As such they were tested for the response to hydrogen between 52 ppm and 50% H_2 in nitrogen with 1% O_2 in nitrogen as the baseline. Furthermore, the selectivity was investigated as the sensitivity to other gases related to syngas production like CO, CO_2, CH_4, and H_2S. No influence of 40% CO or 5% CH_4 was detected, while addition of 40% CO_2 resulted in a 7% change in the baseline depending either on a direct response to CO_2 or an influence of the increase in oxygen concentration, which is a possible reaction product due to this exposure ($CO_2 + H_2 \Leftrightarrow CO + H_2O$ and $CO + 2\ H_2 \Leftrightarrow CH_4 + 1/2O_2$). Furthermore, a concentration of 2,000 ppm H_2S, 10 times higher than the realistic value, did not poison the sensor. This SiC sensor was tested under industrial conditions in the HPR (High Performance Reactor) at the Dept. of Energy's Technology Laboratory (DOE_NETL) in Morgantown, WV, and proved to stand a high temperature gas environment and corrosive gases in stable operation for 5 days with an additional laboratory testing of 480 h for the same sensor. A high quality SiO_2 and a rather thick sputtered Pt film (100 nm), operation at a constant capacitance close to midgap [42] together with sturdy packaging is assigned to the stable operation of the sensor during in total 25 days. The drawback of capacitive measurements is, however, as pointed out earlier, the expensive electronics needed for these measurements.

5.1.2 Cylinder-Specific Monitoring

A high operation temperature also speeds up the chemical reactions on the sensor surface. Resistivity changes of a gas sensor based on $SrTiO_3$ operated at 1,000°C demonstrated individual cylinder deviation monitoring [51]. It was also shown

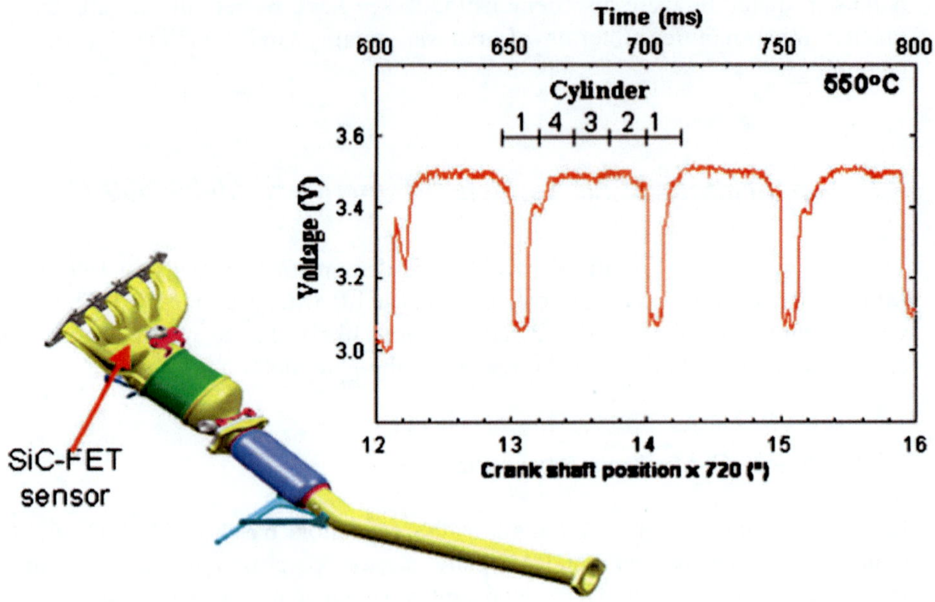

Fig. 6 One SiC-FET sensor operated at 550°C placed in the manifold system where the branches from the cylinders join monitors the deviation of one cylinder, number 1, which has excess fuel while cylinder 4, 3, 2 has a close to stochiometric fuel to air composition [54]. Copyright Elsevier Ltd. Reproduced with permission

possible to monitor deviations of individual cylinders, where the gas atmosphere changes in less than 10 ms, when operating silicon carbide transistor devices at 550°C [52, 53]. The SiC-based FE sensor was placed in the manifold system of a gasoline engine where the different branches from the cylinders join, see Fig. 6. The measurements in Fig. 6. were performed on a stationary engine for which the fuel to air ratio could be varied individually for each cylinder. Here, cylinder number 1 was rich, fed with excess fuel, while the other cylinders had a close to normal fuel to air ratio, that is all oxygen and fuel had reacted and formed carbon dioxide and water.

Special equipment has also been invented to monitor fast gas sensing constants. In one design, two small gas outlets were rapidly moved back and forth just beneath the sensor surface by the aid of a small engine and in this way the gas sensor ambient was switched in milliseconds [54]. Switching between hydrogen and oxygen atmosphere for a SiC sensor with a $TaSi_x$ + Pt gate metal at temperatures $\geq$500°C the response time was shown to be on the order of 10 ms, while the speed of response to ammonia for a sensor with a porous Pt gate at 300°C was estimated to less than 100 ms [55]. In this way, it was for example possible to show that the ammonia sensor fulfills the requirements from car industry of a sensor for control of SCR, selective catalytic reduction, as described later on in this chapter. Other researchers have constructed an ultra small sensor housing (1 μl gas exchange volume), where it was possible to exchange gas ambient in the ms time scale [56], while others used high speed injection valves in front of a small chamber (6 cm^3) [57].

5.1.3 Cold Start and EGR Sensor

At cold start of an engine there is a high risk of condensation and water droplet formation. These droplets can impinge on any surface at a high speed in the engine and the exhaust system. The traditional Lambda sensor used to control the air/fuel ratio in internal combustion engines is normally used in a binary fashion to monitor excess or deficiency of oxygen in order to fine-tune the combustion process, which is why it is needed to keep the sensor at a high operation temperature [58].The zirconia material making up the base of these Lambda, or UHEGO (Universal heated exhaust gas oxygen), sensors is, however, brittle and breaks from the thermo shock induced by impinging droplets when heated to its operation temperature. SiC, a material which is the second hardest after diamond, is believed to solve this problem. The SiC-FET devices also measure the lambda value, represented by a high signal in excess oxygen and a low signal in reducing exhaust conditions, for example, excess hydrocarbons and carbon monoxide, and almost defect-free silicon carbide material was proven to withstand water splash. Two SiC-FET devices were compared to a Lambda sensor in the exhaust system during operation of a 5-cylinder 2.5 l turbo engine, the lambda value varying between 0.94 and 1.07, see Fig. 7 [59].

The lambda value is defined as

$$\lambda = (\text{air}/\text{fuel})_{\text{engine}}/(\text{air}/\text{fuel})_{\text{stoich}} \tag{5}$$

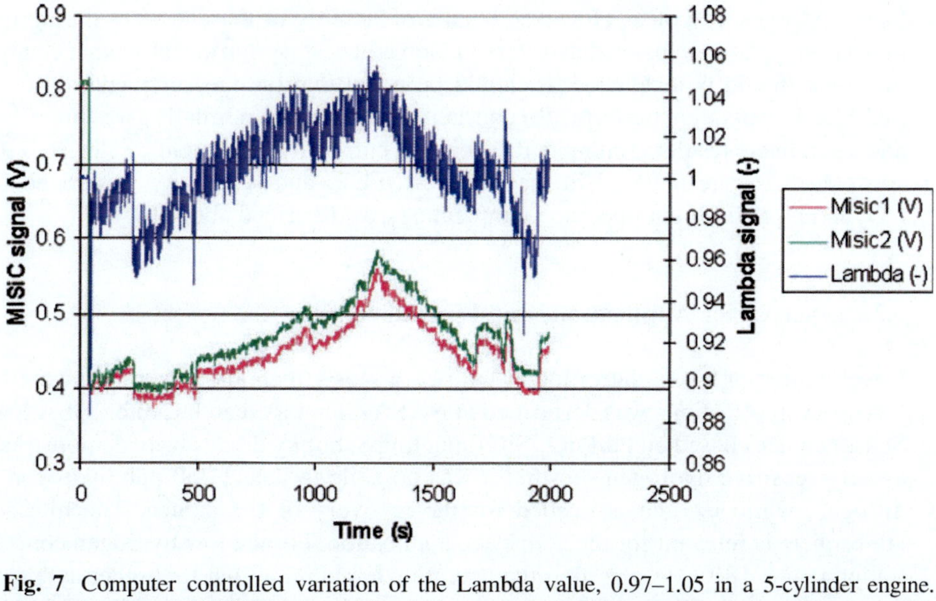

Fig. 7 Computer controlled variation of the Lambda value, 0.97–1.05 in a 5-cylinder engine. Monitoring by two SiC-FET sensors (*the two lower curves*). Requirements on the cold start sensor, resistance to water splash and possible to operate in an ambient temperature range of −20°C to 950°C during in total 240,000 km (5,000 h) [59]. Copyright Elsevier Ltd. Reproduced with permission

and equals one at stoichiometric conditions, while a value above one represent oxidizing conditions in excess air and values below one reducing conditions, with excess amounts of hydrocarbons, hydrogen, and carbon monoxide in the exhaust. However, for long-term operation of SiC-FET sensors at this temperature the contact materials need improvement, see Sect. 6.

Presently an oxygen FET sensor device intended for the control of the exhaust gas recirculation, EGR, system is also being developed; see Sect. 4 [32]. The requirements on this sensor is similar to the cold start sensor regarding, for example, resistance to water splash, since the EGR sensor should monitor the oxygen content in the gas mixture in the engine air intake.

5.2 *Applications at an Ambient Temperature Below 500°C*

5.2.1 Engines, Gas Turbines, and Boiler Plants

Several research groups develop SiC-FET devices with certain applications in mind. Sandvik et al. have processed 4H–SiC-FETs for O_2 and NO_x monitoring in automotive and locomotive engines, gas turbines, and boiler plants [60]. The devices are processed with a 50 nm gate oxide, shorted source and gate contacts and with a 600 nm SiO_2 protection overlayer. The measurements are performed as current change at a constant voltage, typically of 8 V. The response to NO peaks at 350°C for InO_x/SnO_2 as the gate material and at 300°C for InO_x/VO_x, which also seems to have the largest response. However, testing of stability of these devices during five hours lead to the conclusion that drift reduction is necessary and that also other catalyst materials should be evaluated. It should be noted that the measurements by these SiC-FET sensors were performed at constant voltage, which normally gives more drift and less reliable results as compared to constant current measurements. Later, $In_xV_yO_z$ was tested as gate material in these 4H–SiC-FETs and selectivity was found for NO_x and D_2 at different operation temperatures, as described above [47].

5.2.2 Aerospace Applications and Fire Alarms

Development of basic platforms for both core sensor technology and hardware for aerospace applications was performed at NASA Glenn Research Center, USA [61]. Sensor devices based on $Pd/PdO_x/SiC$ tunneling Schottky diodes were demonstrated as very sensitive hydrogen sensors. It was possible to detect 250 ppb hydrogen in nitrogen, while oxygen is needed for the recovery of the device. The nitrogen atmosphere is relevant for the aerospace application. For the low hydrogen concentration range, 180°C was used as the operation temperature but for higher hydrogen concentrations, 0.5% H_2, the device was operated at 450°C in a stable fashion for over 1,400 h. Another application area which has been described by NASA Glenn is sensors for hydrogen and hydrocarbons for fire detection without false alarm in the cargo bay of the aircraft. The Multiparameter Microsensor Fire Detection System, MMFDS was not triggered by environmental parameters such as dust or humidity as tested by the Federal Aviation Administration [61].

5.2.3 NH_3 Sensor for SCR Control

Awareness of the detrimental effect on the environment as well as adverse health effects of nitrogen oxides has led to a tightening of the legislations, for example, for emissions from heavy duty vehicles. With US as the leading country already down to almost zero allowed emissions, the EU countries and Japan are only a few years behind and hopefully the rest of the world will follow shortly. This has increased the efforts to reduce nitrogen oxides and also particulate matter, for which the legislations are equally strong. Several NO_x abatement methods have been tested and are under development but the currently most efficient method is the urea-SCR, where urea is injected in diesel exhaust, see Fig. 8. The urea forms ammonia and carbon dioxide through reaction with water

$$(NH_2)_2CO + H_2O \rightarrow 2NH_3 + CO_2 \tag{6}$$

At high temperatures in the exhaust gas, >150°C is required for this reaction, the ammonia (NH_3) reacts with nitric oxide and nitrogen dioxide (NO_x), which normally coexists, in the catalytic converter and harmless nitrogen (N_2) and water is formed:

$$4NH_3 + 4/xNO_x + O_2 \rightarrow (2 + 2/x)N_2 + 6H_2O \tag{7}$$

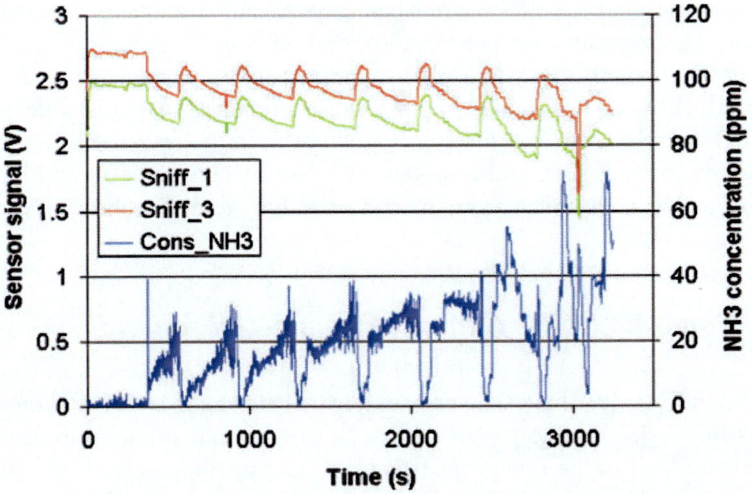

Fig. 8 Measurements by two SiC-FET sensors (*two upper curves*) in a diesel engine equipped with SCR system. Optical reference instrument (*lower curve*) for the NH_3 concentration

The process can be controlled either by a NO_x or NH_3 sensor or both for highest performance. We have developed an NH_3 sensor based on a SiC-FET with a porous Ir gate [44, 45], since this appeared as a very selective sensor to NH_3 provided the sensor is operated at 300°C, see Fig. 3. The performance was mapped also regarding cross sensitivity to NO_2 and N_2O and a background of 50 and 20 ppm, respectively, did not significantly change the response to pulses of 12.5–250 ppm NH_3 in a carrier gas of 10% O_2 in N_2. In diesel exhausts, the NH_3 sensor proved to detect less than 10 ppm of NH_3, see Fig. 8, and mounted in the exhaust system of two diesel trucks with SCR systems, two sensors successfully measured NH_3, with no interference on the sensor signal from NO_x, during a 10-day test session in Spain [2].

5.2.4 NH_3 Sensor for SNCR Control

For stationary combustion, the large catalytic beds needed for urea-SCR are expensive, containing quite large amounts of precious metals, and are therefore not considered for small- and medium-sized combustors, for example, local and domestic heating facilities. Another alternative to SCR, at least for medium-sized boilers, is the thermal $DeNO_x^{TM}$ process patented by Exxon, which is more familiar under the name Selective Non-Catalytic Reduction (SNCR) of NO_x by ammonia. In this process, ammonia (or some source of ammonia) is added directly to the flue gas stream, where at sufficiently high temperatures ammonia undergoes a number of gas phase reactions leading to the conversion of NO_x to N_2 and H_2O. In any of these processes, the amount of ammonia added must not be too large, however, since any excess ammonia, which is emitted to the atmosphere, would again lead to NO_x emissions through the oxidation of ammonia to NO_x by atmospheric oxygen.

From tests on a 5.6 MW wood fuelled and SNCR running boiler good indications for the realization of a field effect sensor-based system for ammonia slip monitoring were obtained. The logarithmic dependence of the FET sensors on ammonia concentration ascertains a good sensitivity to low concentrations of ammonia and the detection limit was clearly shown to be less than 5 ppm. Using an iridium gate transistor device operated at 250°C the selectivity was also fairly good even in the presence of typical flue gas constituents such as CO, hydrocarbons, and NO_x, whereas the same kind of sensor operated at 350°C exhibited more of cross-sensitivity to other substances, see Fig. 9.

5.3 *Domestic Boiler Control, Commercialization*

The SiC-FET devices have also been extensively tested in small- and medium-sized boilers [62–64]. The flue gas of boilers is very similar to diesel exhausts and the synergetic effect of the two projects is obvious. The interest for biofuels like wood logs and pellets has recently increased substantially due to environmental concerns and levies. A sensor system for control of domestic boilers, see Fig. 10, was developed based on two SiC-FET sensors kept at two different temperatures,

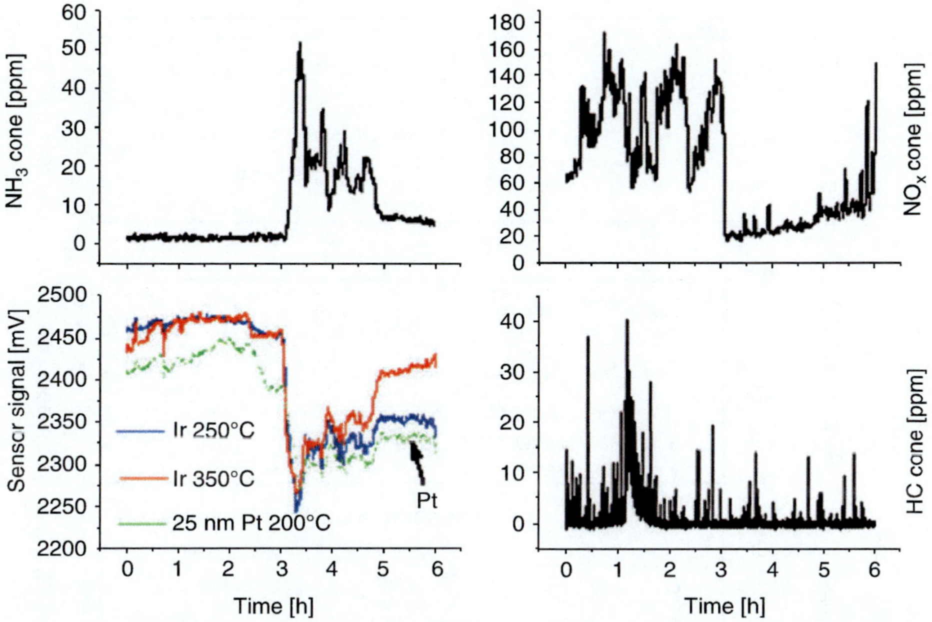

Fig. 9 The ammonia slip encountered during a one-day test of an SNCR system as compared to the signals from three different SiC-based field effect transistor sensors (*lower picture to the left*). The flue gas concentrations of NO_x ($NO + NO_2$) and hydrocarbons are also given for comparison

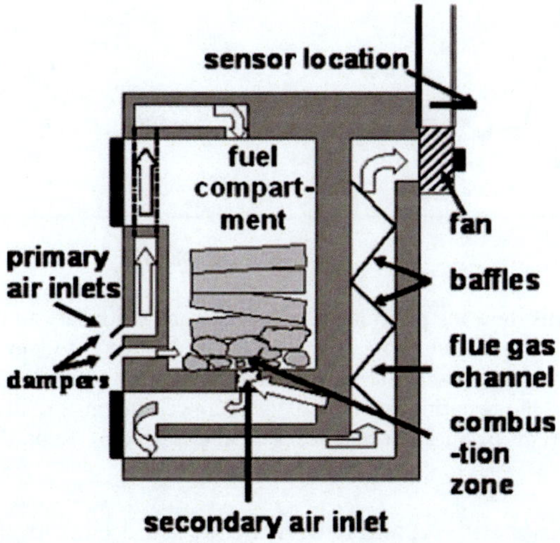

Fig. 10 Schematic cross section of the 40 kW wood fired boiler used during the field tests. The sensor location is also indicated. The primary air is supplied both directly into the combustion zone at the bottom of the fuel compartment and from above at the top of the fuel compartment. The reactions take place in the combustion zone and in the flame, the latter passing through a small hole at the bottom of the fuel compartment into which also the secondary air is supplied

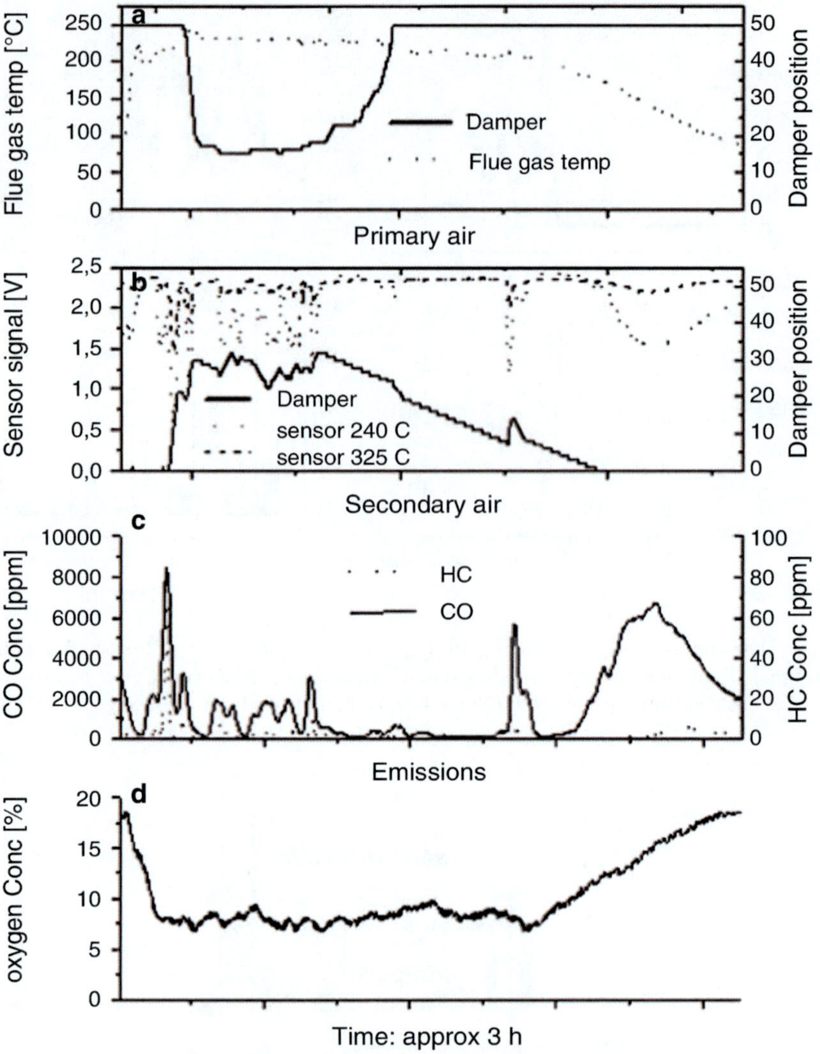

Fig. 11 The diagrams show the position of the primary and secondary air dampers (0 corresponds to zero flow and 50 to the maximum flow of air) related to the flue gas temperature and sensor signals, respectively, when the control scheme was applied to one firing cycle (**a**) and (**b**). The emissions of unburned matter and the resulting flue gas oxygen concentrations are also given (**c**) and (**d**) [62]. Copyright Springer (Part of Springer Science + Business Media). Reproduced with permission

240 and 325°C, respectively, and a thermocouple-type temperature sensor [62]. The primary and secondary air inlets to the boiler were controlled based on the pattern of the sensor signals, see Fig. 11. The control algorithm takes into account the combination of low or high level of the two sensors with some input also from the temperature sensor. The controlled combustion in the wood fuelled boiler increased the efficiency of the combustion and substantially reduced the emissions to the

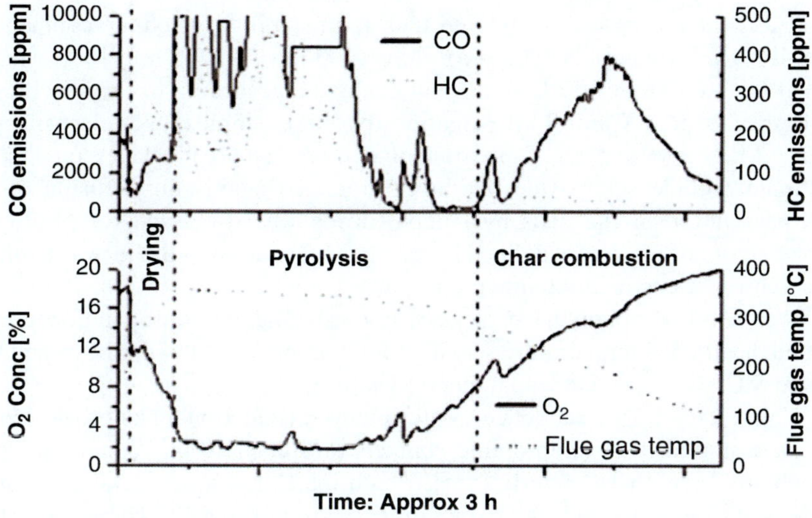

Fig. 12 The diagram shows the characteristics of a typical firing cycle from an uncontrolled 40 kW wood fired boiler for domestic heating, as well as the different phases in the combustion cycle as characterized by the flue gas temperature, oxygen concentration, and the emissions

environment, see Figs. 11 and 12. Furthermore, this still hold for varying quality of the wood, both very dry wood and somewhat humid wood could be used for the combustion with the same good result regarding efficiency and emissions.

The successful control of the domestic boilers resulted in a commercialization activity of this sensor system [8], which was only possible through the close collaboration between researchers, engineers, and end users.

6 Future Trends and Development in SiC-FET Sensors

6.1 Improved Contact Material to SiC

Today, the SiC-FET devices are very stable and perform excellent with very high long-term stability for operation temperature up to 400°C. However, for temperatures above 400°C, the contacts fail, both ohmic pad and gate contacts, due to inferior quality of the contact materials. The thermal stability was, for example, studied using the linear transfer length method, TLM, by Virshup et al. [65] for the system Pt (150 nm)/$TaSi_x$ (50 nm)/Ni (100 nm)/SiC, which remained ohmic for 800 h at 300°C but failed after 240 h at 500°C and already after 36 h at 600°C. Our goal is to develop high quality contacts to SiC in order to reach new applications such as EGR sensors, see Sect. 5. The SiC material fulfills the requirements on these sensors, however, also the contacts must survive, for example, the water splash and epitaxially grown Ti_3SiC_2 is an interesting candidate as described below.

NASA Glenn reported improved long-term stability for ohmic contacts to 6H– and 4H–SiC based on Ti (200 nm)/$TaSi_2$ (400 nm)/Pt (300 nm) [66], which were tested for 1,000 h at 600°C in air and evaluated according to the specific contact resistance (at RT). Careful investigations by, for example, auger electron spectroscopy (AES) revealed the formation of a protective oxide layer on the formed platinum silicide. The oxide overlayer protected against in-diffusion of oxygen and passivation of the pads by silicon nitride was suggested for future studies. A new figure of merit was defined based on the deviation of the forward voltage at a fixed current density and temperature.

Chen and Mehregany have successfully tested heavily nitrogen doped poly-SiC as both heater and temperature sensing film in microhotplate platforms and the high doped SiC also served as metal electrodes [67].

Within the VINN Excellence Center FunMat (Functional Nanoscale Materials), we have started to develop ohmic contacts to SiC based on conducting ceramics, which are three phase materials based on a metal and a nonmetal combined with carbon, nitrogen, oxygen, and sometimes phosphorus [68]. There are around 60 identified materials with different properties but all belong to the category conducting ceramic, so-called MAX materials, which makes them extremely interesting as contacts for high temperature applications. Since Ti_3SiC_2 grows spontaneously on SiC when Ti is present, this has been suggested as an interesting contact material [43, 69]. We have started to develop this contact system for SiC [67]. Capping layers will be needed for bonding but also as protection against in-diffusion of oxygen, since the layer oxidizes at about 400°C [70, 71].

6.2 Packaging

The mounting of SiC-FET devices will constitute the main part of the unit price per sensor and therefore the packaging will be dealt with a lot more in the future. Several interesting prototype platforms were presented by NASA Glenn [61] and by Sandvik et al. [72]. The gas distribution around the sensor surface is a critical parameter for each packaging design. Useful simulations were performed for the design of the housing for the ammonia sensors for SCR control on heavy duty tracks, see Sect. 5 [73].

6.3 3C SiC

3C SiC has gained a large interest recently in the SiC research community [74, 75]. There is hope that this material will show up as a lot less expensive material but still with high performance regarding high temperature applications. The band gap is lower, 2.3 eV, than, for example, that of 4H SiC material, 3.2 eV, but still more than double that of Si, 1.1 eV. We have shown that 4H SiC retains semiconductor properties even at 1,000°C, and it will be interesting to check the high temperature performance of the 3C–SiC material.

6.4 WBG Materials for FETs

Other wide band gap, WBG, materials have been used for FET devices such as Pt–GaN Schottky diodes operated as hydrogen sensors for 21 months in transformer oil [72]. Hydrogen sensors were also demonstrated based on Pt/AlGaN/GaN heterojunction field effect transistors, HFETs [76], for which the hydrogen response was tested in a nitrogen atmosphere, which gives a very high sensitivity (10 ppb of hydrogen was detected at $\geq$450°C) but limits the possible applications. The devices could be operated up to 800°C. The main part of the work on WBG devices is related to biosensors. Chemical functionalization and detection of biomolecules have been frequently reported for AlGaN/GaN [77–79], a ZnO/ZnMgO HFET [80], SiC [81] and diamond [82, 83]. A new and challenging development is represented by nanostructures like carbon nanotubes operated as FET devices proposing single biomolecule detection [84]. Recently also graphene grown in epitaxial monolayers on SiC have been demonstrated as sensors for ppb detection of NO_2 with FET sensors as the preferred device [85, 86]. These technologies will likely find application areas and open up exciting possibilities for new sensing mechanisms and further miniaturization of sensor devices.

7 Conclusions

The close collaboration between researchers, engineers, and end users is important for the possibility to commercialize sensor systems for gas sensing applications, which always includes a large number of influencing parameters. Here, we have mentioned two possibilities to achieve such collaboration, forming spin-off companies and university-based excellence centers in collaboration with industry. Both have in fact contributed to the commercialization of a sensor system based on SiC-FET gas sensors for domestic boiler control. The SiC material is now possible to buy in high quality 4″ wafers and processing for mass production is established. We hope that one successful commercial SiC-based sensor system in the market will speed up the introduction of SiC-based sensor systems for a variety of applications.

Acknowledgment Financial support is acknowledged from VINNOVA, The Swedish Agency for Innovation Systems, Industrial Partners and Linköping University through the VINN Excellence Center FunMat, Functional Nanoscale Materials, at Linköping University, Sweden and from S-SENCE, Swedish Sensor Center, VINNOVA Excellence Center at Linköping University 1995–2005, and from the Swedish Research Council.

References

1. Lundström I, Shivaraman MS, Svensson C (1975) A hydrogen-sensitive MOS field effect transistor. J Appl Phys 26:55–57
2. Lundström I, Sundgren H, Winquist F, Eriksson M, Krantz-Rülcker C, Lloyd Spetz A (2007) Twenty-five year of field effect gas sensors research in Linköping. Sens Actuators B 121:247–262
3. Löfdahl M, Utaiwasin C, Carlsson A, Lundström I, Eriksson M (2001) Gas response dependence on gate metal morphology of field-effect devices. Sens Actuators B 80:183–192
4. Spetz A, Armgarth M, Lunström I (1988) Hydrogen and ammonia response of metal- silicon dioxide-silicon structures with thin platinum gates. J Appl Phys 64(3):1274–1283
5. Spetz A, Helmersson U, Enquist F, Armgarth M, Lundström I (1989) Structure and ammonia sensitivity of thin platinum or iridium gates in metal- oxide -silicon capacitors. Thin Solid Films 177:77–93
6. Lloyd Spetz A, Baranzahi A, Tobias P, Lundström I (1997) High temperature sensors based on metal-insulator-silicon carbide devices. Phys Stat Sol A 162:493–511
7. Wright NG, Horsfall AB (2007) SiC sensors: a review. J Phys D Appl Phys 40:6345–6354
8. SenSiC AB, Isafjordsgatan 39B, SE-164 40 Kista, Sweden
9. Fogelberg J, Eriksson M, Dannetun H, Petersson L-G (1995) Kinetic modeling of hydrogen adsorption in thin films on hydrogen-sensitive field effect devices. Observation of large hydrogen-induced dipoles at the Pd-SiO_2 interface. J Appl Phys 78(2):988–996
10. Wallin M, Grönbeck H, Lloyd Spetz A, Skoglundh M (2004) Vibrational study of ammonia adsorption on Pt/SiO_2. Appl Surf Sci 235:487–500. doi:10.1016/j.apsusc.2004.03.225
11. Eriksson M, Petersson L-G (1998) Real time measurements of hydrogen desorption and absorption during CO exposures of Pd: hydrogen sticking and dissolution. Appl Surf Sci 133:89–97
12. Schalwig J, Kreisl P, Ahlers S, Müller G (2002) Response mechanism of SiC-based MOS field-effect gas sensors. IEEE Sens J 2(5):394–402
13. Dannetun HM, Petersson L-G, Söderberg D, Lundström I (1985) The H_2-O_2 reaction on palladium studied over a large pressure range: independence of the microscopic sticking coefficients on surface condition. Surf Sci 152(153):559–568
14. Salomonsson A, Eriksson M, Dannetun H (2005) Hydrogen interaction with platinum and palladium metal – insulator – semiconductor devices. J Appl Phys 98: 014505-1–014505-9
15. Åbom AE, Haasch RT, Hellgren N, Finnegan N, Hultman L, Eriksson M (2003) Characterization of the metal-insulator interface of field-effect chemical sensors. J Appl Phys 93 (12):9760–9768
16. Åbom AE, Persson P, Hultman L, Eriksson M (2002) Influence of gate metal film growth parameters on the properties of gas sensitive field-effect devices. Thin Solid Films 409:233–242
17. Kahng YH, Lu W, Tobin RG, Loloee R, Ghosh R (2009) The role of oxygen in hydrogen sensing by a platnium-gate silicon carbide gas sensor: an ultrahigh vacuum study. J Appl Phys 105:0645111-1–064511-7
18. Kahng YH, Tobin RG, Loloee R, Ghosh R (2007) Sulfur surface chemistry on the platinum gate of a silicon carbide based hydrogen sensor. J Appl Phys 102:064505-1–064505-9
19. Eisele I, Doll T, Burgmair (2001) Low power gas detection with FET sensors. Sens Actuators B 78:19–25
20. Ostrick B, Fleischer M, Meixner H (2003) The influence of interfaces and interlayers on the gas sensitivity in work function type sensors. Sens Actuators B 95:271–274
21. Moos R, Sahner K, Fleischer M, Guth U, Barsan N, Weimar U (2009) Solid state gas sensor research in Germany – a status report. Sensors 9:4323–4365
22. Andersson M, Ljung P, Mattsson M, Löfdahl M, Lloyd Spetz A (2004) Investigations on the possibilities of a MISiCFET sensor system for OBD and combustion control utilizing different catalytic gate materials. Top Catal 30(31):365–368

23. Lloyd Spetz A, Skoglundh M, Ojamäe L (2008) FET gas sensing mechanism, experimental and theoretical studies. In: Comini E, Faglia G, Sberveglieri G (eds) Solid state gas sensing, chap 4. Springer, Norwell, MA, USA, pp 153–179, ISBN: 978-0-387-09664-3
24. Salomonsson A, Petoral RM Jr, Uvdal K, Aulin C, Käll P-O, Ojamäe L, Strand M, Sanati M, Lloyd Spetz A (2006) Nanocrystalline ruthenium oxide and ruthenium in sensing applications – an experimental and theoretical study. J Nanoparticle Res 8:899–910. doi:10.1007/s11051-005-9058-1
25. Wallin M, Byberg M, Grönbeck H, Skoglundh M, Eriksson M, Lloyd Spetz A (2007) Vibrational analysis of H2 and NH3 adsorption on Pt/SiO2 and Ir/SiO2 model sensors. In Proceedings of IEEE Sensors 2007, Atlanta, USA, 28–31 Oct 2007, pp 1315–1317
26. Wallin M, Grönbeck H, Lloyd Spetz A, Eriksson M, Skoglundh M (2005) Vibrational analysis of H_2 and D_2 adsorption on Pt/SiO_2. J Phys Chem B 109:9581–9588
27. Nakagomi S, Tobias P, Baranzahi A, Lundström I, Mårtensson P, Lloyd Spetz A (1997) Influence of carbon monoxide, water, and oxygen on high temperature catlaytic metal–oxide–silicon carbide structures. Sens Actuators B 45(3):183–191
28. Andersson M, Lloyd Spetz A (2010) Tailoring of SiC based field effect gas sensors for improved selectivity to non-hydrogen containing species. In: Proceedings of IMCS13, Perth, Australia, 12–14 July 2010, p 369
29. Andersson M, Wingbrant H, Lloyd Spetz A (2005) Study of the CO response of SiC based field effect gas sensors. In: Proceedings of IEEE Sensors 2005, Irivine, USA, 31 Oct–2 Nov 2005, pp 105–108
30. Becker E, Skoglundh M, Andersson M, Lloyd Spetz A (2007) In situ DRIFT study of hydrogen and CO adsorption on Pt/SiO_2 model sensors. In: Proceedings of IEEE Sensors 2007, Atlanta, USA, 28–31 Oct 2007, pp 1028–1031
31. Becker E, Andersson M, Eriksson M, Lloyd Spetz A, Skoglundh M (2011) Study of the sensing mechanism towards carbon monoxide of platinum-based field effect sensors. IEEE Sens J 11 (7):1527–1534
32. Andersson M, Lloyd Spetz A (2009) Tailoring of field effect gas sensors for sensing of nonhydrogen containing substances from mechanistic studies on model systems. In: Proceedings of IEEE Sensors 2009, Christchurch, New Zealand, 26–28 Oct 2009, pp 2031–2036
33. Nakagomi S, Sato K, Suzuki S, Kokubun Y (2009) Influence of ambient, gate metal and oxide thickness on interface state density and time constant in MOSiC capacitor. Mater Sci Forum 600–603:735–738
34. Weidemann O, Hermann M, Steinhoff G, Wingbrant H, Lloyd Spetz A, Stutzmann M, Eickhoff M (2003) Influence of surface oxides on hydrogen-sensitive Pd:GaN Schottky diodes. Appl Phys Lett 83(4):773–775
35. Zangooie S, Arwin H, Lundström I, Lloyd Spetz A (2000) Ozone treatment of SiC for improved performance of gas sensitive Schottky diodes. Mater Sci Forum 338–342:1085–1088
36. Neudeck PG, Spry DJ, Truneck AJ, Evans LJ, Chen L-Y, Hunter GW, Androjna D (2009) Hydrogen gas sensors fabricated on atomically flat 4H-SiC webbed cantilevers. Mater Sci Forum 600–603:1199–1202
37. Comini E, Cusma A, Kaciulis S, Kandasamy S, Padeletti G, Pandolfi L, Sberveglieri G, Trinchi A, Wlodarski W (2006) XPS investigation of CoO_x-based MRISiC structures for hydrocarbon gas sensing. Surf Interface Anal 38:736–739
38. Nakagomi S, Lloyd Spetz A (2006) Gas sensor device based on catalytic Metal – Metal Oxide – SiC structure. In: Grimes CA, Dickey EC (eds) Encyclopedia of sensors, vol 4. American Scientific, Stevenson Ranch, CA, USA, pp 155–170
39. Trinchi A, Kandasamy S, Wlodarski W (2008) High temperature field effect hydrogen and hydrocarbon gas sensors based on SiC MOS devices. Sens Actuators B 133:705–716
40. Zhu W, chen XF, Tan OK, Deng J (2002) Hydrogen-sensitive amorphous ferroelectric thin film capacitive devices. Integr Ferroelectron 44:25–75

41. Di Natale C, Buchholt K, Martinelli E, Paolesse R, Pomarico G, D'Amico A, Lundström I, Lloyd Spetz A (2009) Investigation of quartz microbalance and chemfet transduction of molecular recognition events in a metalloporphyrin film. Sens Actuators B 135:560–567
42. Ghosh RN, Tobias P (2005) SiC field effect devices operating at high temperatures. J Electron Mater 34:345–350
43. Racault C, Langlais F, Naslain R (1994) Solid state synthesis and characterization of the ternary phase Ti_3SiC_2. J Mater Sci 29:3384–3392
44. Wingbrant H, Svenningstorp H, Kubinski DJ, Visser JH, Andersson M, Unéus L, Löfdahl M, Lloyd Spetz A (2006) MISiC-FET NH_3 sensors for SCR control in exhaust and flue gases. In: Grimes CA, Dickey EC (eds) Encyclopedia of sensors, vol 6. American Scientific, Stevenson Ranch, CA, USA, pp 205–218
45. Wingbrant H, Svenningstorp H, Salomonsson P, Kubinski D, Visser JH, Löfdahl M, Lloyd Spetz A (2005) Using a MISiC-FET sensor for detecting NH_3 in SCR systems. IEEE Sens J 5 (5):1099–1105
46. Kim CK, Lee JH, Choi SM, Noh IH, Kim HR, Cho NI, Hong C, Jang GE (2001) Pd- and Pt-SiC Schottky diodes for detection of H_2 and CH_4 at high temperature. Sens Actuators B 77:455–462
47. Ali M, Cimalla V, Lebedev V, Stauden Th, Wang Ch, Ecke G, Tilak V, Sandvik P, Ambacher O (2007) Reactively sputtered $In_xV_yO_z$ films for detection of NO_x, D_2 and O_2. Sens Actuators B 123:779–783
48. Karthigeyan A, Gupta RP, Burgmair M, Zimmer M, Sulima T, Venkataraj S, Sharma SK, Eisele I (2004) Iridium oxide as low temperature NO2-sensitive material for work function-based gas sensors. IEEE Sens J 4(2):189–194
49. Ostrick B, Pohle R, Fleischer M, Meixner H (2000) TiN in work function type sensors: a stable ammonia sensitive material for room temperature operation with low humidity cross sensitivity. Sens Actuators B 68:234–239
50. Loloee R, Chorpening B, Beer S, Ghosh RN (2008) Hydrogen monitoring for power plant applications using SiC sensors. Sens Actuators B 129:200–210
51. Meixner H, Gerblinger J, Lampe U, Fleischer M (1995) Thin-film gas sensors based on semiconducting metal oxides. Sens Actuators B 23:119–125
52. Baranzahi A, Tobias P, Lloyd Spetz A, Lundström I, Mårtensson P, Glavmo M, Göras A, Nytomt J, Salomonsson P, Larsson H (1997) Fast responding air/fuel sensor for individual cylinder control. SAE Technical Paper Series 972940, Combustion and Emisson Formation in SI Engines, SP-1300: 231–240
53. Larsson O, Göras A, Nytomt J, Carlsson C, Lloyd Spetz A, Artursson T, Holmberg M, Lundström I, Ekedahl L-G, Tobias P (2002) Estimation of air fuel ratio of individual cylinders in SI engines by means of MISiC sensor signals in a linear regression model. In: Proceedings of SAE2002, Detroit, USA, 4–7 Mar 2002 (2002-01-0847, also in SAE 2002 Transactions J Engines)
54. Tobias P, Lloyd Spetz A, Mårtensson P, Baranzahi A, Göras A, Lundström I (1999) Moving gas outlets for the evaluation of fast gas sensors. Sens Actuators B 58(1–3):389–393
55. Wingbrant H, Svenningstorp H, Salomonsson P, Tengström P, Visser J, Ekedahl L-G, Lundström I, Lloyd Spetz A (2003) The speed of response of MISiCFET devices. Sens Actuators B 93(1–3):286–294
56. Ghosh RN, Tobias P, Hu H, Koochesfahani M (2005) Fast solid state gas sensor characterization with millisecond resolution. In: 4th IEEE conference on sensors, Irvine, CA, 31 Oct–3 Nov 2005
57. Gerblinger J, Lampe U, Meixner H (1995) Sensitivity mechanism of metal oxides to oxygen detected by means of kinetic studies at high temperatures. Sens Actuators B 25:639–642
58. Visser J, Soltis RE (2001) Automotive exhaust gas sensing systems. IEEE Trans Instrum Meas 50(6):1543–1550
59. Wingbrant H, Svenningstorp H, Salomonsson P, Tengström P, Moldin D, Ekedahl L-G, Lundström I, Lloyd Spetz A (2003) Using a MISiCFET device as a cold start sensor. Sens Actuators B 93(1–3):295–303

60. Sandvik P, Ali M, Tilak V, Matocha K, Stauden T, Tucker J, Deluca J, Ambacher O (2006) SiC-based MOSFETS for harsh environment emissions sensors. Mater Sci Forum 527–529:1457–1460
61. Hunter GW, Xu JC, Dungan LK, Ward BJ, Rowe S, Williams J, Makel DB, Liu CC, Chang CW (2008) Smart sensor systems for aerospace applications: from sensor development to application testing. ECS Trans 16(11):333–344
62. Andersson M, Everbrand L, Lloyd Spetz A, Nyström T, Nilsson M, Gauffin C, Svensson H (2007) A MISiCFET based gas sensor system for combustion control in small-scale wood fired boilers. In: Proceedings of IEEE Sensors 2007, Atlanta, USA, 28–31 Oct 2007, pp 962–965
63. Andersson M, Wingbrant H, Petersson H, Unéus L, Svenningstorp H, Löfdahl M, Holmberg M, Lloyd Spetz A (2006) Gas sensor arrays for combustion control. In: Grimes CA, Dickey EC (eds) Encyclopedia of sensors, vol 4. American Scientific, Stevenson Ranch, CA, USA, pp 139–154
64. Unéus L, Artursson T, Mattsson M, Ljung P, Wigren R, Mårtensson P, Holmberg M, Lundström I, Lloyd Spetz A (2005) Evaluation of on-line flue gas measurements by MISiCFET and metal-oxide sensors in boilers. IEEE Sens J 5(1):75–81
65. Virshup A, Porter LM, Lukco D, Buchholt K, Hultman L, Lloyd Spetz A (2009) Investigation of thermal stability and degradation mechanism in Ni-based ohmic contacts to n-type SiC for high-temperature gas sensors. J Electron Mater 38(4):569–573
66. Okojie RS, Lukco D, Chen YL, Spry DJ (2002) Reliability assessment of Ti/$TaSi_2$/Pt ohmic contacts on SiC after 1000 h at 600°C. J Appl Phys 91(10):6553–6559
67. Buchholt K, Ghandi R, Domeij M, Zetterling C-M, Lu J, Eklund P, Hultman L, Lloyd Spetz A (2011) Ohmic contact properties of magnetron sputtered Ti_3SiC_2 on *n*- and *p*-type 4 H-silicon carbide. Appl Phys Lett 98(4):042108. doi:10.1063/1.3549198
68. Eklund P, Beckers M, Jansson U, Högberg H, Hultman L (2010) The $M_{n+1}AX_n$ phases: materials science and thin film processing. Thin Solid Film 518:1851–1878
69. Pecz B, Toth L, di Forte-Poisson MA, Vacas J (2003) Ti_3SiC_2 formed in annealed Al/Ti contacts to p-type SiC. Appl Surf Sci 206:8–11
70. Sun Z, Zhou Y, Li M (2001) High temperature oxidation behavior of Ti_3SiC_2 based material in air. Acta Mater 49:4347–4353
71. Wenzel R, Goesmann F, Schmid-Fetzer R (1998) Diffusion barriers in gold-metallized titanium-based contact structures on SiC. J Mat Sci Mat in Electr 9(2):109–113
72. Sandvik P, Babes-Dornea E, Trudel AR, Georgescu M, Tilak V, Renaud D (2006) GaN-based Schottky diodes for hydrogen sensing in transformer oil. Phys Stat Sol C 6:2283–2286
73. Belov I, Wingbrant H, Lloyd Spetz A, Sundgren H, Thunér B, Svenningstorp H, Leisner P (2006) CFD analysis of packaging and mounting solutions for SiC-based gas sensors in automotive applications. Sens Lett 4:29–37
74. Eriksson J, Roccaforte F, Reshanov S, Leone S, Giannazzo F, LoNigro R, Fiorenza RP, Raineri V (2011) Nanoscale characterization of electrical transport at metal/3 C-SiC interfaces. Nanoscale Res Lett 6:120–124
75. Lebedev AA, Abramov PL, Bogdanova EV, Lebedev SP, Nelson DK, Oganesyan GA, Tregubova AS, Yakimova R (2008) Highly doped p-type 3C-SiC on 6H-SiC substrates. Semicond Sci Technol 23:075004
76. Song J, Lu W (2008) Operation of Pt/AlGaN/GaN-heterojunction Field Effect-Transistor hydrogen sensors with low detection limit and high sensitivity. IEEE Electron Dev Lett 29 (11):1193–1195
77. Baur B, Howgate J, von Ribbeck H-G, Gawlina Y, Bandalo V, Steinhoff G, Stutsmann M, Eickhoff M (2006) Catalytic activity of enzymes immobilized on AlGaN/GaN solution gate field-effect transistors. Appl Phys Lett 89:183901
78. Baur B, Steinhoff G, Hernando J, Purrucker O, Tanaka M, Nickel B, Stutzmann M, Eickhoff M (2005) Chemical functionalization of GaN and AlN surfaces. Appl Phys Lett 87: 26391-1–26391-3

79. Steinhoff G, Baur B, Wrobel G, Ingebrant S, Offenhäuser A, Dadgar A, Krost A, Stutzmann M, Eickhoff M (2005) Recording of cell action potentials with AlGaN/GaN field-effect transistors. Appl Phys Lett 86:033901 (with G. Steinhoff G, Baur B, Wrobel G, Ingebrandt S, Offenhäusser A, Stutzmann M, Eickhoff M (2006) Erratum:[Recording of Cell Action Potentials with AlGaN/GaN Field Effect Transistors], Appl Phys Lett 89: 019901)
80. Koike K, Takagi D, Kawasaki M, Hashimoto T, Inoue T, Ogata K-I, Sasa S, Inoue M, Yano M (2007) Ion sensitive characteristics of an electrolyte-solution-gate ZnO/ZnMgO heterjunction field-effect transistor as a biosensing transducer. Jpn J Appl Phys 46(36):L865–L867
81. Yakimova R, Petoral RM Jr, Yazdi GR, Vahlberg C, Lloyd Spetz A, Uvdal K (2007) Surface functionalization and biomedical applications based on SiC. J Phys D Appl Phys 40:6435–6442
82. Rubio-Retama J, Hernando J, Lopez-Ruiz B, Härtl A, Steinmüller D, Stutzmann M, Lopez-Cabarcos E, Garrido JA (2006) Synthetic nanocrystalline diamond as a third-generation biosensor support. Langmuir 22:5837–5842
83. Helwig A, Müller G, Garrido JA, Eickhoff M (2008) Gas sensing properties of hydrogen-terminated diamond. Sens Actuators B 133:156–165
84. Roy S, Gao Z (2009) Nanostructure based electrical biosensors. Nano Today 4:318–334
85. Pearce R, Iakimov T, Andersson M, Hultman L, Lloyd Spetz A, Yakimova R (2011) Epitaxially grown graphene based gas sensors for ultra sensitive NO_2 detection. Sens Actuators B 155 (2):451–455
86. Emtsev KV, Bostwick A, Horn K, Jobst J, Kellogg GL, Ley L, McChesney JL, Ohta T, Reshanov SA, Röhrl J, Rotenberg E, Schmid AK, Waldman D, Weber HB, Seyller T (2009) Towards wafer-size graphene layers by atmospheric pressure graphitization of silicon carbide. Nat Mater 8:203–207

Development of Planar Potentiometric Gas Sensors for Automotive Exhaust Application

C. Pijolat and J.P. Viricelle

Abstract In the past, we have deeply investigated the development of a β-alumina potentiometric sensor with two coplanar metallic electrodes. Recently, many papers have demonstrated an interest for similar non-Nernstian devices based on stabilised zirconia (*YSZ*) electrolyte. The common particularity of such sensors is related to the use a dissymmetry of electrodes on an electrolyte and to work in a single gas cell. Our main experimental results obtained with β-alumina sensors are reported with the special attention to develop gas sensors for exhaust applications. The sensors are evaluated under CO, HC and NO_x CO. The selectivity depends on the sensor temperatures. The performances appear sufficient to satisfy in the future automotive applications like the control of NOxTrap or DeNOx devices. The measure of NH_3 concentrations seems also possible. The comparison of performances of sensors replacing β-alumina electrolyte by YSZ is also presented. Similar behaviours are obtained for both types of sensors. These comparisons reinforce our proposal of a capacitive model mainly based on the difference of reactivity of the electrodes. It is pointed out that mixed potentials cannot explain the behaviour of our sensors under oxygen. Very good agreements are obtained between this capacitive model and the experimental results, not only versus oxygen, but also with mixture of oxygen and CO.

Keywords Automotive exhaust, Non-Nernstian gas sensors, YSZ, β-Alumina

C. Pijolat (✉) and J.P. Viricelle
Ecole Nationale Supérieure des Mines, SPIN-EMSE, CNRS:FRE3312, Laboratoire des Procédés en Milieux Granulaire, F-42023 Saint-Etienne, 158 cours Fauriel, 42023, Saint-Etienne cedex 02, France
e-mail: cpijolat@emse.fr; viricelle@emse.fr

M. Fleischer and M. Lehmann (eds.), *Solid State Gas Sensors – Industrial Application*, Springer Series on Chemical Sensors and Biosensors (2012) 11: 215–254,
DOI 10.1007/5346_2011_10, © Springer-Verlag Berlin Heidelberg 2012,
Published online: 17 January 2012

Contents

1 Introduction

1.1 State of the Art

World-wide recommendations and laws for limiting environmental pollution become stricter for both industrial and domestic sources of pollution. In particular, European emission regulation standards for vehicles are redefined periodically and the standards named “Euro” are much more stringent [1]. Hence, on-board diagnosis (OBD) is implemented and reinforced. Various equipments have been developed by car manufacturers in order to reduce emissions. The OBD of the efficiency of these systems and the loop control of their running require information from gas sensors able to measure exhaust gas emissions, such as HC (hydrocarbons), CO, NO_x, (NO, NO_2), and also NH_3 for vehicles using Selective Catalytic Reduction (SCR) systems. Hence, there is a strong demand to selective sensors from car manufacturers.

For automotive exhausts, a big effort of research in the 1960s years has led to the development of lambda oxygen sensors based on yttrium stabilised zirconia

(YSZ), These sensors are realised with tubular zirconia devices using a physical separation of gaseous atmosphere between the reference atmosphere (generally atmospheric air, $P_{O_{2ref}}$) and the measuring cell. The electrical potential delivered by the YSZ sensor (Device: exhaust gas, Pt/YSZ/Pt_{ref}, Air_{ref}) is directly correlated with the partial pressure of oxygen in the measuring cell (P_{O_2}) through the Nernst law:

$$E = (RT/4F)\ln(P_{O_2}/P_{O_{2ref}}).$$

Due to the past successful development of lambda oxygen sensors based on YSZ, many researches were focused on this kind of sensor in order to adapt them to monitor other gases than oxygen, especially nitrogen oxides. In particular, the most common concept was to create a diffusion area to limit gas introduction in a first pumping cell, which oxidise reducing gases. Then a sensing cell measures the NO_x concentration [1–22]. Such systems present good performances, but their main drawback is their complexity both for production and for use due to the requirement to monitor two cells.

It has been noticed very early that such Zirconia oxygen sensor can present a deviation to the Nernst law in exhaust gases. J. Fleming has reported first this non-ideal behaviour in 1977 [23] with an explanation of CO oxidation at the anode. According to the different possible reactions of CO, especially the electrochemical reaction with the oxygen of YSZ, the final voltage depends on both P_{O2} and P_{CO} partial pressures.

This non-ideal behaviour has been confirmed by Okamoto in 1981 [24] and the authors introduced then the term of "mixed electrode potential". They confirm that the open circuit potential (noted "EMF" or "OCV") of zirconia galvanic cells, $O_2(P_{O_2})$, $CO(P_{CO})$, $Pt/YSZ/Pt_{ref}$, $O_2(P_{O_2ref})$ can deviate from the Nernst's equation at low temperature, for example 350°C. At high temperature, the following gas reaction occurs, 2 CO + O_2 = 2 CO_2, and the sensor shows an EMF gap at the stoichiometric ratio of this reaction when the equilibrium is established. At low temperature the EMF is considered to be caused by mixed potential according to the two electrochemical reactions:

$$O_{ads} + 2e^- = O^{2-}$$

$$CO_{ads} + O^{2-} = CO_2 + 2e^-$$

We will see further that this notion of "mixed potential" has been used by many authors to explain the behaviour of several electrochemical gas sensors, and especially for NO_x sensors.

In regards of conventional YSZ sensors, a new concept has been proposed by Williams in 1983 [12], which consists to have no physical separation between the two cells (measuring and reference ones). The two electrodes are in only one cell and the EMF is created by a dissymmetry of the electrodes. The device is realised

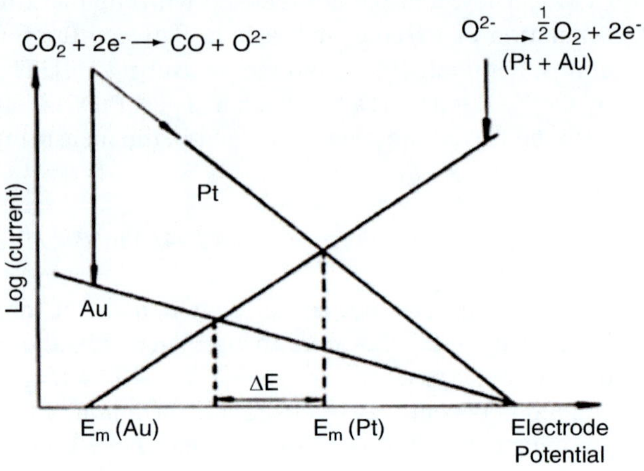

Fig. 1 Mixed potential proposed by Moseley

with electrolytes with mobile ion other than oxygen (β-Alumina for example) and with electrodes made with different sizes and/or materials (Pt and Au for example). This device is sensitive to CO or H_2 in air. P. Moseley has later introduced the general term of "Non-Nernstian" potentiometric sensors [25]. He included two types of dissymmetric devices: on the one hand Seebeck sensors using two different metallic electrodes on a semi-conductor material, and on the other hand mixed potential sensors realised with Au/YSZ/Pt cell (Fig. 1). The EMF depends on the kinetics of the possible reactions at each electrode and is a result of non-equilibrium conditions. At high temperature, equilibrium can be reached and the mixed potential disappears. On the contrary, at low temperature, the mixed potential may be limited by kinetics of reaction or diffusion.

The mixed potential was described by Velasco and Moseley [26] with more details later in 1993. Mixed potentials arise at electrodes where at least two reactions are available simultaneously. At open circuit, the sum of the cathodic currents must be equal to the sum of the anodic currents. As the catalytic activity of the electrodes becomes primordial for such processes, these authors propose to use different semi-conductive oxides as electrodes. At the same period, a screening of different materials was also proposed for non-Nernstian potentiometric zirconia sensors, but with conventional devices working with two separated cells [6].

Our group at St-Etienne has also proposed at this period, similar sensors based on the difference of catalytic activity of Pt and Au electrodes deposited on thick films of β-alumina [27, 28]. The devices are used for the detection of CO with only one gas cell and after a special treatment with SO_2.

Before to come back later on our own development of sensors, it appears interesting a make a short review of the most significant research centres who have proposed at this period a high activity on such types of non-Nernstian sensors.

The team of Yamazoe and Miura at Kyushu University has proposed a lot of publications on electrochemical sensors. They have first studied in 1993 conventional tubular YSZ sensors, but with sulphate as secondary ionic phase under the external electrode. These sensors are used for SO_2 and the responses are in complete agreement with the Nernst law [14]. The same types of sensors are then proposed in 1995 with carbonate phase between YSZ and porous gold electrodes for CO_2 detection. It is always Nernstian sensors, but planar devices appear with the two electrodes placed in the same atmosphere [29]. The notion of mixed potential appears only in 1996 with YSZ sensor for H_2S detection using WO_3 electrodes. The term of "Sensing electrodes" is then proposed. After this date, Yamazoe and Miura have published many papers on mixed potential sensors with a lot of designs for the sensing electrodes in terms of materials and also arrangements with porous metals above oxides: combination of potentiometric and amperometric type for NOx sensors [30], use of porous catalytic filters for H_2 or HC detections [31] or various mixtures of oxides for N_2O sensors [32]. An important effort was then devoted to NOx detection, with the use of spinel-type oxide for the sensing electrode [33] or more recently with thin NiO electrodes [34]. Several interesting explanations are proposed for the relation between mixed potential and catalytic activities of electrodes, including, for example, the differences of electrochemical and non-electrochemical reactions with temperature and materials. Nevertheless, there is no theoretical complete model, which can be used to calculate the values of the mixed potentials.

The team of Traversa and Di Bartolomeo at Roma has published in 2000 a study on a NOx sensor using Nasicon as electrolyte and various perovskite-type oxides as electrodes. These oxides are placed inside the Nasicon tube for the reference electrode with the goal to work with a single gas cell [35], and planar structures have been also proposed earlier [22]. It is very interesting to note that these authors found that the gas-sensing mechanism of these sensors cannot be always explained by mixed potential theory. When semi-conductors are used as electrodes, the semi-conductor sensing mechanisms dominate the sensor response [36]. They have then published several articles on planar non-Nernstian electrochemical sensors, especially focused on exhaust applications [37], and also using Nb_2O_5 electrodes [38, 39].

The team of Wachsman at University of Florida presents near 2000 his first studies also on NOx sensors using the notion of "differential electrode equilibria" and based on catalytic selectivity of two electrodes placed in a single cell containing mixtures of NOx and O_2 [7, 9]. The more recent publications are focused on the use of several oxides for the sensing electrode as WO_3 [20], Cr_2O_3 or SnO_2 [19], Nb_2O_5 [38, 39], and also various metals for the so-called "counter" electrode in Metal/YSZ/WO_3 devices [21]. In the major part of these publications, the catalytic activity of materials is always studied with a very fine attention, but nevertheless without models directly connected to the EMF values.

The team at Los Alamos (Brosha, Mukundan, Garzon) has also published his first paper in 2000 on mixed potential sensors with a dissymmetry of electrodes in single cells as Au/YSZ/Oxides [40, 41]. They have early pointed out that the gold electrode is responsible of long-term evolution of the sensors [42, 43]. The applications are always focused on the exhaust gases with the objective to detect selectively HC/CO with spinels materials for electrodes ($NiCr_2O_4$ or $ZnFe_2O_4$) and, on the other hand, NOx gases with $LaSrCrO_3$ materials for electrodes [44].

The team of Dutta at Columbus published a study in 2002 with planar YSZ sensors for NO detection. They use two Pt electrodes, but the dissymmetry is obtained by the use of a porous aluminosilicate zeolite film over one Pt electrode. The zeolite promotes the NO/NO_2 equilibrium at 500°C [4]. They have also published several studies using new electrode materials as cobaltite or chromites [5] always focused on CO–HC and NOx sensors for diesel engine exhaust [45]. More recently, they proposed planar sensors with different designs using WO3/YSZ/Pt–Y with good results for NOx detection [16, 17].

The team of Hibino at Nagoya has also worked from 1998 with non-Nernstian devices as Au/YSZ/Pt [46]. For HC detection, he used proton conductor-containing platinum electrodes [47]. After a long period working on single-chamber solid oxide fuel cells [18], he comes back to NOx sensors with special electrode materials, which have been developed for SOFC [48].

It is particularly interesting to underline that the teams described above present several similarities:

- The major part of the groups started their studies on such types of sensors at the same period (near 2000)
- They generally use the terms of "Mixed potential" or "Non-Nernstian potential"
- All of them have worked on exhaust gas sensors, and more specifically on NOx sensors
- Many of them are involved in fuel cell development with studies on materials for electrolytes and electrodes
- There are few tentative of models to explain the mixed potential, but never possibility to calculate EMF values in good agreement with the sensor responses

1.2 Research at St-Etienne and Objectives of the Chapter

Concerning our own activity, as exposed previously we have studied from 1990 the development of a planar potentiometric sensor Au/β-Alumina/Pt without reference chamber [27, 49, 50]. We have focused early our attention on CO/NOx sensors for exhaust applications [51]. On the contrary of studies exposed previously, we observed with β-alumina EMF depending on oxygen partial pressure, although we are in a single cell configuration. A capacitive model has been proposed to explain these responses [52]. For this model, several studies have been realised to understand the differences due to the Au or Pt electrodes: calorimetric study [53], influence of electrode sizes [54], or work function measurements [55]. On the basis of these results, we have improved the capacitive model [56].

During the same period (from 2000), we have also worked on SC-SOFC (Single-Chamber Solid Oxide Fuel Cell) with devices, which are very closed to our Non-Nernstian sensors, but replacing the β-alumina electrolyte first by YSZ and now by

CGO or SDC. The objective is to use planar Anode/Electrolyte/Cathode device in a single cell without physical separation between anode and cathode. These developments are mainly based on the difference of catalytic activity of the two electrodes in regards of a mixture HC – Air [3], and in this sense, these studies are very close to those of Non-Nernstian sensors. It is interesting also to notice that we have a similar demarche as Hibino: from sensors to SOFC, and coming back to sensors using SOFC electrolytes and electrodes materials as illustrated by a recent publication with LSM electrode, which is a well-known cathode material in SOFC [57].

The objective of the present article is to make a review of our studies on potentiometric sensors, on the one hand with experimental results especially in relation to the field of exhaust gas sensors, and on the other hand with a description of our capacitive model.

2 Experimental

2.1 Sensors Preparation

As exposed previously, we have worked on the development of potentiometric sensors for automotive applications using a simple device. For this, the two electrodes are deposited on an electrolyte without a physical separation between the electrodes. We have first used β-alumina as electrolyte, and later YSZ. The results reported in this chapter have been obtained with Au/β-alumina/Pt devices and with Au/YSZ/Pt devices.

To satisfy the industrial requirements, it was decided to produce sensors through thick-film technology. Screen-printing technology was used to develop robust prototypes on alumina substrates. The sensing elements having a size of $6 \times 6\ mm^2$ are deposited on α-alumina substrates ($0.85 \times 5.1\ cm^2$), which include a platinum heater on the opposite side. Gold and platinum electrodes ($2 \times 6\ mm^2$) are then screen-printed on the solid electrolyte (Fig. 1).

For β-alumina-based device, a special development of the screen-printed ink consisting in a mixture of Na-β-alumina and Na-aluminosilicate glass was performed in collaboration with Politecnico di Torino. The thick film based on β-alumina is annealed at 900°C. More details of this process can be found in [49].

YSZ sensors are prepared in similar conditions. The only difference from β-alumina prototypes is the electrolyte: YSZ ink is prepared from a commercial YSZ powder and organic binders. No other mineral binders are added. The ink is annealed at 1,380°C for 2 h.

Typical sensor is presented in Fig. 2. Protective dielectric layers are deposited on the platinum heater and its electrical connexions (side A) and on the electrical connexions of the gold and platinum electrodes (side B).

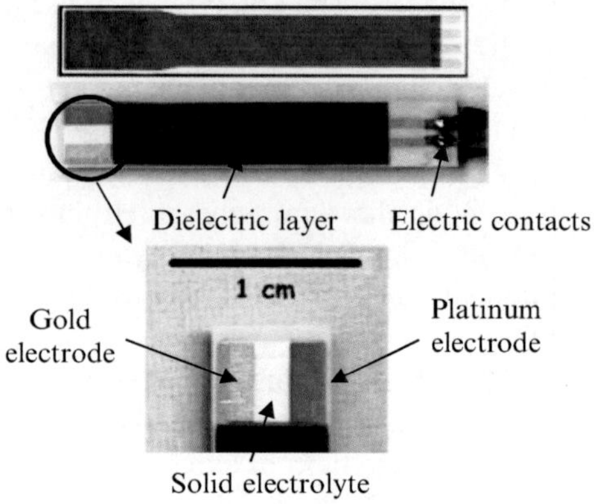

Fig. 2 Photograph of the sensor: (**a**) heating side, (**b**) sensing side with detail of the sensing element

2.2 *Sensor Bench Test*

A laboratory test bench has been developed to test up four sensors simultaneously with controlled gas injections and data acquisition. The automation (computer controlled electro valves and mass flow controllers) allows generating various gases (CO, HC, NO_2, NH_3) alone or mixed with various concentrations in the range 0–1,000 ppm. HC represents a hydrocarbon mixture of 25% methane, 50% propene and 25% butane (volume composition). Oxygen and water vapour concentrations can also be varied but were fixed, respectively, at 12% and 2% vol. in this study. Gas flows were fixed at 5 l/h and the gases were not preheated.

The monitored signal ΔV (mV) is the potential difference between platinum and gold electrode. The sensor response is defined as the difference between the signal under gas and under air ($\Delta V_{gas} - \Delta V_{air}$).

3 Results with β-Alumina Sensors

3.1 *Typical Sensor Responses to Oxidant and Reducing Gases on Laboratory Bench*

The β-alumina sensors have been tested under oxidising and reducing gases, for example with CO and NO_2 in air. As observed with many gas sensors, the responses are in opposite sense between oxidising and reducing gases. Moreover, the intensities of the responses depend on the temperature: the highest sensitivities are obtained at high temperatures for CO and at lower temperatures for NO_2, respectively, 530°C for CO and 330°C for NO_2 for the experiments reported on Fig. 3. On the contrary, we can see on this set of experiments that the sensitivities are very low, close to zero, for CO at 330°C and for NO_2 at 530°C.

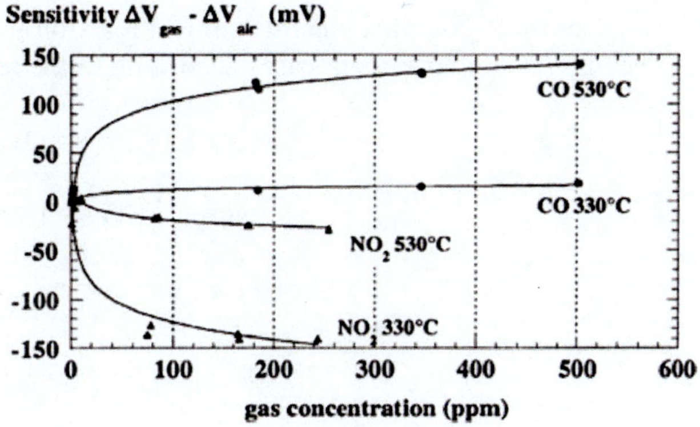

Fig. 3 β-alumina sensor responses versus CO and NO_2 concentrations at 330 and 530°C

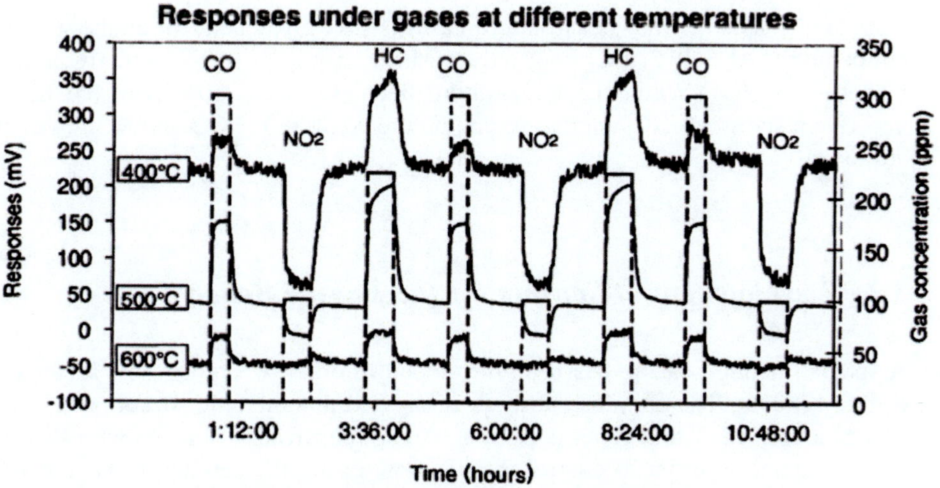

Fig. 4 β-alumina sensor signals on laboratory bench ($O_2 = 20\%$) versus time to cyclic injections of CO (300 ppm), NO_2 (100 ppm) and HC (225 ppm) at 400, 500 and 600°C

Typical responses of β-alumina sensors on laboratory test bench at three temperatures (400°C, 500°C and 600°C) are shown in Fig. 4 ($O_2 = 20\%$). For these experiments, successive introductions of CO (300 ppm), NO_2 (100 ppm) and HC (300 ppm) are repeated during several cycles. A 1-h step under pure air is always realised between each gas injection.

Reducing gases such as CO and HC lead to a positive response compared to air [increase of $(V_{Pt} - V_{Au})$], whereas oxidant gases such as NO_2 decrease the difference of potential. This behaviour is in agreement with the oxido-reduction reactions occurring between the gases and the electrodes, leading to either a production or consumption of chemisorbed oxygen species if we consider simple chemical reactions [reactions (1)–(2)], or a production and consumption of O^{2-} ions if we consider more complex electrochemical reaction [reactions (3)–(4)]. The resulting response is similar in both cases according to the reaction (5),

$$CO + O_{ads} \rightarrow CO_2 \quad (1)$$

$$NO_2 \rightarrow NO + O_{ads} \quad (2)$$

$$CO + O^{2-} \rightarrow CO_2 + 2e^- \quad (3)$$

$$NO_2 + 2e^- \rightarrow NO + O^2 \quad (4)$$

$$O_{ads} + 2e^- \rightarrow O^{2-} \quad (5)$$

It can be seen that the sensitivity to gases is strongly dependent on the working temperature. At 600°C, there are only positive responses to CO and HC and no response to NO_2. When the temperature decreases from 500°C to 400°C, the negative response to NO_2 increases, and on the contrary the CO and HC responses are lower.

3.2 *Development of Sensors for the Engine Bench Tests*

A special metallic casing has been developed in order to evaluate the sensors on engine benches. The alumina substrate is inserted in a metallic socket (part left of Fig. 5) and fixed with a ceramic cement. A cap (part right of Fig. 5) with slits has been designed in order to control the gas flow inside the cap and to have consequently a good control of the temperature of the sensor tip. The cap is fixed on the socket with a threading and another threading allows fixing the sensor on automotive exhausts.

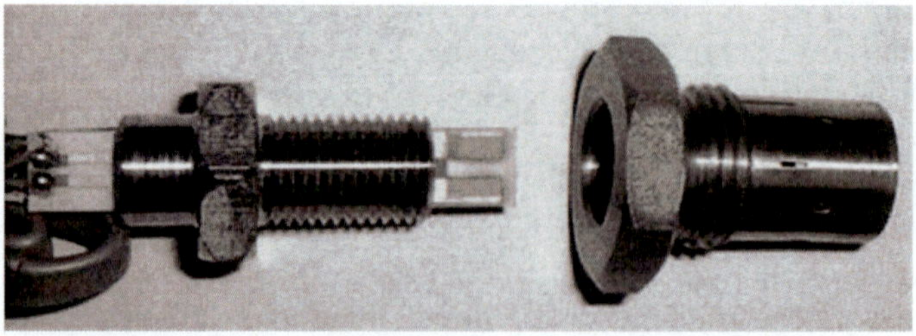

Fig. 5 Photograph of the sensor prototype with its casing (socket and cap)

3.3 Protective Layer for the Sensing Element

In order to perform the tests in automotive exhausts, it was also necessary to protect the sensing element from soot deposition, which may short circuit the electrodes. We can see on Fig. 6 top views of sensors before and after working in an exhaust bench. The black colour on the right photograph is due to the soot deposition after only few minutes of test. Consequently to soot deposition, problems of electrical short circuits appear with normal sensors. To avoid these problems, it was necessary to add a porous protective layer on the sensing element.

Independently of the soot deposition problem, some degradation of the sensors performances was observed in relation to the electrodes. The degradation is due to irreversible deposition of chemical species according to the working conditions. For example in experiments with HC with sensors at medium temperature, we can observe (Fig. 7) the deposition of carbon species on the part of the platinum electrode close to the electrical connexion. These depositions are responsible of long-term drifts, which must be absolutely avoided in automotive applications.

To solve the problem exposed above, a porous α-alumina layer was developed in collaboration with Politecnico of Torino [58]. A new screen-printed ink formulation was developed. The protective layer was annealed at 700°C after deposition. The thick film was made with a mixture of α-alumina and of a transition alumina

Fig. 6 Top view of sensors before (*left*) and after (*right*) tests in exhausts

Fig. 7 Electrode degradation during sensor ageing: carbon deposition on platinum electrode

(γ-type) issued from a gel containing an alumina precursor. The specific surface area of the γ-alumina, evaluated separately by B.E.T. analysis, was about 180 m^2/g. A well-distributed, homogeneous porosity (about 30–40% by volume, as evaluated by image analyses) is present in the thick film (Fig. 8).

Such protected sensors were evaluated during 150 h on laboratory bench and 30 h on engine bench without any degradation of their performances [58]. An example of a long duration test with protected sensors on laboratory test bench is reported on Fig. 9. Four sensors with a porous protective layer are tested during cycles with alternative injections of CO (300 ppm) and NO_2 (100 ppm). We can notice that the porous layer does not affect the behaviour of the sensors, and moreover their long-term stability is improved.

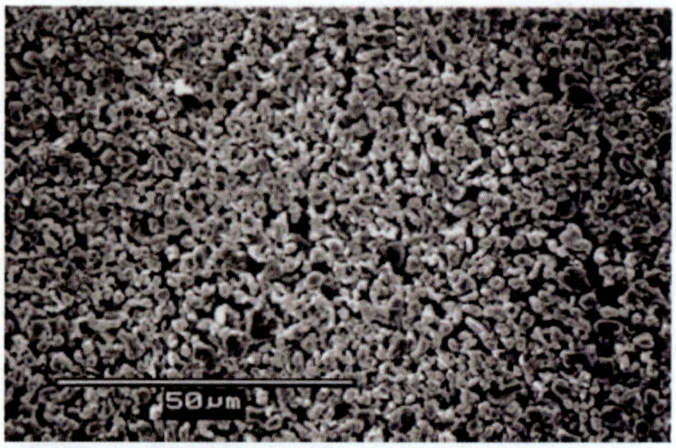

Fig. 8 SEM images of the alumina porous protective layer

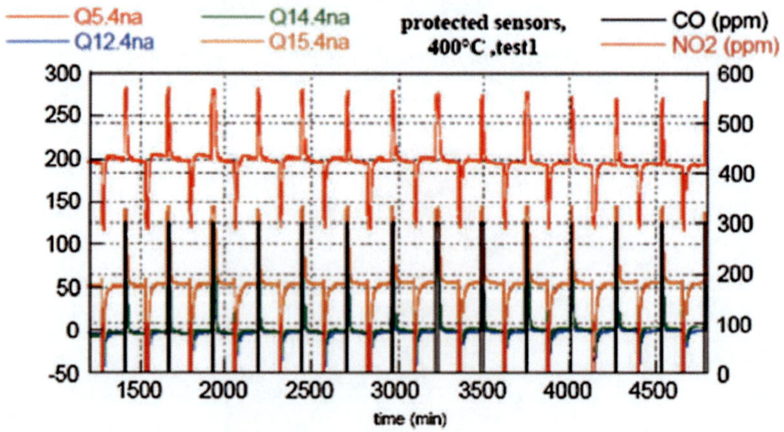

Fig. 9 Stability test on laboratory bench, at 400°C, with cyclic injections of CO (300 ppm) and NO_2 (100 ppm)

3.4 Tests of β-Alumina Sensors on Motor Bench

The β-alumina sensors equipped with their protective layer and metallic casing have been evaluated on engine benches for different applications. The sensors have been directly placed on exhaust pipes on gasoline or diesel engines (Fig. 10). We reported below typical results obtained in such experiments.

The first test reported here was performed on a diesel exhaust with a sensor placed at the outlet of a NO_x trap in order to monitor this device. The working temperature of the sensor has been fixed at a low value near 350°C in order to be more selective for NOx detection. A NO_x analyser was used for comparison. Figure 11 shows that the sensor signal suddenly changes when the engine richness is changed from rich to poor. From this point, NOx gases are produced. The NOx concentration is controlled with the gas analyser. The concentration increases from 0 to 600 ppm after 300 s. The sensor placed in parallel with the analyser also gives a response, with a normal decreasing voltage due to the oxidant gas. This response is in full agreement with the NOx analyser. Consequently, the sensors can be used to detect the point when the trap is full and begins to release NOx gases, and so for the management of NO_x Trap regeneration.

The second example is always in relation with diesel engine, but with the aim to control HC before and after the Light-Off catalyst. The working temperature of the sensor has been fixed in order to detect mainly HC than NOx (high temperature near 600°C). The experiments have been realised during a standard step representative of a start phase of urban cycle. Four signals are reported on Fig. 12: speed (RPM), NOx and HC concentrations (from gas analysers) and the sensors response. We can notice a very good agreement between the HC concentration and the signal delivered by the sensor. In consequence, the use of two sensors placed up stream and down stream the catalyst can allow the optimisation of the working conditions of this catalyst.

Some β-alumina sensors have also been used for NH_3 detection. Ammonia detection is now of prime importance with heavy duty diesel engine for SCR application. The SCR is necessary to reduce the NOx concentration according to the new depollution regulations. The SCR is based on the reduction of the NOx

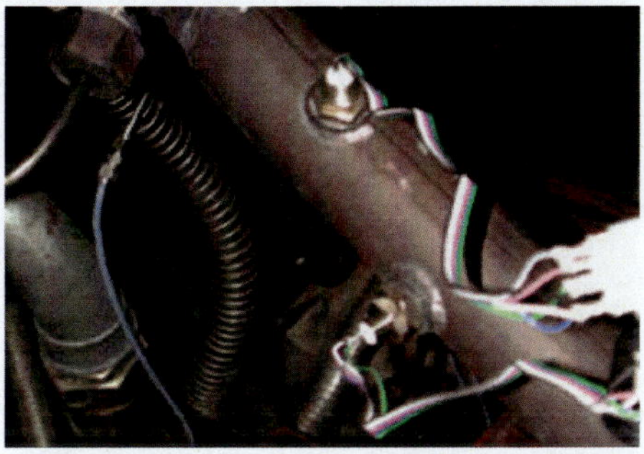

Fig. 10 Exhaust pipe with two plugged sensors on engine test bench

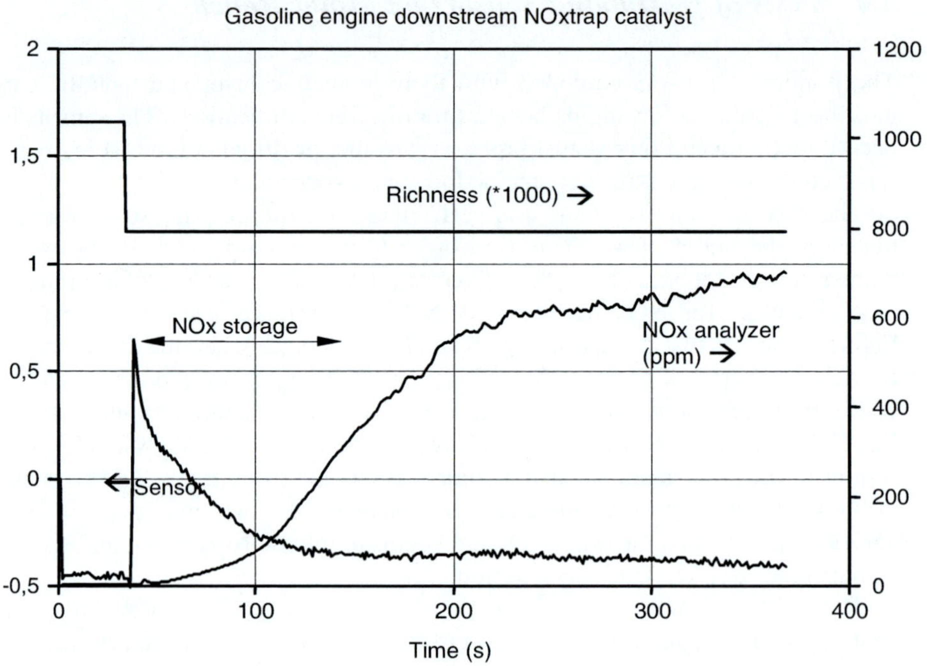

Fig. 11 Comparison of β-alumina sensor and analyser responses to NO_x at the outlet of a NO_x trap

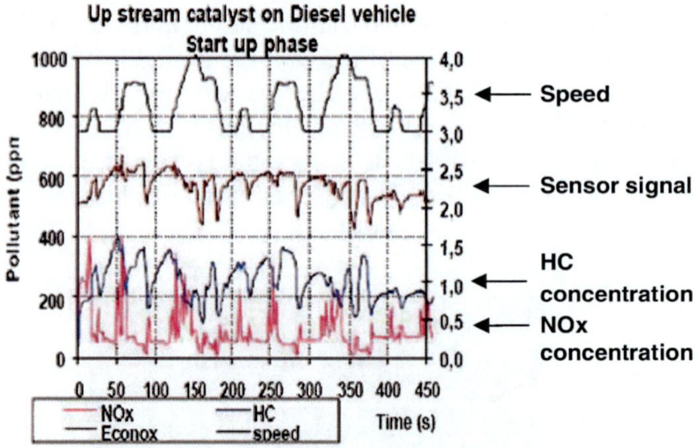

Fig. 12 Sensor response during the start-up of urban driving cycle of a diesel vehicle

gases using urea and with a production of NH_3 in case of urea excess. Consequently, it is now of prime importance to have NH_3 sensors for the control of SCR devices. For these experiments, we have used a β-alumina sensor with working conditions close to those generally used for NOx detection. The sensor was placed at the outlet of a heavy duty diesel engine in parallel with a NH_3 analyser. Figure 13 shows the comparison of the analyser signal and the sensor response to ammonia at one load point for the engine. A good correlation is obtained between both signals allowing the use of the sensor for such SCR control. The map reported on Fig. 14 is the correlation between the sensor signal and the real NH_3 concentration. Even if we do not get a straight line, this result appears quite interesting for future exploitation on SCR systems.

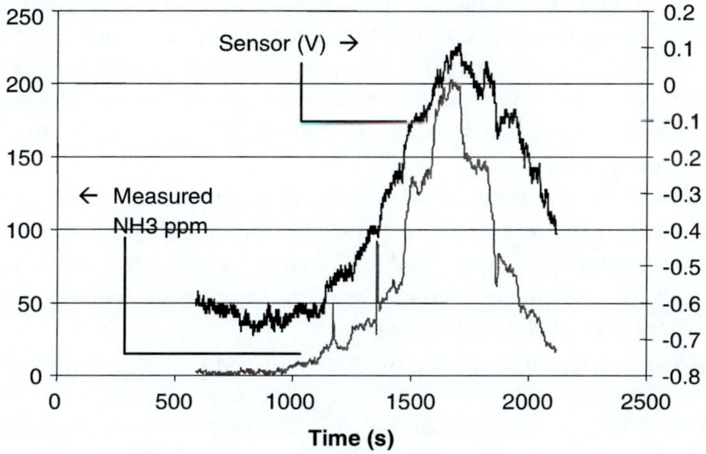

Fig. 13 Comparison of alumina sensor and analyser responses to ammonia at one load point for HD diesel engine

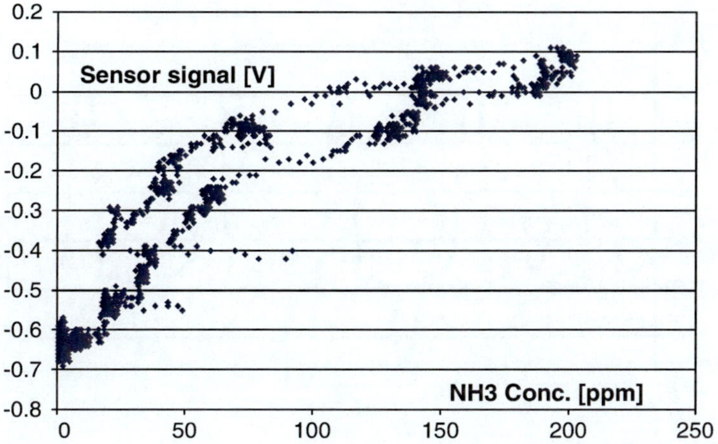

Fig. 14 Sensor signal of Fig. 12, re-plotted as a function of ammonia concentration

4 Comparison of β-Alumina and YSZ Sensor Performances

4.1 General Behaviour

As it has been described in the introduction paragraph, and consequently to the oxygen Lambda sensors development, the majority of the potentiometric sensors studied for the exhaust applications use YSZ as electrolyte material. In our case, we have first developed our exhaust sensors on the base of β-alumina. The major difference is related to the nature of the mobile ion, O^{2-} with YSZ, and Na^+ with β-alumina. Taking into account this difference, we have studied a capacitive model in order to explain the majority of our experimental results. This model is presented later in Sect. 6. Nevertheless, in order to have a more clear view on the phenomena occurring on non-Nernstian sensors, we have decided to make comparisons of both types of electrolytes.

We present in this part a comparison of experimental results obtained with sensors, which differ only by the nature of the electrolyte, on the one hand β-alumina, and on the other hand YSZ. The two devices are Au/β-alumina/Pt and Au/YSZ/Pt. For this part of the study, sensors were prepared without protective layer and tested only on the laboratory bench.

The general behaviours of the two types of sensors are very close one to each other. The responses of two YSZ sensors for successive injections of 300 ppm CO and 100 ppm NO_2 are reported on Fig. 15. The responses are in opposite sense for CO and NO_2 in complete agreement with β-alumina sensors.

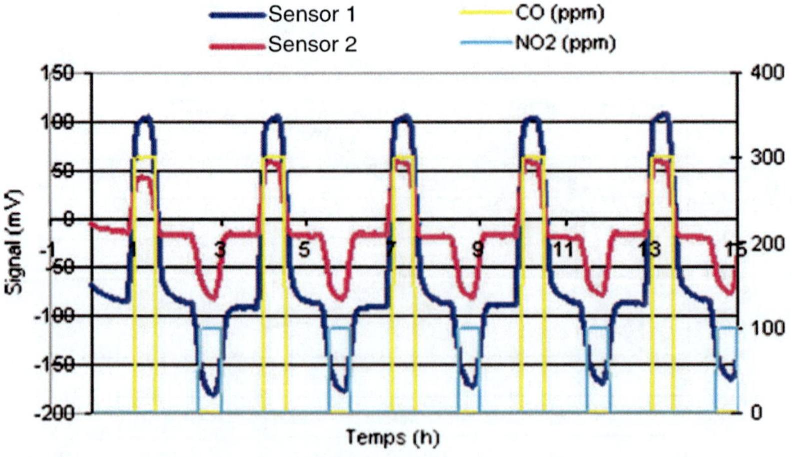

Fig. 15 Dynamic responses of two Pt/YSZ/Au sensors (sensors 1 and 2 varying by the conditions of YSZ preparation) to CO (300 ppm) and NO_2 (100 ppm) at 500°C

4.2 CO, HC and NO_2 Performances

Responses to increasing CO, HC and NO_2 concentrations at 500°C for both types of sensors are shown in Fig. 16. A similar behaviour is observed for reducing gases CO and HC. The responses continuously increase versus gas concentration, with a high rate in the range 0–100 ppm. For NO_2, the response is in opposite sign compared to reducing gases, with similar absolute value, around 20 mV for gas concentration of 100 ppm. This behaviour is coherent with past results obtained with β-alumina sensors (Fig. 5). These results demonstrate that the general behaviours of the two types of sensors are very similar.

The influence on the temperature on the two types of sensors has been evaluated under the reducing and oxidising gases. The comparison for CO is shown in Fig. 17 for a fixed concentration of 600 ppm CO. Results obtained for HC, not reported here, and are similar to those of CO. A maximum of sensitivity is observed at 400°C for both types of sensors. For higher temperatures, the response strongly decreases and is close to zero at 600°C. For NO_2, the behaviour is slightly different between the two sensors. For YSZ sensors, the response is also negative in all studied temperature range as observed with β-alumina sensors. The response is generally maximum (absolute value) at lower temperature, and it continuously decreases with increasing temperature. Nevertheless, the intensities of the responses and the temperature dependency can be different from one sensor to the other one. It seems that the reactivity with NO_2 is strongly affected by the material microstructure. Due to possible NO–NO_2 conversion, the results are also influenced by experimental conditions such as gas flows and temperature. We will come back later on the difficulty of NOx experiments.

4.3 NH_3 Responses

As expressed previously, a strong demand concerns also ammonia detection for vehicles using SCR system. The responses to this gas at low concentrations 5, 10, 20 and 50 ppm and at 500°C are reported in Fig. 18 for both studied sensors. A weak response is observed in the range 0–10 ppm. Then the response increases significantly to reach around 50 mV for 50 ppm.

The interference with 100 ppm NO_2 for the same NH_3 concentrations and same temperature (500°C) was studied and the dynamic signal of the sensor is shown on Fig. 19 in the case of YSZ sensors. NH_3 and NO_2 have been simultaneously injected. For YSZ sensors, NH_3 concentration is increased while maintaining a fixed NO_2 concentration without intermediate air injection. It can be seen on Fig. 19 that the response of YSZ sensors remains negative, thus corresponding to NO_2 predominance, for NH_3 injections of 5, 10 and 20 ppm. However, with 50 ppm of NH_3, the signal increases, indicating a predominant effect of ammonia. With β-alumina sensors (not reported here), we have observed in this set of experiments that the response remains negative compared to sensor baseline whatever the ammonia concentration is. It seems that the cross sensitivity between NO_2 and

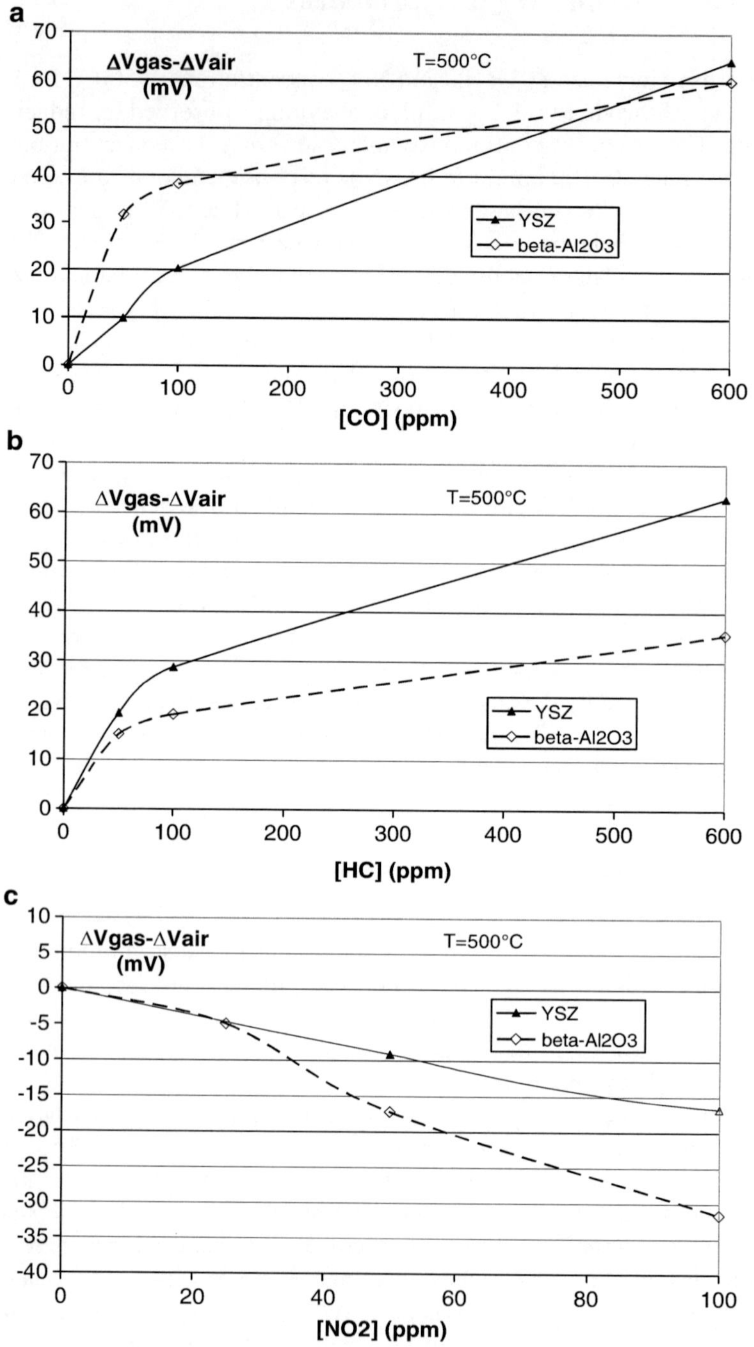

Fig. 16 Comparison of YSZ and β-alumina sensor responses at 500°C: versus (**a**) CO, (**b**) HC, (**c**) NO_2 concentrations

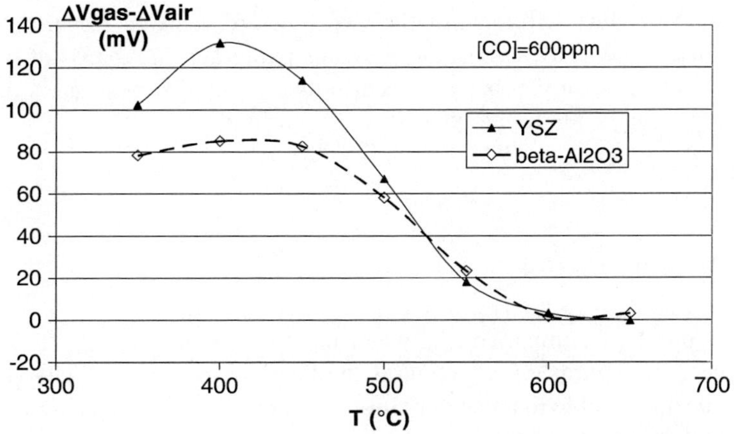

Fig. 17 Comparison of YSZ and β-alumina sensor responses to CO (600 ppm) versus temperature

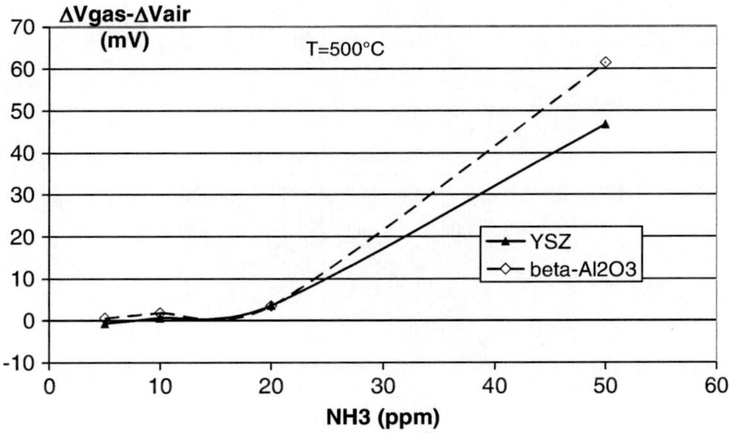

Fig. 18 Comparison of YSZ and β−alumina sensor responses to ammonia at 500°C

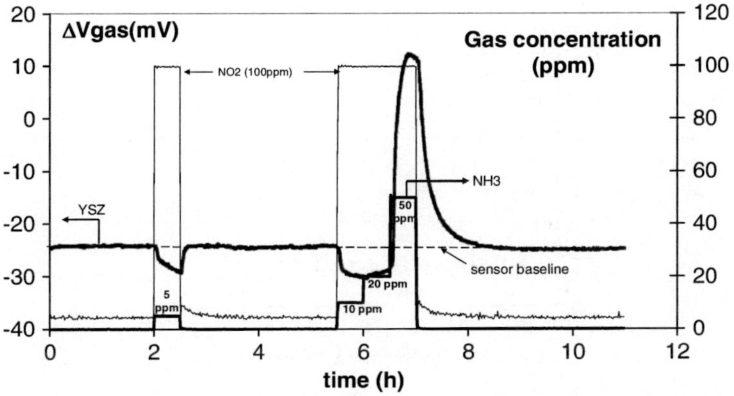

Fig. 19 Interference effect of NO_2 (100 ppm) on ammonia response at 500°C for YSZ sensor

NH_3 could be thus different for the two types of sensors. Nevertheless, we can express the same remark as in the previous paragraph in regards of the important difficulties to control the exact concentrations of the different gases in the case of NOx experiments due to the NO–NO_2 equilibrium.

4.4 Sensor Ageing

Another point of comparison between the two studied devices is their thermal resistance, in regards of a potential application in exhaust. The performances have to be quite stable in harsh conditions. A simplified test in laboratory conditions has been performed to estimate sensitivity degradation due to thermal ageing of the materials. The sensors have been submitted to a rapid thermal cycling between 200°C and 600°C with a period of 2 min (60 times, Fig. 20). An injection of 600 ppm of CO at 500°C was performed after each cycle in order to measure its evolution. The results are reported on Fig. 21. They indicate that no major

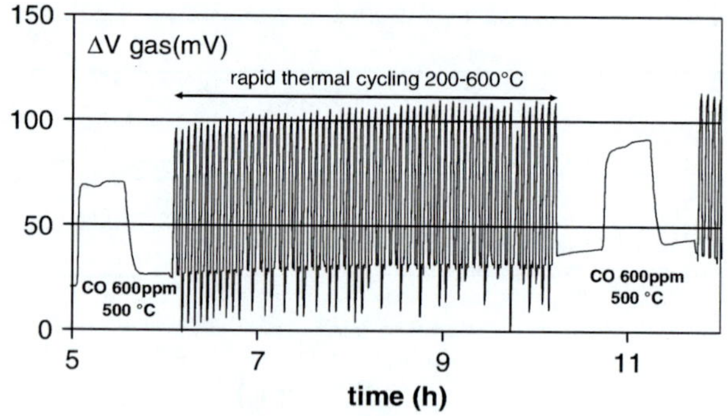

Fig. 20 Sensor signal during thermal cycling 200–600°C at a period of 2 min, and reference CO injections (600 ppm, 500°C)

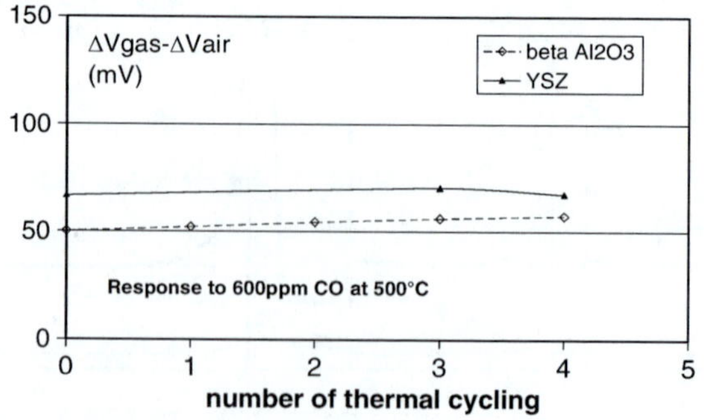

Fig. 21 Sensor response to CO 600 ppm at 500°C after cycles of thermal ageing 200–600°C at a period of 2 min (shown in Fig. 20)

degradation is observed after 4 cycles, for both sensors. This conclusion is important in regard of the question of the stability of β-alumina. It is generally admitted that YSZ is a very good electrolyte due to its very good stability, contrary to β-alumina which is often presented as not very stable at high temperature. Our experiments seem to demonstrate that both electrolytes are very similar in terms of stability.

5 Prospective Studies

5.1 Electrodes Materials

In our previous studies, the electrolyte, either β-alumina or YSZ, was associated with Au and Pt electrodes. However, the principle of this sensor is based on the difference of reactivity of the electrodes towards the target gas. Hence, many couples of electrodes can be used. For example, we obtained interesting results with Au–YSZ–LSM devices (LSM: $La_{0.8}$ $Sr_{0.2}$ MnO_3), which present similar behaviour as Au–YSZ–Pt sensors (Fig. 22 from [57]).

In particular, many oxides or mixed-metal oxides sensing electrodes have been investigated especially for NO_x detection by Miura [1, 30, 59, 60]. The adequate choice of the electrode material allows modifying the sensor selectivity.

5.2 Amperometric Mode: Electrode Polarisation

Although most of the experiments are conducted in potentiometric mode (measurement of open circuit voltage between electrodes), it is also interesting to consider amperometric mode, in particular polarisation curves I–V. Such results obtained in

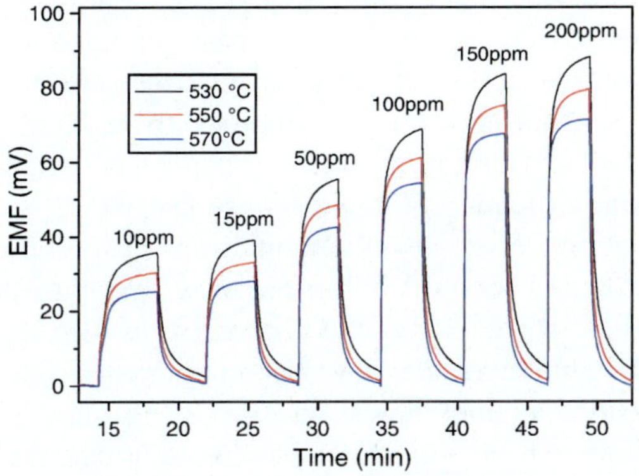

Fig. 22 Dynamic responses of Au–YSZ–LSM sensor to CO

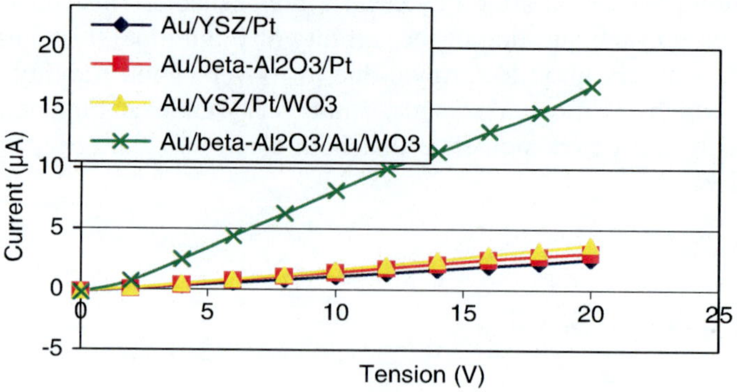

Fig. 23 Polarisation curves *I–V* for various sensors devices, during exposure to 300 ppm CO at 600°C

our laboratory are reported in Fig. 23 for various sensors based on YSZ or β-alumina and varying electrodes couples. In agreement with results published by the group of Traversa and Di Bartolomeo [35], it may also be interesting to couple metallic electrodes with semiconducting oxides to modify significantly the response characteristic, and thus the selectivity.

Another interesting point concerning current–voltage characteristics is the possibility to use polarisation to modify the sensor sensitivity, and thus to improve selectivity. This method was applied to differentiate NO and NO_2, either by applying a bias current with a two electrodes device [11], or by polarisation with a third electrodes [30].

5.3 *Catalytic Filter for Selectivity Modification*

Another way to improve selectivity is the addition of catalytic filters, which can be either added as a filtering layer on the sensing element, or integrated as a separate filter in the sensor packaging (cap). In the past, we have used both methods with various types of sensors (potentiometric or resistive sensors): zeolite filters placed upstream to the sensors showed potentialities to remove alcohol response [2]; rhodium filter deposited directing on the sensing layers allowed to remove NO_2 response without modification of CO response [50, 61].

For NOx detection, recent developments were proposed by the group of Dutta [4–16], where Pt-loaded zeolite Y filters are used upstream YSZ sensors to remove CO and hydrocarbon interferences, and to give a total NOx signal, independently on the ratio NO/NO_2. Similar results have been obtained in our group with platinum-based filters deposited as thick layers on Au/YSZ/Pt sensors (Fig. 24): the sensors with a platinum filter exhibit almost no response to 300 ppm CO and present similar signal to NO and NO_2. This result is explained by the catalytic conversion of CO into CO_2, and the activation of the equilibrium NO/NO_2 by the catalyst: for a fixed

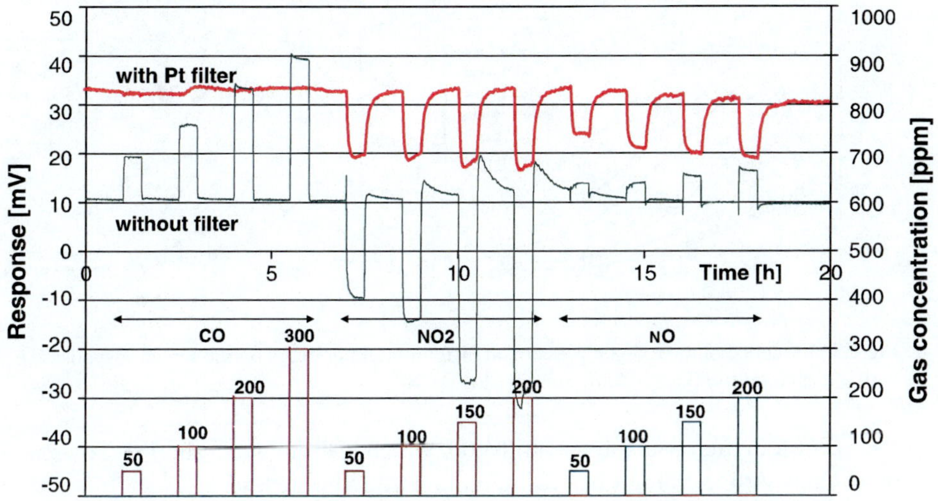

Fig. 24 Effect of a platinum filter on the response of a Au–YSZ–Pt sensor to CO, NO, NO_2 at 450°C

temperature, whatever is the initial NO or NO_2 concentration, the NO/NO_2 equilibrium is reached at the sensor surface and the response corresponds to total NOx concentration.

6 Discussion and Modelling

6.1 Limitation of the Mixed Potential Model

It has been exposed in the introduction paragraph that now many groups of research work on potentiometric sensors with a dissymmetry of electrodes in single cell. We have also demonstrated in the second paragraph that these types of devices can be really used for exhaust sensors with interesting performances. Despite all the interest generated by these new concepts of detection, modelling of the phenomenon does not appear to have been addressed in a systematic way. Most often, the authors just mentioned the existence of a mixed potential, which implies by definition the existence of at least two redox reactions on each electrode. Nevertheless, mixed potential is not sufficient to explain all experimental results and more justifications have to be brought.

The first remark is that some of these devices are particularly sensitive to the action of oxygen alone. The results reported on Fig. 25 reflect the influence of the partial pressure of oxygen and temperature on a dissymmetric sensor using β-alumina as electrolyte (Au/β-alumina/Pt) [24, 25, 27–34, 46–48, 50–52, 56, 57,

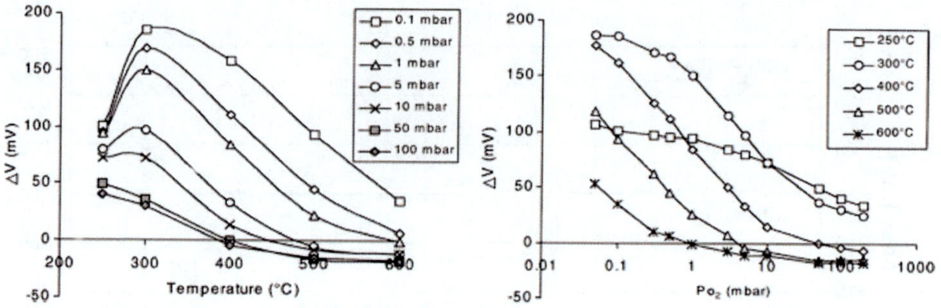

Fig. 25 Influence of the oxygen pressure at different temperatures on the sensor responses (EMF) with a dissymmetric device Au/β-alumina/Pt

59, 61–68]. In this case, the phenomenon, which is limited to a single redox reaction on each electrode, cannot be attributed to a mixed potential.

The second point concerns the expression of mixed potential. This expression is based on the Nernst law applied to the two reactions. But in this case, the reactions are not at equilibrium, and it is important to consider a kinetic process. This naturally implies a good knowledge of the reaction process that will allow to take into account possible intermediate species and to write new elementary steps. In general, these systems are not easy to solve and the solution will depend on the assumptions made about the shifting of equilibriums, and of limiting steps. In most cases, it is the assumption of the steady state, which is predominant.

To illustrate this purpose, we will express the mixed potential for the catalytic oxidation of carbon monoxide by oxygen. The approach consists to use Rideal's hypothesis. Rideal's hypothesis, which supposes the presence of interactions between the adsorbed oxygen and the carbon monoxide, leads to:

$$O_2 + 2e^- + 2s \underset{k_{1Ox}}{\overset{k_{1Red}}{\rightleftarrows}} 2O^- - s$$

and:

$$2CO + 2(O^- - s) \underset{k_{2Red}}{\overset{k_{2Ox}}{\rightleftarrows}} 2CO_2 + 2s + 2e^-$$

where "k_i" represents the kinetic rate constant of either reduction (k_{iRed}) or oxidation (k_{iOx}) reactions (i).

The global reaction is, in the present case, $CO + \frac{1}{2}O_2 \leftrightarrow CO_2$. In contrast to the global reaction, the intermediate reactions reveal a new species, namely adsorbed oxygen "O–s" on adsorption sites "s", whose concentration can be expressed from a degree of recovery θ.

Let us go back to the global system and consider separately the two reactions. They are both at the origin of an electric current characterised by the rate R of the reaction. R is proportional to the number of electron grammes "dn" extracted from the reducing agent during dt and per surface unit, therefore:

$$R = R_{\mathrm{Ox}} - R_{\mathrm{Red}} = \lambda \frac{\mathrm{d}n}{\mathrm{d}t},$$

where R_{Ox} and R_{Red} are, respectively, the oxidation and reduction rate, λ is the proportional coefficient. dn/dt is therefore proportional to the density of the current i, for which the following convention was adopted: the current is considered to be positive in the case of an oxidation, and negative in the case of a reduction, thus leading to:

$$i = q \frac{\mathrm{d}n}{\mathrm{d}t} = i_{\mathrm{Ox}} + i_{\mathrm{Red}}.$$

If we consider that the two reactions may be considered as an elementary step, the Van't Hoff law may be applied, and:

$$i_1 = q\lambda_1 \left(k_{1\mathrm{Ox}} \theta^2 - k_{1\mathrm{Red}} P_{\mathrm{O_2}} (1 - \theta)^2 \right),$$

$$i_2 = q\lambda_2 \left(k_{2\mathrm{Ox}} P_{\mathrm{CO}}^2 \theta^2 - k_{2\mathrm{Red}} P_{\mathrm{CO_2}}^2 (1 - \theta)^2 \right)$$

with $\lambda_1 = \lambda_2 = 2$ in this case. The kinetic constants $k_{i\mathrm{Ox}}$ and $k_{i\mathrm{Red}}$ can be written, respectively:

$$k_{i\mathrm{Ox}} = k_{i\mathrm{Ox}}^0 \exp\left(-\frac{\tilde{E}_{i\mathrm{Ox}}}{RT} \right)$$

and

$$k_{i\mathrm{Red}} = k_{i\mathrm{Red}}^0 \exp\left(-\frac{\tilde{E}_{i\mathrm{Red}}}{RT} \right)$$

where $\tilde{E}_{i\mathrm{Ox}}$ and $\tilde{E}_{i\mathrm{Red}}$ are activation energies for respective oxidation and reduction reactions. These reactions with charged species lead to a potential barrier, which displaces the redox equilibrium as a result of activation energies modification as represented on Fig. 26.

Thus:

$$\tilde{E}_{i\mathrm{Ox}} = E_{i\mathrm{Ox}} - \beta_i q V_{\mathrm{D}}^i$$

and

$$\tilde{E}_{i\mathrm{Red}} = E_{i\mathrm{Red}} + \alpha_i q V_{\mathrm{D}}^i.$$

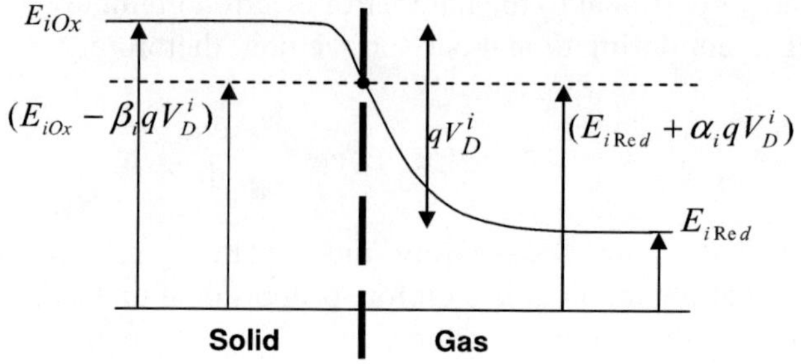

Fig. 26 Modification of redox activation energies E_i due to the creation of potential barrier (due to charged species) at the reactive interface gas/solid; α_i and β_i are, respectively, the reduction and oxidation transfer coefficient of potential ($\alpha + \beta = 1$)

It follows that:

$$k_{i\mathrm{Ox}} = k^0_{i\mathrm{Ox}} \exp\left(-\frac{(E_{i\mathrm{Ox}} - \beta_i qV^i_{\mathrm{D}})}{RT}\right)$$

and

$$k_{i\mathrm{Red}} = k^0_{i\mathrm{Red}} \exp\left(-\frac{(E_{i\mathrm{Red}} + \alpha_i qV^i_{\mathrm{D}})}{RT}\right)$$

which can be written:

$$k_{i\mathrm{Ox}} = k'_{i\mathrm{Ox}} \exp\frac{+\beta_i qV^i_{\mathrm{D}}}{RT}$$

with:

$$k'_{i\mathrm{Ox}} = k^0_{i\mathrm{Ox}} \exp\frac{-E_{i\mathrm{Ox}}}{RT}.$$

And

$$k_{i\mathrm{Red}} = k'_{i\mathrm{Red}} \exp\frac{-\alpha_i qV^i_{\mathrm{D}}}{RT}$$

with:

$$k'_{i\mathrm{Red}} = k^0_{i\mathrm{Red}} \exp\frac{-E_{i\mathrm{Red}}}{RT}.$$

The global current is then expressed as:

$$i_1 = q\lambda_1\left[k'_{1\mathrm{Ox}}\exp\left(\frac{\beta_1 qV_{\mathrm{D}}^1}{RT}\right)\theta^2 - k'_{1\mathrm{Red}}\exp\left(-\frac{\alpha_1 qV_{\mathrm{D}}^1}{RT}\right)P_{\mathrm{O}_2}(1-\theta)^2\right],$$

$$i_2 = q\lambda_2\left[k'_{2\mathrm{Ox}}\exp\left(+\frac{\beta_2 qV_{\mathrm{D}}^2}{RT}\right)P_{\mathrm{CO}}^2\theta^2 - k'_{2\mathrm{Red}}\exp\left(\frac{-\alpha_2 qV_{\mathrm{D}}^2}{RT}\right)P_{\mathrm{CO}_2}^2(1-\theta)^2\right].$$

If both reactions are taken into account, then the current i will represent the summation of both phenomena. However, it should be made clear that there is only one possibility for the electrical diffusion barrier potential and the profile of $V = f(x)$. It results in the same value V_{M} for the potential barrier as well as the α and β coefficients in both cases, so:

$\alpha = \alpha_1 = \alpha_2$ and $\beta = \beta_1 = \beta_2$

V_{M} is referred as the "mixed potential".

We can now express i as $i = i_1 + i_2$, where:

$$i_1 = q\lambda_1\left[k'_{1\mathrm{Ox}}\exp\left(\frac{\beta qV_{\mathrm{M}}}{RT}\right)\theta^2 - k'_{1\mathrm{Red}}\exp\left(-\frac{\alpha qV_{\mathrm{M}}}{RT}\right)P_{\mathrm{O}_2}(1-\theta)^2\right],$$

$$i_2 = q\lambda_2\left[k'_{2\mathrm{Ox}}\exp\left(+\frac{\beta qV_{\mathrm{M}}}{RT}\right)P_{\mathrm{CO}}^2\theta^2 - k'_{2\mathrm{Red}}\exp\left(\frac{-\alpha qV_{\mathrm{M}}}{RT}\right)P_{\mathrm{CO}_2}^2(1-\theta)^2\right].$$

In some cases, one can also assume that for reaction 1, the oxidation current is negligible in front of reduction one. On the opposite, for reaction 2, oxidation current is more important (Tafels approximation). So that $i = i_1 + i_2 = i_{1\mathrm{Red}} + i_{2\mathrm{Ox}}$ and:

$$i = q\left[\lambda_2 k'_{2\mathrm{Ox}}\exp\left(\frac{\beta qV_{\mathrm{M}}}{RT}\right)P_{\mathrm{CO}}^2\theta^2 - \lambda_1 k'_{1\mathrm{Red}}\exp\left(-\frac{\alpha qV_{\mathrm{M}}}{RT}\right)P_{\mathrm{O}_2}(1-\theta)^2\right].$$

This means that the considered equilibrium is $\mathrm{Ox}_1 + \mathrm{Red}_2 \rightarrow \mathrm{Ox}_2 + \mathrm{Red}_1$.The "$i = 0$" condition, which expresses the stationary character of the process, yields the expression for V_{M}. If we take into account the fact that $\alpha + \beta = 1$, we obtain, for the present case:

$$V_{\mathrm{M}} = \frac{RT}{q(\alpha+\beta)}\ln\frac{\lambda_1 k_{1\mathrm{Red}}}{\lambda_2 k'_{2\mathrm{Ox}}}\frac{P_{\mathrm{O}_2}(1-\theta)^2}{P_{\mathrm{CO}}^2\theta^2} = A + \ln\frac{P_{\mathrm{O}_2}(1-\theta)^2}{P_{\mathrm{CO}}^2\theta^2}$$

with:

$$A = \frac{RT}{q(\alpha+\beta)}\ln\frac{\lambda_1 k'_{1\mathrm{Red}}}{\lambda_2 k'_{2\mathrm{Ox}}}.$$

Coming back to a sensor where we have two different electrodes (gold and platinum for example), the difference of the potential is then given by:

$$\Delta V = V_{\mathrm{Au}} - V_{\mathrm{Pt}}.$$

Assuming that the kinetic process is identical on both electrodes, it leads to:

$$\Delta V = A_{\mathrm{Au}} + \ln\frac{P_{\mathrm{O_2}}(1-\theta_{\mathrm{Au}})^2}{P_{\mathrm{CO}}^2\theta_{\mathrm{Au}}^2} - A_{\mathrm{Pt}} - \ln\frac{P_{\mathrm{O_2}}(1-\theta_{\mathrm{Pt}})^2}{P_{\mathrm{CO}}^2\theta_{\mathrm{Pt}}^2}.$$

As the two electrodes are located in the same gaseous atmosphere, the expression becomes:

$$\Delta V = A + \ln\frac{\theta_{\mathrm{Pt}}^2(1-\theta_{\mathrm{Au}})^2}{\theta_{\mathrm{Au}}^2(1-\theta_{\mathrm{Pt}})^2}$$

with:

$$A = A_{\mathrm{Au}} - A_{\mathrm{Pt}}.$$

In this case, the difference of potential ΔV measured on the device is not directly related to the partial pressure of gas, and it is difficult to express θ as a function of P. The only assumption that we can still do is to assume that the value of θ is small in regards of 1, and in which case, the potential is entirely independent of the partial pressures of oxygen and carbon monoxide.

If it is not impossible to imagine other kinetics models, we may think that the solutions obtained will be complex or identical to those presented in the previous example. These remarks, and the one made about oxygen alone in particular, suggest that other models should be explored. We will see, with some results obtained with some sensors, how this problem has been addressed by imagining other physical and chemical aspects related to capacitive effects.

6.2 *Complementary Experiments with Au/β-Alumina/Pt Device*

In this part, most significant results concerning the physicochemical behaviour of the sensor devices in regard of oxygen and carbon monoxide are presented.

For oxygen, results have been previously reported in Fig. 25 during discussion on mixed potential. The complex responses of sensors to oxygen have to be considered for the study of interactions with carbon monoxide.

For better understanding of all these phenomena, we have started a physical chemical study on β-alumina device associated with a platinum electrode and a gold electrode (Fig. 27). Figure 28 shows the sensor's response to carbon monoxide, depending on oxygen presence and temperature.

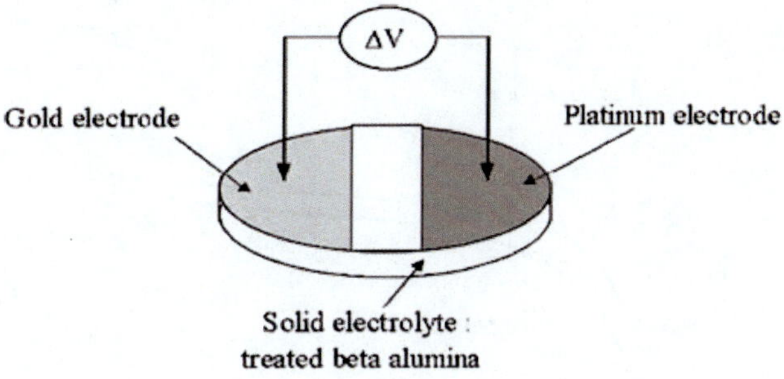

Fig. 27 Schematic of the studied β-alumina device

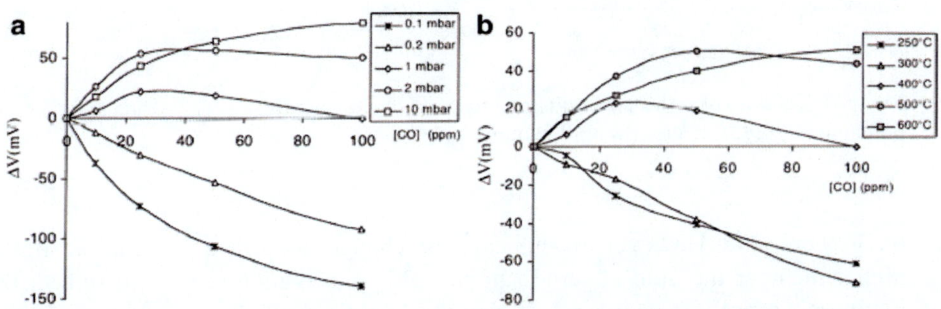

Fig. 28 Response of β-alumina sensor to carbon monoxide at different concentrations: (**a**) influence of oxygen pressure for tests at 400°C; (**b**) influence of temperature for PO_2 = 100 Pa = 1 mbar

The reported curves appear to be relatively complex regarding their evolution according to different parameters of the system (partial pressure of gas, temperature). To understand and control such behaviour, it seemed interesting to conduct investigations, in order to better characterise the system. Thus, we first started to evaluate the charge potential effects related to the adsorption of gas on the electrodes, and/or the ionic material. To do this, we used two different techniques.

6.2.1 Vibrating Capacitor Method (PUP)

This method consists to measure the effect, which may exist between a reference electrode R, deemed insensitive to the action of gas, and the solid material *S* on which the chemical adsorption of gases is able to vary the electric charge of this electrode and consequently its potential surface. The results in Fig. 29 show an increase of surface potential value when the oxygen pressure rises on a β-alumina–gold structure. This logarithmic evolution is visible for experiments performed at 300 and 400°C.

The results obtained with the platinum covered β-alumina device are relatively similar to those of gold-covered β-alumina: the surface potential increases with

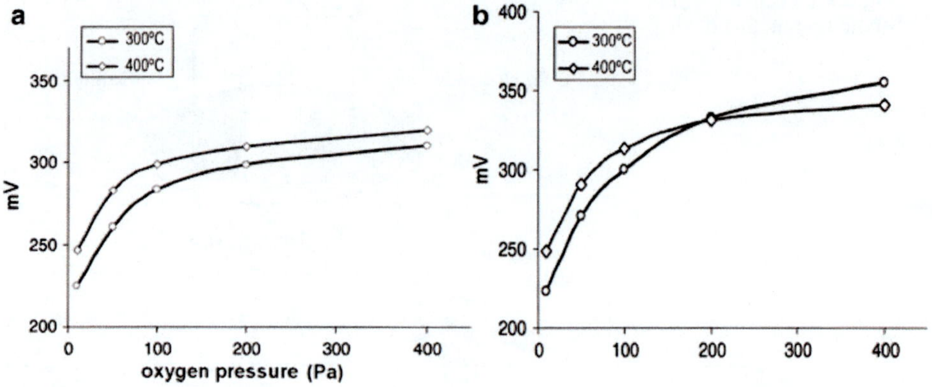

Fig. 29 Evolution of surface potential as a function of oxygen pressure, at 300 and 400°C: (**a**) on β-alumina—gold structure; (**b**) on β-alumina—platinum structure

oxygen pressure. However, in this case the change in surface potential is slightly higher than for the gold-covered sample. This observation seems to confirm that platinum's affinity for oxygen is higher than gold's one.

Thus, the results we obtained using gold or platinum confirm the formation of additional charged species that induce the surface electrical effects.

6.2.2 Impedance Spectroscopy Technique

The electrical characteristics of heterogeneous interfaces render them very good candidates to be used in gas sensor applications. Indeed, we often find that the electrical or electrochemical nature of the data supplied by such sensors depends on the nature and intensity of the processes taking place at the interfaces of the device. In particular, to characterise electrode/electrolyte interface, impedance spectroscopy consisting in electrical measurement with AC current at various frequency is commonly used.

In the event of ideally polarisable electrodes, it is preferable to work at low frequencies, which is to say at a few Hertz. The accumulation of charges near the electrodes has to be associated with ion diffusion phenomena inside the material. The lower the measurement frequency, the higher the depth reached inside the material of the disturbance caused by the alternating current, and the greater is the amplification of the electrode polarisation process. The mobilisation of these charge carriers, which are no longer necessarily available for current conduction, then leads to an increase in the material's resistance at low frequencies, as represented in Fig. 30 by the 55° slope of the straight line (45° in the ideal case). This line will be considered as representing an electrode polarisation process.

Experiments are performed here on sintered pellets associated with two gold electrodes. This result confirms the activated nature of the ionic conductivity.

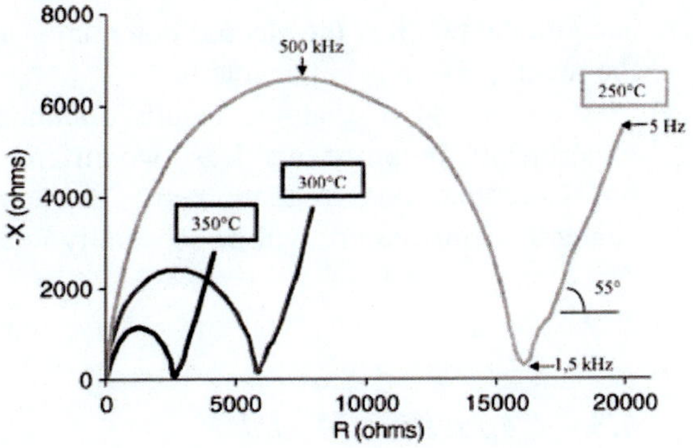

Fig. 30 Nyquist representation of the impedance spectrum obtained with sintered β-alumina associated with two gold electrodes, at various temperatures, and under air

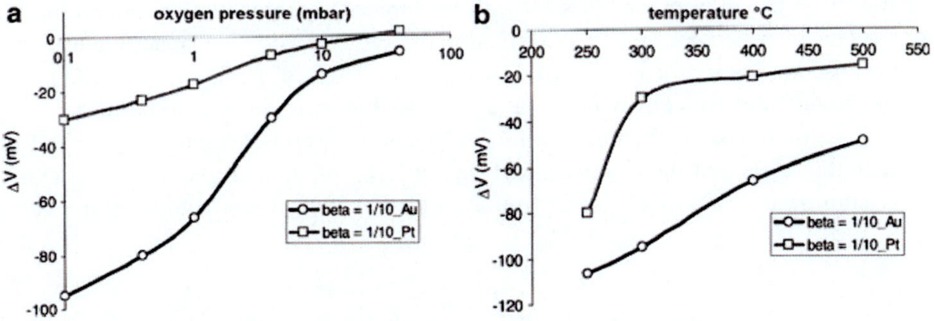

Fig. 31 Electric responses of sensors composed of two electrodes of same nature with a ratio $\beta = 1/10$: (**a**) as a function of the oxygen pressure at 300°C; (**b**) as a function of the temperature for an oxygen pressure of 0.1 mbar

6.2.3 Electrodes Sizes

The last aspect of this study concerns the influence of the size of electrodes on the device's response to the action of gas. We know that the functioning of such a sensor is necessarily associated with the dissymmetry of its structure. According to the role of electrodes, it seemed interesting to create an asymmetrical condition, playing on the size ratio "β" of two electrodes made here by the same metal, namely gold or platinum.

They are in the literature only few studies, which deal on electrode size influences [66, 69].

The results, displayed in Fig. 31, show that the electric responses of the sensors with surface ratio, $\beta = 0.1$, are quite large, especially if gold electrodes are used.

First, we can notice that the electric response of the sensors is negative (the reference electrode was the smallest one). This means that the potential of a large electrode is less than that of a small electrode. We also notice that the potential difference is much greater with gold electrodes than with platinum electrodes. The results involving the size of the electrodes are relatively unexpected if we take into account the fact that the electric potential is an intensive variable. This implies that the metal plays a peculiar role here.

We can add to that the results obtained from calorimetry experiments that underline the presence of at least two different oxygen species chemically adsorbed on β-alumina, one of them being characterised by a bond with endothermic character. This feature will be especially relevant to the modelling of the process.

6.3 Capacitive Model

Based on the previous results [56], the chemical and physical processes inducing oxygen adsorbed species are described in the Figs. 32 and 33. Capacitive effects are proposed to justify the presence of chemically adsorbed species at the triple point of the device metal/electrolyte/gas, and the possible migration of sodium ions.

This model, described in detail in [63], has two parallel capacitors, capacitor C_1 with the surface of the electrode and capacitor C_2 on the periphery of the electrode. Assuming that C_2 to C_1 is negligible, and that the effects achieved are directly

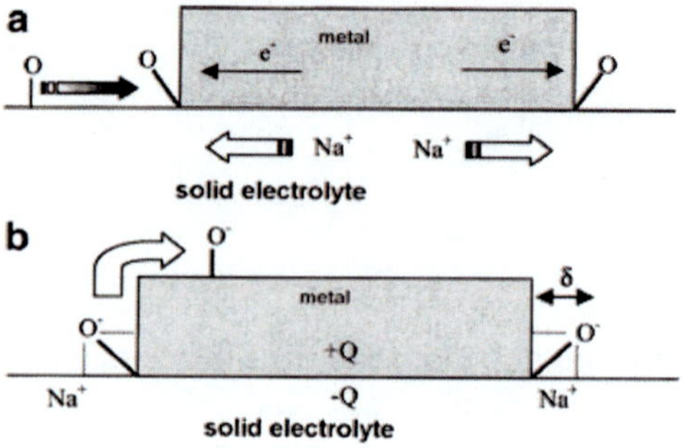

Fig. 32 Scheme explaining the electrostatic phenomena inducing equilibrium between the charges in the system

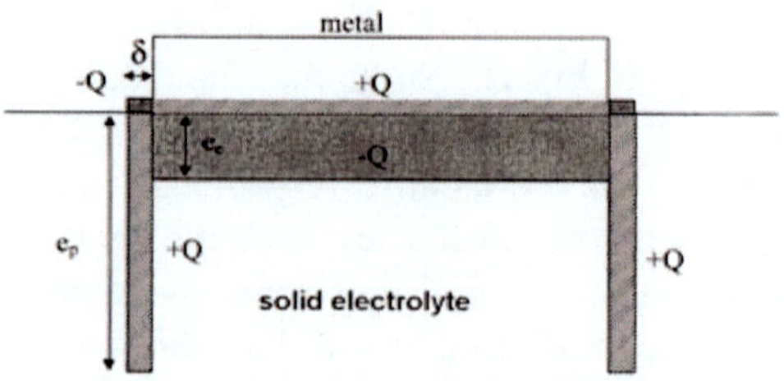

Fig. 33 Scheme displaying the location of space charge areas present in the system

related to species adsorbed at the triple point, the potential V can be expressed from the relationship $Q = C_1V$ as:

$$V = \frac{qN_0^2\theta^2}{[\mathrm{Na}^+]\varepsilon}\frac{(2\pi R\delta)^2}{(\pi R)^2} = \frac{4qN_0^2\theta^2}{[\mathrm{Na}^+]\varepsilon}\left(\frac{\delta}{R}\right)^2$$

$$V = \alpha \times \theta^2\left(\frac{\delta}{R}\right)^2$$

where: N_0: Number of adsorption sites
θ: Coverage fraction of adsorbed oxygen
R: Electrode radius
δ: Width of periphery spacc charge area
ε: Dielectric permittivity of electrolyte
$[\mathrm{Na}^+]$: Ion concentrations in β-alumina electrolyte

If we admit that the value of δ is not a function of R, and that it keeps a relatively constant value, we note that the produced potential depends of the electrode's radius, and consequently, its surface.

Besides, the ratio δ/R being without unit, the expression of the potential keeps its specificity: it is an intensive parameter. The consequence is that the difference of potential recorded at the poles of two electrodes 1 and 2 will depend on the coverage degree θ_1 and θ_2, respectively, and on their radius R_1 and R_2:

$$\Delta V = V_1 - V_2 = \alpha\delta^2\left[\left(\frac{\theta_1}{R_1}\right)^2 - \left(\frac{\theta_2}{R_2}\right)^2\right].$$

This expression will allow us to represent all of the results observed on such a device, that is, the influence of the nature and the geometry of the electrodes. The two cases considered during this study are the following:

– In the first case, the two electrodes have the same size but have different natures, which imply that R_1 and R_2 are equal, and that θ_1 and θ_2 are different, if we admit that the concentration of the adsorbed species is linked to the catalytic activity of each metal. The expression of ΔV becomes:

$$\Delta V = V_1 - V_2 = \frac{\alpha\delta^2}{R^2}\left[\theta_1^2 - \theta_2^2\right].$$

– In the second case, the two electrodes have the same nature but different sizes. Given that the coverage degrees are not a function of the area, we will have $\theta_1 = \theta_2 = \theta$, with:

$$\Delta V = V_1 - V_2 = \alpha_i \delta^2 \theta^2 \left[\frac{1}{R_1^2} - \frac{1}{R_2^2} \right].$$

For oxygen adsorption process, we consider all the following reactions:

$$O_2 + 2\beta \Leftrightarrow 2(O - \beta)$$

$$(O - \beta) + \mathrm{s} + \mathrm{metal} \Leftrightarrow (O - \mathrm{s})^{\delta^+_-} + \beta$$

$$(\mathrm{O} - \mathrm{s})^{\delta^+_-} + \mathrm{e}^- \Leftrightarrow (\mathrm{O} - \mathrm{s})^-$$

$$(O - \mathrm{s})^- + \alpha \Leftrightarrow (O - \alpha)^- + \mathrm{s}$$

with K_1, K_2, K_3, and K_4 the equilibrium constants and θ_1 the coverage degree of the specie $(\mathrm{O} - \mathrm{s})^{\delta^+_-}$. This species, fixed at the metal/solid electrolyte interface, will be considered in strong interaction with these two materials. This double interaction could imply polarisation effects, noted by δ^+_-, of the adsorbed oxygen. This polarisation, favoured by the presence of large quantities of electrons in the metal, will create inductive effects and, consequently, an accumulation of charges in the solid electrolyte. In such a condition, the expression of the coverage degree θ_1 will then be:

$$\theta_1 = \frac{K_2 \sqrt{K_1} \sqrt{P_{\mathrm{O}_2}}}{1 + K_2 \sqrt{K_1} \sqrt{P_{\mathrm{O}_2}} (1 + K_3)}.$$

And the expression of the theoretical potential difference $V\Delta$ at the poles of the device is:

$$\Delta V = \gamma \left[\theta_{\mathrm{Pt}}^2 - \theta_{\mathrm{Au}}^2 \right],$$

$$\Delta V = \gamma \left[\left(\frac{K_\alpha^{\mathrm{Pt}} \sqrt{P_{\mathrm{O}_2}}}{1 + K_\alpha^{\mathrm{Pt}} \sqrt{P_{\mathrm{O}_2}} (1 + K_3^{\mathrm{Pt}})} \right)^2 - \left(\frac{K_\alpha^{\mathrm{Au}} \sqrt{P_{\mathrm{O}_2}}}{1 + K_\alpha^{\mathrm{Au}} \sqrt{P_{\mathrm{O}_2}} (1 + K_3^{\mathrm{Au}})} \right)^2 \right].$$

With $K_{\alpha_i} = K_2 \sqrt{K_1}$

For carbon monoxide, and assuming Langmuir–Hinshelwood hypothesis, that is accepting the additional reaction: $\mathrm{CO} + \mathrm{s} \Leftrightarrow \mathrm{CO} - \mathrm{s}$ compared to the previous model, it becomes:

$$\theta_1 = \frac{K_2 \sqrt{K_1} \sqrt{P_{\mathrm{O}_2}}}{1 + K_2 \sqrt{K_1} \sqrt{P_{\mathrm{O}_2}} (1 + K_3) + K_4 P_{\mathrm{CO}}}.$$

And consequently:

$$\Delta V = \gamma\left[\left(\frac{K_{\alpha_i}^{\mathrm{Pt}}\sqrt{P_{\mathrm{O}_2}}}{1 + K_{\alpha_i}^{\mathrm{Pt}}\sqrt{P_{\mathrm{O}_2}}(1 + K_3^{\mathrm{Pt}}) + K_4^{\mathrm{Pt}} P_{\mathrm{CO}}}\right)^2 - \left(\frac{K_{\alpha_i}^{\mathrm{Au}}\sqrt{P_{\mathrm{O}_2}}}{1 + K_{\alpha_i}^{\mathrm{Au}}\sqrt{P_{\mathrm{O}_2}}(1 + K_3^{\mathrm{Au}}) + K_4^{\mathrm{Au}} P_{\mathrm{CO}}}\right)^2\right].$$

To validate previous relationships and first the expression of ΔV as a function of oxygen pressure and temperature, we have to take four parameters for each metal (Au and Pt): K_α, ΔH_α, K_3, ΔH_3 (ΔH_i is the enthalpy variation associated with the reaction with an equilibrium constant K_i).

Physicochcmical considerations concerning oxygen adsorption, and in particular calorimetry results impose some conditions to previous parameters: $\Delta H_\alpha^{\mathrm{Au}} > \Delta H_\alpha^{\mathrm{Pt}} > 0$, $K_\alpha^{\mathrm{Au}} > 0$, $K_\alpha^{\mathrm{Pt}} > 0$ and $K_3^{\mathrm{Au}} > 0$, $K_3^{\mathrm{Pt}} > 0$.

Numerical simulations allowed calculating values reported in Table 1, which give the best agreement with all experimental curves with complex shape describing variations of the potential difference ΔV versus oxygen pressure and temperature (Fig. 34). Once these values have been determined considering oxygen experimental results, same methodology was applied to calculate the parameters

Table 1 Values of the different parameters exploited for the simulation

	Platinum	Gold
K_a^0 (u.a.)	4.6×10^5	1.38×10^5
ΔH_α (kJ/mole)	58	80
K_3 (u.a.)	9.7×10^{-13}	7.4×10^{-12}
ΔH_3 (kJ/mole)	−172	−130
K_4^0 (u.a.)	13.6	1.61
ΔH_4 (kJ/mole)	−14.3	−35.8

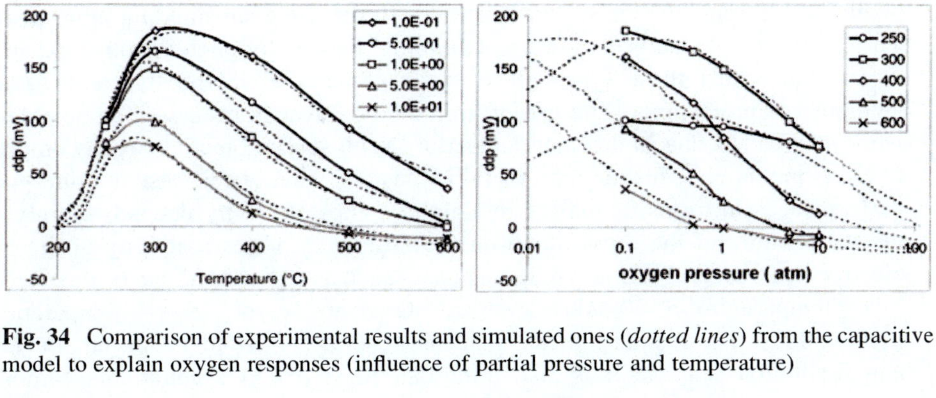

Fig. 34 Comparison of experimental results and simulated ones (*dotted lines*) from the capacitive model to explain oxygen responses (influence of partial pressure and temperature)

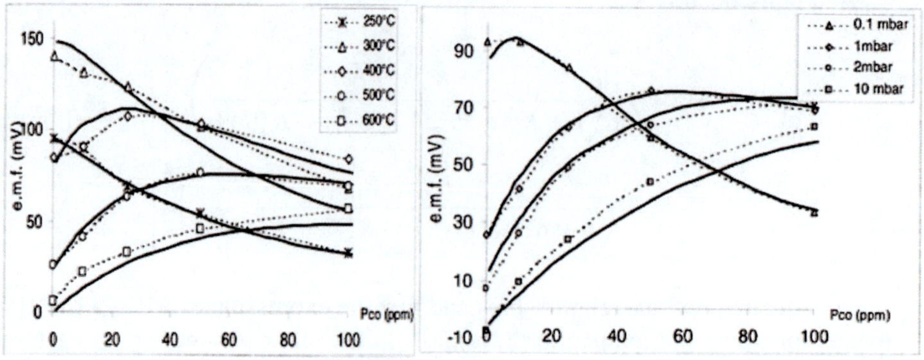

Fig. 35 Comparison of experimental results and simulated ones (*dotted lines*) from the capacitive model to explain carbon monoxide response (influence of partial pressure of CO, O_2 and of temperature)

K_4, ΔH_4 for both metals, to simulate CO response. Calculated values are reported in Table 1 and comparison of simulated curves and experimental ones is presented in Fig. 35. Given the extreme complexity of the obtained experimental results, such accordance appears to be an excellent argument in favour of this model.

7 Conclusion

The objective of this chapter was to make of review of our past and current sensor developments for automotive exhausts. We have first worked with β-alumina electrolyte, and then with YSZ. In both cases, the sensors are very similar using a dissymmetry of electrodes: Au/β-alumina/Pt and Au/YSZ/Pt. The prototypes of sensors have been developed using screen-printing technology. Their interest compared to other developments consists in their real simplicity both in terms of production due to the absence of reference chamber and of use as the signal is only a potential difference between two electrodes in the same surrounding atmosphere. The experimental results obtained on laboratory benches or on automotive exhausts demonstrated that these types of sensors can be used in the future to satisfy automotive applications. The use of a protective layer is necessary to be able to work in exhausts due to the soot deposition. With such protective layer, complete devices including sensors and metallic casings have been tested with good performances in terms of long-term stability. The selectivity depends mainly on the temperature of the sensors and on the difference of gas reactivity of the two electrodes. In general manner, reducing gases such as CO and HC can be detected at high temperature (500–600°C) and NOx mostly at lower temperatures (350–450°C). The results indicate that the general behaviour is similar for YSZ and β-alumina sensors. It is also important to take into account this common behaviour in terms of understanding and models. By another way, the improvement of selectivity will be especially dependent in the future on the use of superficial catalytic films, and of course on the nature of the two electrodes. Some of our recent experiments are in complete agreement with such performance improvements. Our current objectives and studies are focused in these directions.

A capacitive model has been developed to explain our results with the device Au/β-alumina/Pt. The main difference with other non-Nernstian models is related to the necessity in our case to explain EMF variations with PO_2. β-alumina is an ionic conductor by Na+ ions, and we cannot use only a model based on electrochemical reactions exchanging at the triple point O^{2-} with the electrolyte as in the case of YSZ. We have seen that a capacitive model allows explaining many of our experimental results. Nevertheless, our recent comparison of the two types of devices, Au/β-alumina/Pt and Au/YSZ/Pt demonstrates that the performances of both sensors are very similar. Consequently, even if many points of the capacitive model can be debated, we think it must be partly taken into account in regards of the lack of concrete models with mixed potential. Generally, the authors propose the term of "mixed potential", and then develop a discussion including all parameters, which can influence the EMF values (adsorption–desorption steps, catalytic activity, electro or non-electrochemical reaction…). But many experimental results cannot be explained only by mixed potential. In consequence, the term "non-Nernstian" has been introduced, but never with kinetic model allowing calculating the EMF values. It is particularly important to notice that our mathematical simulations are in very good agreement with our experimental data, not only under O_2, but also with CO using the same values for the different parameters. Moreover, our last experiments with similar results obtained with β-alumina or YSZ reinforce the fact that a model close to a capacitive model, or including partly some of its major points, could be studied in the future for dissymmetric sensors using electrolytes as YSZ.

References

1. Zhuiykov S, Miura N (2007) Development of zirconia-based potentiometric NOx sensors for automotive and energy industries in the early 21st century: what are the prospects for sensors. Sens Actuators B 121:639–651
2. Riegel J, Neumann H, Wiedenmann HM (2002) Exhaust gas sensors for automotive emission control. Solis State Ionics 152–153:783–800
3. Rotureau D, Viricelle JP, Pijolat C, Caillol N, Pijolat M (2005) Development of planar SOFC device using screen-printing technology. Electroceramics IX. J Eur Ceram Soc 25:2633–2636
4. Szabo FN, Du H, Akbar Sheikh A, Soliman A, Dutta KP (2002) Microporous zeolite modified yttria stabilized zirconia (YSZ) sensors for nitric oxide (NO) determination in harsh environments. Sens Actuators B 82:142–149
5. Szabo FN, Dutta KP (2004) Correlation of sensing behaviour of mixed potential sensors with chemical and electrochemical properties of electrodes. Solid State Ionics 171:183–190
6. Vogel A, Baier G, Schule V (1993) Non-Nerstian potentiometric zirconia sensors: screening of potential working electrode materials. Sens Actuators B 15–16:147–150

7. Wachsman ED, Jayaweera P (2000) Selective detection of NO_X by differential electrode equilibria. Solid-state ionic-devices II-Ceramic sensors. Ed. Electrochemical Society Proceedings: 298–304
8. Wachsman DE, Jayaweera P, Krishnan G, Sanjurjo A (2000) Electrocatalytic reduction of NO_X on $La_{1-X}A_XB_{1-Y}B'_YO_{3-\delta}$: evidence of electrically enhanced activity. Solid State Ionics 136–137:775–782
9. Wachsman ED, Azad AM (2003) Solid state potentiometric gaseous oxide sensor. US Patent/ 0066519 A1
10. West LD, Montgomery CF, Armstrong RT (2005) "NO-selective" NO_x sensing elements for combustion exhausts. Sens Actuators B 111–112:84–90
11. West LD, Montgomery CF, Armstrong RT (2006) "Total NO_x" sensing elements with compositionally identical oxide electrodes. J Electrochem Soc 153:H23–H28
12. Williams DE (1983) Brevet original UK Patent Application$_{(19)}$ GB$_{(11)}$ 2 119 933 A
13. Xiong W, Kale MG (2006) High-selectivity mixed-potential NO_2 sensor incorporating Au and $CuO+CuCr_2O_4$ electrode couple. Sens Actuators B 119:409–414
14. Yan Y, Shimizu Y, Miura N, Yamazoe N (1993) Characteristics and sensing mechanism of SO_x sensor using stabilized zirconia and metal sulphate. Sens Actuators B 12:77–81
15. Yan Y, Shimizu Y, Miura N, Yamazoe N (1994) High-performance solid-electrolyte SO_X sensor using MgO-stabilized zirconia tube and Li_2SO_4-$CaSO_4$-SiO_2 auxiliary phase. Sens Actuators B 20:81–87
16. Yang J-C, Dutta KP (2007) Promoting selectivity and sensitivity for a high temperature YSZ-based electrochemical total NO_X sensor by using a Pt-loaded zeolite Y filter. Sens Actuators B 125:30–39
17. Yang J-C, Dutta KP (2009) Solution-based synthesis of efficient WO_3 sensing electrodes for high temperature potentiometric NO_X sensors. Sens Actuators B 136:523–529
18. Yano M, Tomita A, Sano M, Hibino T (2007) Recent advances in single-chamber solid oxide fuel cells: a review. Solid State Ionics 177:3351–3359
19. Yoo J, Wachsman DE (2007) NO_2/NO response of Cr_2O_3- and SnO_2- based potentiometric sensors and temperature-programmed reaction evaluation of the sensor elements. Sens Actuators B123:915–921
20. Yoo J, Chatterjee S, Wachsman DE (2007) Sensing properties and selectivities of a WO_3/YSZ/ Pt potentiometric NO_x sensor. Sens Actuators B 122:644–652
21. Yoo J, Oh D, Wachsman DE (2008) Investigation of WO_3-based potentiometric sensor performance (M/YSZ/WO_3 M=Au, Pd, and TiO_2) with varying counter electrode. Solid State Ionics 179:2090–2100
22. Yoon JW, Grilli ML, Dibartolomeo E, Polini R, Traversa E (2001) The NO_2 response of solid electrolyte sensors made using nano-sized $LaFeO_3$ electrodes. Sens Actuators B 76:483–488
23. Fleming William J (1977) Physical principles governing nonideal behaviour of the zirconia oxygen sensor. J Electrochem Soc Electrochem Sci Technol 21–28
24. Okamoto H, Obayashi H, Kudo T (1981) Non-ideal EMF behaviour of zirconia oxygen sensors. Solid State Ionics 3(4):453–456
25. Moseley PT (1987) Non-Nernstian, potential-generating gas sensors. Solid State Gas Sensors 139–150, Ed. Adam Hilger
26. De Arias Velasco AA, Moseley PT, Peat R, Pelaez JG (1993) Atmosphere-dependent potentials at oxide interfaces. Sens Actuators B 15–16:55–62
27. Lalauze R, Visconte E, Montanaro L, Pijolat C (1993) A new type of mixed potential sensor using a thick film of beta-alumina. Sens Actuators B 13–14:241–243
28. Pijolat C, Visconte E, Lalauze R (1990) Détecteur de la présence d'un gaz dans une atmosphère. Brevet n°90 12:413
29. Miura N, Yan Y, Sato M, Yao S, Nonaka S, Shimizu Y, Yamazoe N (1995) Solid-state potentiometric CO_2 sensors using anion conductor and metal carbonate. Sens Actuators B 24–25:260–265

30. Miura N, Lu G, Yamazoe N (1998) High-temperature potentiometric/amperometric NO_x sensors combining stabilized zirconia with mixed-metal oxide electrode. Sens Actuators B52:169–178
31. Miura N, Lu G, Yamazoe N (2000) Progress in mixed-potential type devices based on solid electrolyte for sensing redox gases. Solid State Ionics 136–137:533–542
32. Kanazawa E, Kugishima M, Shimanoe K, Kanmura Y, Teraoka Y, Miura N, Yamazoe N (2001) Mixed potential type N_2O sensor using stabilized zirconia- and SnO_2-based sensing electrode. Sens Actuators B 75:121–124
33. Miura N, Shuiykov S, Ono T, Hasel M, Yamazoe N (2002) Mixed potential type sensor using stabilized zirconia and $ZnFe_2O_4$ sensing electrode for NO_x detection at high temperature. Sens Actuators B 83:222–229
34. Plashnitsa Vladimir V, Ueda T, Elumalai P, Miura N (2008) NO_2 sensing performances of planar sensor using stabilized zirconia and thin-NiO sensing electrode. Sens Actuators B 130:231–239
35. Dibartolomeo E, Traversa E, Baroncini M, Kotzeva V, Kumarr V (2000) Solid state ceramic gas sensors based on interfacing ionic conductors with semiconducting oxides. J Eur Ceram Soc 20:2691–2699
36. Dibartolomeo E, Grilli ML, Traversa E (2004) Sensing mechanism of potentiometric gas sensors based on stabilized zirconia with oxide electrodes. J Electrochem Soc 151(5): H133–H139
37. Grilli ML, Dibartolomeo E, Lunardi A, Chevallier L, Cordiner S, Traversa E (2005) Planar non-nernstian electrochemical sensors: field test in the exhaust of a spark ignition engine. Sens Actuators B 108:319–325
38. Chevallier L, Dibartolomeo E, Grilli ML, Traversa E (2008) High temperature detection of CO/HCs gases by non-Nernstian planar sensors using Nb_2O_5 electrode. Sens Actuators B 130:514–519
39. Chevallier L, Dibartolomeo E, Grilli ML, Mainas M, White B, Wachsman DE, Traversa E (2008) Non-Nernstian planar sensors based on YSZ with a Nb_2O_5 electrode. Sens Actuators B 129:591–598
40. Garzon Fernando H, Mukundan R, Brosha Eric L (2000) Solid-state mixed potential gas sensors: theory, experiments and challenges. Solid State Ionics 136–137:633–638
41. Brosha Eric L, Mukundan R, Brown David R, Garzon Fernando H, Visser JH, Zanini M, Zhou Z, Logothetis EM (2000) CO/HC sensors based on thin films of $LaCoO_3$ and $La_{0.8}Sr_{0.2}CoO_{3-\delta}$ metal oxides. Sens Actuators B 69:171–182
42. Brosha Eric L, Mukundan R, Brown David R, Garzon Fernando H (2002) Mixed potential sensors using lanthanum manganate and terbium yttrium zirconium oxide electrodes. Sens Actuators B 87:47–57
43. Brosha Eric L, Mukundan R, Garzon Fernando H (2002) The role of heterogeneous catalysis in the gas-sensing selectivity of high-temperature mixed potential sensors. Solid State Ionics Devices III:261–271
44. Brosha Eric L, Mukundan R, Lujan R, Fernando H (2006) Mixed potential NO_x sensors using thin film electrodes and electrolytes for stationary reciprocating engine type applications. Sens Actuators B 119:398–408
45. Figueroa LO, Lee C, Akbar Sheikh A, Szabo FN, Trimboli AJ, Dutta KP, Sawaki N, Soliman Ahmed A, Verweij H (2005) Temperature-controlled CO, CO_2 and NO_x sensing in a diesel engine exhaust stream. Sens Actuators B 107:839–848
46. Hibino T, Wang S, Kakimoto S, Sano M (1998) Detection of propylene under oxidizing conditions using zirconia-based potentiometric sensor. Sens Actuators B 50:149–155
47. Hashimoto A, Hibino T, Mori K, Sano M (2001) High-temperature hydrocarbon sensors based on a stabilized zirconia electrolyte and proton conductor-containing platinum electrode. Sens Actuators B 81:55–63
48. Nagao M, Yoshii T, Namekata Y, Teranishi S, Sano M, Tomita A, Hibino T (2008) De-NO_X reactor and NO_X sensor usint In^{3+}-doped SnP_2O_7 with PtRhBa/C electrode. Solid State Ionics 179:1655–16661

49. Guillet N, Lalauze R, Viricelle JP, Pijolat C, Montanaro L (2002) Development of a gas sensor by thick film technology for automotive applications: choice of materials-realization of a prototype. Mater Sci Eng C 21:97–103
50. Pijolat C, Pupier C, Sauvan M, Tournier G, Lalauze R (1999) Gas detection for automotive pollution control. Sens Actuators B 59:195–202
51. Pijolat C, Pupier C, Testud C, Lalauze R, Montanaro L, Negro A, Malvicino C (1998) Electrochemical sensors for CO/N_x detection in automotive applications. J Electroceram 2–3:181–191
52. Pupier C, Pijolat C, Marchand JC, Lalauze R (1999) Oxygen role in the electrochemical response of a gas sensor using ideally polarizable electrodes. J Electrochem Soc 146:2360–2364
53. Guillet N, Pijolat C, Lalauze R (2002) Calorimetric study of interaction between oxygen and β-Al_2O_3, β-Al_2O_3 + Au and β-Al_2O_3 + Pt. J Thermal Anal Calorimetry 68:15–23
54. Guillet N, Lalauze R, Viricelle J-P, Pijolat C (2002) The influence of the electrode size on the electrical response of a potentiometric gas sensor to the action of oxygen. IEEE Sens J 2:349–353
55. Guillet N, Rives A, Lalauze R, Pijolat C (2003) Study of adsorption of oxygen on β-Al_2O_3 + Au and β-Al_2O_3 + Pt work function measurements-proposition of a model. Appl Surf Sci 210:286–292
56. Guillet N, Lalauze R, Pijolat C (2004) Oxygen and carbon monoxide role on the electrical response of a non-Nernstian potentiometric gas sensor; proposition of a model. Sens Actuators B 98:130–139
57. Morata A, Viricelle JP, Tarancon A, Dezanneau G, Pijolat C, Peiro F, Morante JR (2008) Development and characterisation of a screen-printed mixed potential gas sensor. Sens Actuators B130:561–566
58. Billi E, Viricelle J-P, Montanaro L, Pijolat C (2002) Development of a protected gas sensor for exhaust automotive applications. IEEE Sens J 2:342–348
59. Park CO, Miura N (2006) Absolute potential analysis of the mixed potential occurring at the oxide/YSZ electrode at high temperature in NO_x-containing air. Sens Actuators B 113:316–319
60. Zhuiykov S, Nakano T, Kunimoto A, Yamazoe N, Miura N (2001) Potentiometric NOx sensors based on stabilized zirconia and $NiCr_2O_4$ sensing electrode operating at high temperatures. Electrochem Commun 3:97–101
61. Hugon O, Sauvan M, Benech P, Pijolat C, Lefebvre F (2000) Gas separation with a zeolite filter, application to the selectivity enhancement of chemical sensors. Sens Actuators B 67:235–243
62. Holt TC, Azad AM, Swartz SL, Rao RR, Dutta KP (2002) Carbon monoxide sensor for PEM fuel cell systems. Sens Actuators B 87:414–420
63. Lalauze R (2008) Physical chemistry of solid-gas interfaces. ISTE and Wiley, New York
64. Lalauze R, Pijolat C, Couput JP (1981) Procédé, capteur et dispositif de détection de gaz dans un milieu gazeux. Brevet n°81 19:536
65. Li X, Kale MG (2007) Influence of sensing electrode and electrolyte on performance of potentiometric mixed-potential gas sensors. Sens Actuators B 123:254–261
66. Menil F, Daddah BO, Tardy P, Debeda H, Lucat C (2005) Planar LiSICON-based potentiometric CO_2 sensors: influence of the working and reference electrodes relative size on the sensing properties. Sens Actuators B 107:695–707
67. Miura N, Yan Y, Lu G, Yamazoe N (1996) Sensing characteristics and mechanism of hydrogen sulfide sensor using stabilized zirconia and oxide sensing electrode. Sens Actuators B 34:367–372
68. Pijolat C, Viricelle JP, Tournier G, Montmeat P (2005) Applications of membranes and filtering for gas sensor improvement. Thin Solid Films 490:7–16
69. Zosel J, Tuchtenhagen D, Ahlborn K, Guth U (2008) Mixed potential gas sensor with short response time. Sens Actuators B 130:326–329

Atmospheric Humidity Measurements Using Gas Sensors

R. Philipona

Abstract Atmospheric humidity is strongly related to precipitation and more general to weather and weather prediction and as the most important greenhouse gas water vapour is strongly related to climate. For weather forecasting purposes, boundary layer and lower troposphere humidity observations are of prime interest. With climate change, accurate humidity measurements in the upper troposphere and lower stratosphere became very important. Upper air observations started in the second part of the nineteenth century and regular radiosonde balloon soundings are made since nearly 70 years. Humidity sensors on radiosondes had large uncertainties for many years but experienced a rapid evolution and were strongly improved over the last three decades. This article presents an overview of operational and scientific sensors used in upper air humidity measurements and then focuses on thin-film capacitive gas sensors, which are the most used hygrometers in present-day operational radiosondes. Information on international developments and future needs for upper air humidity observations are given as a motivation for further improvements of such humidity gas sensors.

Keywords Capacitive polymer sensors, Radiosondes, Upper air measurements

Contents

R. Philipona (✉)
MeteoSwiss, Aerological station, 1530 Payerne, Switzerland
e-mail: rolf.philipona@meteoswiss.ch

M. Fleischer and M. Lehmann (eds.), *Solid State Gas Sensors – Industrial Application*, Springer Series on Chemical Sensors and Biosensors (2012) 11: 255–264,
DOI 10.1007/5346_2011_9,
Published online: 10 December 2011

Abbreviations

ARM	Atmospheric Radiation Measurements
BSRN	Baseline Surface Radiation Network
ECMWF	European Centre for Medium-Range Weather Forecasts
EUCOS	EUMETNET Composite Observing System
GAW	Global Atmosphere Watch
GCOS	Global Climate Observing System
GPS	Global Positioning System
GRUAN	GCOS Reference Upper Air Network
GTS	Global Telecommunication System
GUAN	GCOS Upper Air Network
MOZAIC	Measurements of OZone, water vapour, carbon monoxide and nitrogen oxides by in-service AIrbus airCraft
NDACC	Network for the Detection of Atmospheric Composition Change
WMO	World Meteorological Organisation

1 Introduction

1.1 Upper Air Observations

Upper air observations are meteorological measurements made in the free atmosphere, either directly or indirectly. Classic direct observations are made by soundings determining one or several upper air meteorological variables by means of instruments carried aloft by balloon, aircraft, kite, glider, rocket and so on. Indirect measurements use remote-sensing methods from the surface or from satellites and were developed only in recent decades. Balloon-borne radiosondes are used since nearly 70 years and have been the primary source for vertical profiles of atmospheric parameters used in operational weather forecasting and, more recently for climate studies and long-term monitoring. Radiosondes normally measure pressure, temperature and relative humidity, and are provided with a radio transmitter for sending this information to the observing station. At most operational upper air- or aerological stations, the radiosonde system is also used for upper-wind determination. In addition, some radiosondes are flown with sensing systems for atmospheric constituents, such as ozone concentration or radioactivity.

1.2 Weather Prediction and Environmental Monitoring

Upper air measurements of temperature and relative humidity are two of the basic measurements used in the initialization of the analyses of numerical weather prediction models for operational weather forecasting. Worldwide, about 860 upper air stations on land, generating a vertical sounding twice per day, are organised in several regional basic synoptic networks of the World Meteorological Organisation (WMO). Nearly 1.5 million radiosondes are used per year of which probably 50% for military purposes. Civil observations are exchanged globally on the Global Telecommunication System (GTS) of the WWW, and are available freely and unrestricted according to WMO data policy. The observations are archived in numerous leading national meteorological centres and in the two WMO World Data Centres. Each Regional Association reviews and agrees on its selection of stations every 4 years in the light of changing requirements, technological progress and economic possibilities. The performance of these stations with respect to data availability and quality is regularly monitored and reports are periodically compiled and published by the WMO Secretariat and designated national centres (in Europe ECMWF and EUCOS).

Radiosondes provide most of the in situ temperature and relative humidity measurements over land, while radiosondes launched from remote islands or ships provide a limited coverage over the oceans. Surface remote-sensing instruments such as microwave radiometers, lidars, wind profilers and GPS receivers are more and more used at designated stations to determine temperature, humidity and wind profiles as well as column integrated water vapour. Such instruments usually have limited altitude range and spatial resolution but much higher temporal resolution.

Temperature and relative humidity with resolution in the vertical similar to radiosondes can also be observed by aircraft either during ascent, descent, or at cruise levels. The aircraft observations are used to supplement the radiosonde observations, particularly over the sea. Satellite observations of temperature and water vapour distribution have lower vertical resolution than radiosonde or aircraft measurements. Satellite observations have greatest impact on numerical weather prediction analyses over the oceans and other areas of the globe where radiosonde and aircraft observations are sparse or unavailable.

Accurate measurements of the vertical structure of temperature and water vapour fields in the troposphere are extremely important for all types of forecasting, especially regional and local forecasting. The measurements indicate the existing structure of cloud or fog layers in the vertical. Furthermore, the vertical structure of temperature and water vapour fields determines the stability of the atmosphere and, subsequently, the amount and type of cloud that will be forecast.

High-resolution measurements of the vertical structure of temperature and relative humidity are also important for environmental pollution studies (for instance, identifying the depth of the atmospheric boundary layer). High resolution in the vertical is also necessary for forecasting the effects of atmospheric refraction on the propagation of electromagnetic radiation or sound waves [1].

1.3 Aviation and Ballistic Applications

Civil aviation, artillery and other ballistic applications, such as space vehicle launches, have operational requirements for measurements of the density of air at given pressures derived from radiosonde temperature and relative humidity measurements. Such applications also strongly rely on accurate upper air wind profiles.

1.4 Climate Change and Atmospheric Research Studies

Radiosonde observations are vital for studies of upper air climate change. Hence, it is important to keep adequate records of the systems used for measurements and also of any changes in the operating or correction procedures used with the equipment. In this context, it has proved necessary to establish the changes in radiosonde instruments and practises that have taken place since radiosondes were used on a regular basis. Climate change studies based on radiosonde measurements require extremely high stability in the systematic errors of the radiosonde measurements. However, the errors in early radiosonde measurements of some meteorological variables, particularly relative humidity, were too high to provide acceptable long-term references at all heights reported by the radiosondes. Thus, improvements to and changes in radiosonde design were and are still necessary. Furthermore, expenditure limitations on meteorological operations require that radiosonde consumables remain affordable if widespread radiosonde use is to continue. Therefore, certain compromises in system measurement accuracy have to be accepted by users, taking into account that radiosonde manufacturers are producing systems that need to operate over an extremely wide range of meteorological conditions:

1,050 to 5 hPa for pressure
50 to −90°C for temperature
100 to 1% for relative humidity

Radiosonde systems must comply with the rules of electromagnetic compatibility and in the same time being able to sustain continuous reliable operation when operating in heavy rain, in the vicinity of thunderstorms, and in severe icing conditions.

2 Historical and Present-Day Upper Air Humidity Sensors

2.1 *Standard Humidity Sensors on Operational Radiosondes*

In the early decades of upper air soundings goldbeater's skin (the outer membrane of cattle intestine, which varies in length with changes in relative humidity), hair hygrometers and films of lithium chloride on strips of plastic (whose electrical resistance varies with relative humidity) were most commonly used. These sensor types perform poorly at temperatures below −20°C and suffer from significant hysteresis effects and biases [1]. Later, plastic or glass strip coated with a hygroscopic film containing carbon particles that changes electrical resistance with relative humidity, called carbon hygristors, were used. However, carbon hygristors suffer from a moist bias at relative humidities above 60% [2] and reveal a pronounced hysteresis after passing and exiting clouds. In general, its performance is unreliable at temperatures below −40°C or at low relative humidities [1]. Thin-film capacitive sensors, consisting of a hygroscopic polymer film between two electrodes are faster and more reliable than carbon hygristors and the capacity measured is directly proportional to relative humidity. Thin-film capacitive polymer sensors are by far the most common type of humidity sensors on operational radiosondes today.

2.2 *Research and Reference Humidity Sensors*

Dewpoint or frostpoint hygrometers have also been used on balloon-borne radiosondes for many years [3], but up to present rather for scientific than operational use. An important feature of this technique is that it is an absolute measurement. This means that a property of water vapour can be related to another fundamental quantity—in this case temperature—which is easier to measure. The technique therefore does not require a calibration to a laboratory standard of water vapour, rather it only requires an accurate temperature calibration, which is much easier to achieve than a water vapour calibration. Dewpoint or frostpoint hygrometry has its foundation in equilibrium thermodynamics of a two-phase system. If the temperature can be measured at which the vapour phase of water and the condensed phase are in equilibrium, then the partial pressure can be calculated based on the measured frostpoint (or dewpoint) temperature using the Clausius Clapeyron equation, which describes the relation between the vapour pressure and the temperature of a two-phase system in thermodynamic equilibrium (WMO uses the equation established by Prof. Sonntag). Frostpoint hygrometers use active cooling and heating to maintain a thin layer of frost on a small polished metal mirror. To achieve this, the mirror temperature must be changed as the amount of water vapour in air varies. The thickness of the frost layer is monitored by shining a small beam of light at the mirror and measuring how much of the light is scattered by the frost. The controlled frost point temperature of the mirror is a direct measure of how much water vapour is in the air. Frostpoint hygrometers have certain advantages over thin-film capacitors particularly in the upper troposphere and lower stratosphere but they are much more technical and expensive and therefore less practical for everyday use. Nevertheless, the rather low-cost chilled mirror sensor SnowWhite® from Meteolabor (Switzerland) is often used as test and reference hygrometer [4].

2.3 *Fluorescence Hygrometers for Stratospheric H_2O Measurements*

Even more sophisticated are fluorescence Lyman-α hygrometers that are used in recent years to measure very small amounts of water vapour particularly in the upper troposphere and lower stratosphere. The method used to measure H_2O by a fluorescence technique was developed by Kley and Stone [5] and Bertaux and Delannoy [6]. The photodissociation of H_2O molecules by radiation at wavelengths <137 nm produces electronically excited OH. The electronically excited OH relaxes to the ground state by fluorescence or by collisions with other molecules. By measuring the intensity of the emitted fluorescence, the H_2O abundance can be determined.

3 Thin-Film Capacitive Sensors

3.1 *Principle of Operation*

As mentioned above, thin-film capacitive sensors are the most common type of humidity sensors used on operational radiosondes today. Achievements in micro technology have encouraged the development of a large variety of very small humidity sensors for miscellaneous applications to measure the water vapour content in gaseous systems. Today, more than 75% of these miniaturised humidity sensors in the market use a capacitive technique [7]. Most of these capacitive sensors are based on dielectric changes of thin films upon water vapour uptake as a measure of the water vapour content. The porous polymer material acts as a hydroactive sponge, whereby the water molecules within the polymer material are in thermodynamic equilibrium with the gas phase, that is the rate of adsorption of molecules onto the surface is exactly counterbalanced by the rate of desorption of molecules into the gas phase [8]. The water adhesion is characterised by physical hydrogen bonds through the "weak" Van der Waals interaction of water molecules with the hydrophilic groups of the polymer molecules (e.g., [9]).

The capacitive thin film moisture sensor responds to changes of relative, rather than absolute humidity in the surrounding air as well as to changes of temperature. It is, therefore, commonly calibrated in terms of relative humidity. The thin film polymer either absorbs or releases water molecules as the relative humidity of the ambient air rises or drops. The dielectric properties of the polymer film depend on the amount of water contained in it. Since the dielectric constant of water vapour is much larger than the dielectric constant of the polymer film, even small changes of relative humidity change the dielectric properties of the film, and hence the capacitance of the polymer sensor. The electronics of the instrument measure the capacitance of the sensor and convert it into a humidity reading.

The response time of the humidity sensor is dependent on the polymer's ability to adsorb and desorb water vapour and on the sensor design, whereby it is strongly dependent on the temperature of the sensor [7]. The response time of capacitive humidity sensors is on the order of a second at ambient temperature but can reach hundreds of seconds at very cold temperature. The sensor is sensitive to chemical contamination by either additional bonding of the non-water molecules or reducing the ability of the polymer to adsorb water molecules, which may cause either a dry bias or reduce the sensitivity of the sensor.

3.2 Upper Air Applications

A prominent meteorological application of thin-film capacitors is the Humicap sensing element developed by Vaisala (Finland) in the 1970's [10]. Humicap sensing elements are used on their RS80 radiosondes since 1980 and they integrated an improved version later into their RS90 and RS92 radiosondes. Based on thin film technology, the sensor consists of a hydroactive polymer film as dielectric between two electrodes applied on a glass substrate. Two types of polymer materials have been developed, the A-type and the H-type polymer. The newer H-polymer (introduced in 1990) is more stable and less hydrophilic compared to the A-type polymer.

Capacitive polymer sensors are nowadays used by most of the manufacturers of operational radiosondes. In fact, the quality of radiosondes deploying thin-film capacitive sensors largely improved and they get close to the ability to provide humidity data that may be used for long-term climate studies. Nevertheless, correction of systematic biases and time lag effects, which are necessary and done individually using different approaches, easily lead to differences on relative humidity with relative uncertainty of $\pm 5\%$ at ambient temperature above -40°C for different radiosondes. At lower temperature, relative uncertainty is increasing to $\pm 10\%$ or more. Night-time measurements show lower uncertainty than daytime measurements mainly due to solar radiation effects that are difficult to correct [11].

Another airborne application of capacitive sensors is the deployment of humidity sensors on board of commercial aircrafts to measure water vapour concentration between the surface and up to 13 km altitude in the context of the MOZAIC (Measurement of Ozone and Water Vapour by AIRBUS In Service Aircraft) project. The sensing element consists of a capacitive sensor (Humicap-H) with a hydroactive polymer film as dielectric whose capacitance depends on the relative humidity [8] plus a platinum resistance sensor (PT100) for the direct measurement of the temperature at the humidity sensing surface. The humidity and temperature signals are linearized by a microprocessor controlled transmitter unit (HMP-230, Vaisala, Finland), which passes the relative humidity and the temperature signal to the automated data acquisition system of MOZAIC located in the avionic bay of the aircraft [12].

4 Intercomparisons and Reference Upper Air Observations

4.1 International Radiosonde Intercomparisons

The World Meteorological Organization (WMO) periodically organises international intercomparisons inviting all leading radiosonde manufacturers to present and test their radiosonde systems. The intercomparison, organised every 4–5 years, is a continuous 4-week sounding campaign, designed to be an objective venue for assessing the qualities and performance of different radiosonde systems. The last two intercomparisons were held in Mauritius (2005) and Brazil (2001). The most recent campaign, hosted by the China Meteorological Administration, was held in July 2010 at the Yangjiang Observing Station, about 200 km from Hong Kong. More radiosondes than ever before were entered into the intercomparison—eleven models from eight different countries were flown in the quality (operational) radiosondes group, and five models from four countries participated in the scientific sensor intercomparison. About 62 quality radiosonde test flights were performed and 10 scientific sounding system test flights. The choice of the site was justified by the excellent vertical structure in relative humidity obtained in nearly all test flights.

All quality radiosondes used thin-film capacitor humidity sensors. New sensors were tested during the campaign and humidity measurement accuracy was improved by taking into account more details than before. New time lag corrections were added, which improve performance especially in cold temperatures, where the time constant of the humidity sensor is greater. This improved measurement accuracy in both daytime and night-time soundings. Also, a new algorithm was tested to correct the effects of solar radiation during daytime soundings. Both solar radiation and time lag correction algorithms have the biggest impact in measurements taken in high humidity conditions in the upper troposphere, at altitudes of approximately 10–15 km.

4.2 *Reference Upper Air Network*

While for many decades operational radiosonde measurements have primarily been used for weather prediction, upper air observations have recently become very important for climate change research and monitoring. To reduce the estimated uncertainties of quality radiosondes as well as for future reference instruments, large efforts have been started in collaboration with WMO in the scope of the Global Climate Observing System (GCOS) Reference Upper Air Network (GRUAN), [13]. The resulting GRUAN will be a network of several dozen stations, made operational in a phased process, and has been identified by the climate community as being required to meet the following key needs:

- Provide long-term high-quality upper air climate records, with complete estimates of measurement error;
- Constrain and adjust data from more spatially comprehensive global observing systems (including satellites and current radiosonde networks);
- Fully characterise the properties of the atmospheric column and their changes; and
- Ensure that potential gaps in satellite programs do not invalidate the long-term climate record.

GRUAN has a strong lead centre providing scientific leadership and oversight, managing the network, training operators, and ensuring proper data archival and free dissemination: The Meteorological Observatory Lindenberg, Richard Aßmann Observatory, will serve as GRUAN Lead Centre, based on the Deutscher Wetterdienst's offer of substantial financial, scientific and technical support, in response to a call for interest issued by the GCOS Secretariat. GRUAN is further building on existing observational networks, such as NDACC, ARM, GUAN, GAW and BSRN.

5 Conclusions

Humidity sensors for upper air observations experienced rapid changes and improvements over the last three decades. Achievements in micro technology encouraged the development of very small humidity sensors to measure the water vapour content in the atmosphere. Today, more than 75% of these miniaturised humidity sensors in the market use a capacitive technique and almost all operational radiosondes use thin-film capacitive polymer sensors. In the lower and middle troposphere, capacitive sensors measure relative humidity with relative uncertainty of $\pm 5\%$ at ambient temperature above -40°C. At lower temperatures, uncertainties are larger and are strongly dependent on solar radiation and time lag correction algorithms, which have the biggest impact on measurements taken in high humidity conditions in the upper troposphere, at altitudes of 8–15 km and temperatures down to -90°C.

Scientific reference measurements in the upper troposphere and lower stratosphere are often made with other sensor techniques like frostpoint or fluorescence hygrometers, but climate change issues are a real challenge for low-cost thin-film capacitive gas sensors, which allow operational radiosonde measurements with low uncertainty up to the tropopause in large upper air networks and on board of commercial aircrafts.

References

1. WMO (2008) Measurements of upper air temperature, pressure, and humidity. Guide to Meteorological Instruments and Methods of Observation, Chap. 12, WMO No. 8, Geneva, seventh edition
2. Schmidlin FJ, Ivanov A (eds) (1998) Radiosonde relative humidity sensor performance: the WMO intercomparison Sept. 1995. Preprint volume. 10th Symposium on Meteorological Observations and Instrumentation, Amer Meteor Soc, 68–71
3. Suomi VE, Barrett WE (1952) An experimental radiosonde for the investigation of the distribution of water vapor in the stratosphere. Rev Sci Instr 23:272–292
4. Fujiwara M, Shiotani M, Hasebe F, Vömel H, Oltmans SJ, Ruppert PW, Horinouchi T, Tsuda T (2003) Performance of the meteolabor "Snow White" chilled mirror hygrometer in the tropical troposphere: comparisons with the Vaisala RS80 A/HHumicap sensors. J Atmos Oceanic Technol 20:1534–1542
5. Kley D, Stone E (1978) Measurement of water vapour in the stratosphere by photodissociation with Ly -a (1216_a) light. Rev Sci Instrum 49:691–874
6. Bertaux JL, Delannoy A (1978) Vertical distribution of H_2O in the stratosphere as determined by UV fluorescent in situ measurements. Geophys Res Lett 5(12):1017–1020
7. Rittersma ZM (2002) Recent achievements in miniaturised humidity sensors – a review of transduction techniques. Sensors Actuators A 96:196–210
8. Anderson PS (1995) Mechanism for the behaviour of hydroactive materials used in humidity sensors. J Atmos Oceanic Technol 12:662–667
9. Matsuguchi M, Umeda S, Dadaoka Y, Sakai Y (1998) Characterization of polymers for a capacitive type humidity 1 sensor based on water absorption behavior. Sensor Actuator 49:179–185
10. Salasmaa E, Kostamo P (1975) New thin film humidity sensor. In: Proc. of Third Symposium on Meteorological Observations and Instrumentation. Amer Meteor Soc, Washington, DC, pp 33–38
11. Miloshevich LM, Vömel H, Whiteman DN, Leblanc T (2009) Accuracy assessment and correction of Vaisala RS92 radiosonde water vapor measurements. J Geophys Res 114: D11305. doi:10.1029/2008JD011565
12. Marenco A, Thouret V, Nedelec P, Smit H, Helten M, Kley D, Karcher F, Simon P, Law K, Pyle J, Poschmann G, Wrede RV, Hume C, Cook T (1998) Measurement of ozone and water vapour by airbus in-service aircraft: the MOZAIC airborne program, an overview. J Geophys Res 103:25631–25642
13. Seidel D, Berger F, Diamond H, Dykema J, Goodrich D, Immler F, Murray W, Peterson T, Sisterson D, Sommer M, Thorne P, Vömel H, Wang J (2009) Reference upper-air observations for climate: rationale, progress, and plans. Bull Am Meteorol Soc 90:361–369. doi:10.1175/2008BAMS2540.1

Concluding Remarks

Sensing of physical quantities is a technology which has been available for a long time. Sensing of chemical quantities is much more complex due to the incomparable larger amount of targets and thus the higher amount and complexity of required detection principles. Therefore, this field requires especially intensive collaboration between the stakeholders who need to mutually understand the capabilities and needs of the others.

This was the main reason to bring together the views of all the different groups needed to produce new gas sensing applications. This cannot be done comprehensively in one book; yet we believe that the different viewpoints on this complex matter are well expressed in the chapters. This will hopefully contribute to a better mutual understanding of how the job has to be done and let gas sensors play an even more important role in the near future.

Munich, Germany — Maximilian Fleischer
Wattwil, Switzerland — Mirko Lehmann

M. Fleischer and M. Lehmann (eds.), *Solid State Gas Sensors – Industrial Application*, Springer Series on Chemical Sensors and Biosensors (2012) 11: 265–266,
DOI 10.1007/5346_2011, © Springer-Verlag Berlin Heidelberg 2012

Index